Plant Ionomics

Plant Ionomics

Sensing, Signaling, and Regulation

Edited by

Vijay Pratap Singh
University of Allahabad
Prayagraj
India

Manzer H. Siddiqui
King Saud University
Riyadh
Saudi Arabia

WILEY

Registered Offices
John Wiley & Sons, Inc., 111 River Street, Hoboken, NJ 07030, USA
John Wiley & Sons Ltd, The Atrium, Southern Gate, Chichester, West Sussex, PO19 8SQ, UK

For details of our global editorial offices, customer services, and more information about Wiley products visit us at www.wiley.com.

Wiley also publishes its books in a variety of electronic formats and by print-on-demand. Some content that appears in standard print versions of this book may not be available in other formats.

Library of Congress Cataloging-in-Publication Data applied for
Hardback ISBN 9781119803010

Cover Design: Wiley
Cover Image: © Pasotteo/Shutterstock

Set in 9.5/12.5pt STIXTwoText by Straive, Pondicherry, India

Contents

List of Contributors

Saghir Abbas
Department of Botany
Faculty of Life Sciences
Government College University
Faisalabad, Pakistan

Zeeshan Ahmed
Xinjiang Key Laboratory of Desert Plant Roots Ecology and Vegetation Restoration
Xinjiang Institute of Ecology and Geography
Chinese Academy of Sciences
Urumqi, China

State Key Laboratory of Desert and Oasis Ecology
Xinjiang Institute of Ecology and Geography

Chinese Academy of Sciences
Urumqi, China

Cele National Station of Observation and Research for Desert-Grassland Ecosystems
Cele, China

and

University of Chinese Academy of Sciences
Beijing, China

Amna
Department of Plant Sciences
Faculty of Biological Sciences
Quaid-i-Azam University
Islamabad, Pakistan

Jesús Amo
Departamento de Nutrición Vegetal
Centro de Edafología y Biología Aplicada del Segura-CSIC
Murcia, Spain

Muhammad Arslan Ashraf
Department of Botany
Government College University Faisalabad
Faisalabad, Pakistan

Savita Bhardwaj
Department of Botany
School of Bioengineering and Biosciences
Lovely Professional University
Phagwara, Punjab, India

Nikita Bisht
Microbial Technologies Division
Council of Scientific and Industrial Research-National Botanical Research Institute (CSIR-NBRI)
Lucknow, Utter Pradesh, India

Puneet Singh Chauhan
Microbial Technologies Division
Council of Scientific and Industrial Research-National Botanical Research Institute (CSIR-NBRI)
Lucknow
Utter Pradesh, India

Neda Dalir
Department of Soil Science
Tarbiat Modares University
Tehran, Iran

Salar Farhangi-Abriz
Department of Plant Eco-physiology
Faculty of Agriculture
University of Tabriz
Tabriz, Iran

Kazem Ghassemi-Golezani
Department of Plant Eco-physiology
Faculty of Agriculture
University of Tabriz
Tabriz, Iran

Corina Graciano
Instituto de Fisiología Vegetal

Consejo Nacional de Investigaciones Científicas y Técnicas

Universidad Nacional de La Plata
Buenos Aires, Argentina

Iqbal Hussain
Department of Botany
Government College University Faisalabad
Faisalabad, Pakistan

Mudassir Iqbal Shad
Department of Botany
Government College University Faisalabad
Faisalabad, Pakistan

Muhammad Iqbal
Department of Botany
Government College University Faisalabad
Faisalabad, Pakistan

Khasan Ismoilov
University of Chinese Academy of Sciences
Beijing, China

and

CAS Key Laboratory of Biogeography and Bioresource in Arid Land
Chinese Academy of Sciences
Urumqi, China

Hassan Javed Chaudhary
Department of Plant Sciences
Faculty of Biological Sciences
Quaid-i-Azam University
Islamabad, Pakistan

Harshita Joshi
Microbial Technologies Division
Council of Scientific and Industrial Research-National Botanical Research Institute (CSIR-NBRI)
Lucknow, Utter Pradesh, India

Dhriti Kapoor
Department of Botany
School of Bioengineering and Biosciences
Lovely Professional University
Phagwara, Punjab, India

Rajeev Kumar
Department of Botany
School of Bioengineering and Biosciences
Lovely Professional University
Phagwara, Punjab, India

Vicente Martínez
Departamento de Nutrición Vegetal
Centro de Edafología y Biología Aplicada del Segura-CSIC, Murcia, Spain

Almudena Martínez-Martínez
Departamento de Nutrición Vegetal
Centro de Edafología y Biología Aplicada del Segura-CSIC, Murcia, Spain

Nebojša Milošević
Department of Pomology and Fruit Breeding
Fruit Research Institute Čačak
Čačak, Republic of Serbia

Tomo Milošević
Department of Fruit Growing and Viticulture
Faculty of Agronomy
University of Kragujevac
Čačak, Republic of Serbia

Ghulam Murtaza
Faculty of Environmental Science and Engineering
Kunming University of Science and Technology, Kunming, PR China

Manuel Nieves-Cordones
Departamento de Nutrición Vegetal
Centro de Edafología y Biología Aplicada del Segura-CSIC
Murcia, Spain

Rizwan Rasheed
Department of Botany
Government College University Faisalabad
Faisalabad, Pakistan

Francisco Rubio
Departamento de Nutrición Vegetal
Centro de Edafología y Biología Aplicada del Segura-CSIC, Murcia, Spain

Sehrish Sadia
Department of Biological Sciences
University of Veterinary and Animal Sciences-Lahore, Pattoki, Pakistan

Muhammad Shahid
Department of Bioinformatics and Biotechnology
Government College University
Faisalabad, Pakistan

Muhammad Sohail Akram
Department of Botany
Faculty of Life Sciences
Government College University
Faisalabad, Pakistan

Kashif Tanwir
Department of Botany
Faculty of Life Sciences
Government College University
Faisalabad, Pakistan

Muhammad Tariq Javed
Department of Botany
Faculty of Life Sciences
Government College University
Faisalabad, Pakistan

and

Department of Plant Sciences
Faculty of Biological Sciences
Quaid-i-Azam University
Islamabad, Pakistan

Akash Tariq
Xinjiang Key Laboratory of Desert Plant Roots Ecology and Vegetation Restoration
Xinjiang Institute of Ecology and Geography

Chinese Academy of Sciences
Urumqi, China

State Key Laboratory of Desert and Oasis Ecology

Xinjiang Institute of Ecology and Geography
Chinese Academy of Sciences
Urumqi, China

Cele National Station of Observation and Research for Desert-Grassland Ecosystems
Cele, China
and University of Chinese Academy of Sciences, Beijing, China

Monika Thakur
Division Botany
Department of Bio-Sciences
Career Point University
Hamirpur, Himachal Pradesh, India

Abd Ullah
Xinjiang Key Laboratory of Desert Plant Roots Ecology and Vegetation Restoration

Xinjiang Institute of Ecology and Geography, Chinese Academy of Sciences
Urumqi, China

State Key Laboratory of Desert and Oasis Ecology

Xinjiang Institute of Ecology and Geography

Chinese Academy of Sciences
Urumqi, China

Cele National Station of Observation and Research for Desert-Grassland Ecosystems
Cele, China

and

University of Chinese Academy of Sciences
Beijing, China

Tunisha Verma
Department of Botany
School of Bioengineering and Biosciences
Lovely Professional University
Phagwara, Punjab, India

Fanjiang Zeng
Xinjiang Key Laboratory of Desert Plant Roots Ecology and Vegetation Restoration
Xinjiang Institute of Ecology and Geography

Chinese Academy of Sciences
Urumqi, China

State Key Laboratory of Desert and Oasis Ecology

Xinjiang Institute of Ecology and Geography

Chinese Academy of Sciences
Urumqi, China

and

Cele National Station of Observation and Research for Desert-Grassl and Ecosystems
Cele, China

Zhihao Zhang
Xinjiang Key Laboratory of Desert Plant Roots Ecology and Vegetation Restoration

Xinjiang Institute of Ecology and Geography

Chinese Academy of Sciences
Urumqi, China

State Key Laboratory of Desert and Oasis Ecology

Xinjiang Institute of Ecology and Geography

Chinese Academy of Sciences
Urumqi, China

Cele National Station of Observation and Research for Desert-Grassland Ecosystems
Cele, China

and

University of Chinese Academy of Sciences
Beijing, China

Muhammad Zubair
Discipline of Zoology
University of Veterinary and Animal Sciences-Lahore, Pattoki, Pakistan

Preface

Ionome includes the role of mineral nutrients, namely phosphorus (P), nitrogen (N), calcium (Ca), potassium (K), sulfur (S), magnesium (Mg), etc., and trace metals namely iron (Fe), copper (Cu), manganese (Mn), molybdenum (Mo), cobalt (Co), zinc (Zn), etc. in plant growth and development. Although all mineral nutrients and trace elements are essential for growth and developmental processes of plants, concentration greater than the required level becomes toxic to the plants. Apart from posing toxicity at higher concentration, nutrients under safe limit play important role in alleviating toxicity induced by various stresses.

Appropriate and adequate amount of any nutrient is required at specific stage of development and time to sustain the life of the plant. Hence, in the past decade, much progress has been made in understanding the regulation of nutrients homeostasis under normal and stressed conditions. Further, on one hand an approach has been developed for generating biofortified grains and on other hand nutrients have been successfully employed for managing stress challenges in plants. Taking into account the progress made in ionomics and plants, this book has compiled recent knowledge on sensing, signaling, and regulation of nutrients uptake in plants under changing environment.

A total of 13 chapters have been compiled in this book. Chapter 1 has comprehensively complied information on the regulation of metabolites by nutrients in plants. Chapter 2 deals with agricultural production relation with nutrient applications. Chapter 3 deals with the role of nutrients in ROS metabolism in plants. Chapter 4 deals with polyamines metabolism and their regulatory mechanism in plant development and in abiotic stress tolerance. Chapter 5 narrates the mycorrhizal symbiosis and nutrients uptake in plants. Chapter 6 deals with nutrient availability and root system behavior. Chapter 7 describes in detail the potassium transport systems at the plasma membrane of plant cells and their potential roles in improving potassium use efficiency of crops. Chapter 8 narrates the role of nutrients in modifications of fruit quality and antioxidant activity. Chapter 9 deals with the nutrients use efficiency in plants. Chapter 10 describes the nutrients uptake and their transport in plants. Chapter 11 comprehensively deals with the regulation of phytohormonal signaling by nutrients in plants. Chapter 12 gives details on the nutrients regulation and abiotic stress tolerance in plants. Chapter 13 deals with nutrient management and stress tolerance in crops. Overall, we believe that this book will serve as an important source of information for students, researchers, and academicis.

Vijay Pratap Singh
Manzer H. Siddiqui

1

Regulation of Metabolites by Nutrients in Plants

Akash Tariq[1,2,3,4], Fanjiang Zeng[1,2,3], Corina Graciano[5], Abd Ullah[1,2,3,4], Sehrish Sadia[6], Zeeshan Ahmed[1,2,3,4], Ghulam Murtaza[7], Khasan Ismoilov[4,8], and Zhihao Zhang[1,2,3,4]

[1] *Xinjiang Key Laboratory of Desert Plant Roots Ecology and Vegetation Restoration, Xinjiang Institute of Ecology and Geography, Chinese Academy of Sciences, Urumqi, China*
[2] *State Key Laboratory of Desert and Oasis Ecology, Xinjiang Institute of Ecology and Geography, Chinese Academy of Sciences, Urumqi, China*
[3] *Cele National Station of Observation and Research for Desert-Grassland Ecosystems, Cele, China*
[4] *University of Chinese Academy of Sciences, Beijing, China*
[5] *Instituto de Fisiología Vegetal, Consejo Nacional de Investigaciones Científicas y Técnicas, Universidad Nacional de La Plata, Buenos Aires, Argentina*
[6] *Department of Biological Sciences, University of Veterinary and Animal Sciences, Lahore, Pakistan*
[7] *Faculty of Environmental Science and Engineering, Kunming University of Science and Technology, Kunming, PR China*
[8] *CAS Key Laboratory of Biogeography and Bioresource in Arid Land, Chinese Academy of Sciences, Urumqi, China*

Introduction

Mineral nutrients are required for plant growth and metabolism. Seventeen nutrients that are essential for plant growth have been identified to date. Macronutrients or essential elements such as nitrogen (N), phosphorus (P), potassium (K), sulfur (S), and magnesium (Mg) are required in large amounts, whereas micronutrients or trace elements including chloride (Cl), copper (Cu), manganese (Mn), iron (Fe), zinc (Zn), cobalt (Co), molybdenum (Mo), and nickel (Ni) are required in smaller amounts. Some researchers also consider sodium (Na) and silicon (Si) as essential micronutrients because of their role in certain crop plants; however, whether they are essential for plant growth and metabolism remains unclear (Maathuis 2009; Marschner 2012). Plants acquire these nutrients from the soil through the roots and carbon (C) and oxygen (O_2) from the air. The concentrations of the available forms of these nutrients in the soil solution are low; for example, N can be readily absorbed by plant roots in nitrate (NO_3^-) and ammonium (NH_4^+) forms and P in orthophosphate (HPO_4^{2-} and H_2PO_4) forms (Soetan et al. 2010; Mazid et al. 2011). Their availability can be further altered in both time and space because of several environmental factors, such as precipitation, temperature, wind, and soil properties. Consequently, plants have evolved various flexible physiological and morphological mechanisms to facilitate the

Plant Ionomics: Sensing, Signaling, and Regulation, First Edition.
Edited by Vijay Pratap Singh and Manzer H. Siddiqui.

uptake of nutrients. Each mineral carries equal importance and a deficiency of any of them can affect growth and disrupt the plant life cycle.

Metabolites are chemical compounds produced during metabolism and catalyzed by different naturally occurring enzymes in the cell. Plants can produce a variety of metabolites ranging from 0.1 to 1 million, which help plants survive in diverse environments, as these metabolites play important roles in structure, defense, energy, signaling, and interactions with other organisms (Rai et al. 2017). Plant metabolites are divided into two categories (primary and secondary) according to their functions. Primary metabolites (PMs) are directly involved in growth and cellular functions; they are thus also referred to as vital metabolites because they are generally common in all plant species. These are the intermediate products of anabolic metabolism and are formed in the growth phase. PMs are produced in large amounts and can be easily extracted. Examples of PMs include carbohydrates, protein, nucleic acids, lipids, and vitamins. Secondary metabolites (SMs) are the derivatives of PMs, and they are not directly involved in growth and development, but they are ecologically relevant. SMs are formed in the stationary phase; they are also called "natural products" and are different in every plant species because of environmental and genetic differences. There are more than 30 000 SMs, but they are produced in small concentrations, and extracting them can be difficult (Baranauskiene et al. 2003). Plant SMs have diverse applications and have been used in pharmaceuticals, flavoring compounds, antimicrobial agents, dyes, antioxidants, and agrochemicals. Examples of SMs include alkaloids, terpenoids, phenolic compounds, and glycosides.

There is a strong linkage between plant metabolites and mineral nutrients, but understanding the large diversity of plant metabolites requires knowledge of their biosynthesis, degradation, regulation, and transportation processes. Their biosynthesis involves various metabolic pathways such as absorption, assimilation of C, photosynthesis, protoplast formation, respiration, transpiration, translocation, and storage, and these pathways are regulated by mineral nutrients available in different forms in the soil solution (Soetan et al. 2010). Mineral nutrient availability and composition have a profound effect on the type and amount of PMs and SMs produced. Plants growing in soil with an unbalanced composition and limited nutrient concentrations experience serious morphological, biochemical, physiological, and molecular disruptions (Amtmann and Armengaud 2009). There is an increased interest in unraveling the regulatory roles of mineral nutrients in the biosynthesis of plant metabolites. The aim of this chapter is to improve our understanding regarding the regulatory roles of mineral nutrients in the biosynthesis of plant metabolites.

Nitrogen (N)

N is a key plant macronutrient that accounts for nearly 4% of the total dry matter of plants. It is an important structural component of several organic compounds, including amino acids, proteins, chlorophyll pigments, and nucleic acids. N is also involved in the biochemistry of several nonprotein compounds such as polyamines, co-enzymes, SMs, and photosynthetic pigments. N is an important plant macronutrient that is necessary for the biosynthesis of various PMs (i.e. amino acids, chlorophyll, nucleic acids, lipids, proteins, and enzymes) and SMs (i.e. flavonoids and phenolic compounds) in plants (Kováčik and Klejdus 2013; Shiwakoti et al. 2016). N availability controls plant metabolism and alters the N-based metabolites into C-based plant metabolites (Strissel et al. 2005; del Mar Rubio-Wilhelmi et al. 2012a). C-based plant metabolites mostly contain phenolic acids, anthocyanins, flavonoids, and coumarins (Kováčik et al. 2011). Phenolics consist of thousands of nonenzymatic antioxidants that are ubiquitous in plants and an integral part of human nutrition (Balasundram et al. 2006). The concentrations of flavonoids produced in the secondary metabolism of plants are affected by N, which regulates flavonoid biosynthesis by affecting the C flow allocation between primary and secondary metabolism. For example, excessive N application reduces the content of flavonoids in plants (Ballizany et al. 2012; Shiwakoti et al. 2016), and low N increases the content of carbohydrates (e.g. starch and fructose) but decreases the content of N-rich metabolites (e.g. proteins, free amino acids, and polyamines) in plants (Paul and Driscoll 1997; Coruzzi and Bush 2001).

Soil solution carries N mainly in the form of nitrate (NO_3^-) and ammonium (NH_4^+), but, NO_3^--N is readily available for plants uptake (at high pH). Plants absorb NO_3^-, which is then reduced to NH_4^+ through enzymatic action of glutamine synthetase/glutamate synthase (GS/GOGAT) enzymes. Scheible et al. (2004) reported downregulation of most genes responsible for chlorophyll and plastid protein synthesis, photosynthesis, and upregulation of many genes involved in secondary metabolism following two days of N deprivation in *Arabidopsis thaliana*. It is noted that N limitation mainly governs the shift from N-based to C-based plant metabolites (Kováčik et al. 2011; del Mar Rubio-Wilhelmi et al. 2012b). This effect of N limitation on phenolic metabolites is thought to be the result of reductions in other macronutrients, such as the depletion of K, and does not increase phenolic metabolites and vice versa (Nguyen et al. 2010).

Leaf N limitation generally alters the pattern of N allocation in photosynthetic organs and directly regulates the content and enzymatic activity of ribulose 1,5-bisphosphate carboxylase/oxygenase (Rubisco) and chlorophyll, thereby affecting the photosynthetic rate. Kováčik and Klejdus (2013) noted that concentrations of phenolic metabolites (especially chlorogenic acid) and free amino acids (mainly proline and arginine) increased in *Matricaria chamomilla* plants upon NH_4^+ application. Coumarin-related metabolites showed a similar increase under N limitation, and herniarin mainly accumulated in plants under NO_3^- addition and umbelliferone under NH_4^+ addition.

Phosphorus (P)

P does not exist in elemental form but is firmly fixed (>90%) in soil minerals, therefore it is unavailable to plants. P is assimilated by the plants as phosphate ($H_2PO_4^-$) at a pH of 5, and it is unevenly distributed because it is relatively immobile in soils. In addition, inorganic phosphate (Pi) is an important part of several biological compounds, such as sugar phosphates, phospholipids, DNA, RNA, and nucleotides. P has a crucial role in energy metabolism and activation of various intermediates in the C cycle of photosynthesis. These are the reasons why plant growth and yield depend on the availability of P.

Aside from its structural roles, P is involved in reversible protein phosphorylation (facilitated by kinases and phosphatases) to modulate protein activity. In this mechanism, terminal Pi from ATP or GTP nucleotide is transferred to Ser or the protein residues. ATP is an important energy compound with a pyrophosphate bond. Pyrophosphate is an energy-rich bond and participates in active ion uptake and the synthesis of various PMs and SMs during hydrolysis (30 kJ mol^{-1}). GTP, CTP, and UTP are energy-rich phosphonucleotides which also regulate the nucleic acid metabolism. CTP provides energy during phospholipid biosynthesis, and UTP is the main structural constituent of sucrose, starch, and cellulose (Maathuis and Diatloff 2013). Pi also regulates various enzymes involved in carbohydrate metabolism (Ashihara and Stupavska 1984; Avigad and Bohrer 1984). P is a constituent of energy metabolism, and the phospho groups are responsible for the activation of enzymes and metabolic intermediates. For example, P activates intermediates in the photosynthetic carbon cycle and during the entire primary metabolism. P limitations have a substantial effect on plants at the morphological, physiological, biochemical, and molecular levels (Secco et al. 2013; Pant et al. 2015). Morcuende et al. (2007) determined the metabolite levels, transcripts, and maximum enzyme activity in the seedlings of Arabidopsis grown under P-added and P-depleted liquid culture. Reduction occurred in the levels of ATP and hexose phosphates (Glc6P, Fru6P, and Glc1P), owing to P limitation, recovered immediately after restocking with a knock-on effect on UDP-glucose, which took few hours for its completion. Tricarbonic acids such as 3-phosphoglyceric acid (3PGA) and phosphoenolpyruvate (PEP), accumulated due to P limitation, are rapidly used after restocking and lead to the quick initiation of glycolysis. However, reduction in organic acids (malate, 2-oxoglutarate, citrate, and fumarate) and storage carbohydrates (starch and sucrose) due to P limitation did not get back to their normal state within 24 hours of restocking. These findings clearly show that in response of P restocking, plants do modifications to retrieve their energy stores and the biosynthesis of phosphorylated substrates for glycolysis.

Soil contains extremely low concentrations of Pi (~0.1–1 mM), and P limitation is likely widespread in agriculture. Several studies revealed that P deficiency affects numerous PMs and SMs in soybean and Arabidopsis (Hurry et al. 2001; Morcuende et al. 2007; Hernandez et al. 2009). However, how the metabolite profile during P limitation and in mutants is affected during the P-deprivation response remains unclear.

Morcuende et al. (2007) noted that P limitation in *A. thaliana* can induce or suppress more than1000 genes involved in many metabolic processes. P limitation leads to the accumulation of carbohydrates, amino acids, and organic acids. P-limited plants also show significant variations in the expression of many genes associated with secondary metabolism and carbon assimilation. Upon P deficiency, plants generally recycle P from organic molecules (Yuan and Liu 2008). P limitation can induce substantial variations in gene expression in plants, but the diversity of these metabolic alterations and their regulation procedure have been poorly studied. Pant et al. (2015) profiled nearly 350 PMs and SMs of P-added and P-deficient *A. thaliana*, wild-type, and mutants of the central P-signaling components (PHR1 and PHO2, and microRNA399 overexpressor). P availability has a significant effect on the metabolite profile in the roots and shoots of *A. thaliana* wild-type, phr1, pho2, and miR399OX seedlings. The concentrations of many glucosinolates, phenylpropanoids, flavonoids, flavonoid glycosides, esculetin, ferulic acid, caffeic acid, kaempferol glucuronide, kaempferol 3-*O*-glucoside, kaempferol 3-*O*-rutinoside, and many other glucosinolates markedly increased in wild-type *A. thaliana* under P limitation.

Potassium (K)

K is a vital macro-mineral element (inorganic cation) that is necessary for the optimal growth and development of plants. K being an abundant cation in plant tissues plays fundamental role in numerous biochemical and biophysical processes, such as osmoregulation, ion homeostasis, enzyme activation, and protein formation. Nonstomatal inhibition of photosynthesis, such as the inhibition of chlorophyll synthesis, is often observed during unfavorable conditions, and the K supply can positively affect chlorophyll biosynthesis (Tiwari et al. 1998), the chloroplastic CO_2 level, and the electron transport system. The free ionic form of K (K^+) plays a fundamental role in regulating many physiological processes, such as membrane polarization, protein synthesis, and osmotic adjustments, and thus controls osmotically driven processes such as phloem transport and stomatal function (Zorb et al. 2014). When K^+ is sufficient, carboxylation and the overall photosynthetic rate lead to normal growth and yield. K-limitation reduces photosynthetic rates because of the low concentration of CO_2 stemming from stomatal closure (Jin et al. 2011). In contrast, a sufficient K level can improve the photosynthetic rate by regulating stomatal and gas exchange attributes and enzymes such as Rubisco (Erel et al. 2015). Furthermore, K-induced increases in photosynthesis might also stem from a reduction in reactive oxygen species. For example, the K supply can decrease oxidative damage by increasing the activities of key enzymes of the antioxidant system, such as SOD, POD, and CAT (Ahanger and Agarwal 2017a, 2017b). Plants have developed efficient mechanisms for K uptake to guarantee normal growth under K-limited conditions. For example, main strategies for efficient K uptake capacity include potassium redistribution between cytosol and vacuole for cellular homeostasis and adaptation of the root system. Execution of these strategies needs accurate regulatory signaling pathways and mechanisms. The accumulation of sugars and amino acids increased in both leaves and roots, but organic acids showed reduction upon low K stress. Moreover, amino acids with negative charges (Asp and Glu) and most organic acids were decreased while amino acids carrying positive charge (Lys and Gln) increased in all three genotypes (Ashley et al. 2006). The selective reductions in acidic amino acids can facilitate the maintenance of the charge balance upon low K stress. Armengaud et al. (2009) noted the responses of shoot and roots of two-week old Arabidopsis (*A. thaliana*) plant species to K availability. The concentrations of nitrate, pyruvate, organic acids (2-oxoglutarate and malate), glutamate, and aspartate (negatively charged amino acids) were significantly reduced in the roots, but the concentrations of soluble carbohydrates (e.g. glucose, sucrose, and fructose) and numerous amino acids such as glycine, arginine, and glutamine (positively charged amino acids) and those with a high N:C ratio were significantly increased in the roots. In addition, metabolite alterations under K depletion were observed later in the shoots than in the roots. Carbohydrates (sucrose and reducing sugars) and organic acids were increased in the shoots. Moreover, N-rich and basic amino acids like arginine, glutamine, asparagine, lysine, and histidine largely increased in shoots relative to the roots.

Although K is not metabolized, but there are numerous metabolic processes that are highly K-dependent. For example, (i) direct enzyme activation of enzymes: vacuolar PPase isoforms, which are responsible for the accumulation of protons in the vacuolar lumen, are largely dependent on the K^+ level; (ii) translation: ribosome-mediated protein synthesis also depends on a sufficient K^+ concentration; and (iii) transportation: both across membranes and over long distances, among other important roles (Wyn Jones and Pollard 1983; Marschner 1995). These processes assist in observing the metabolic impairments especially under K-deficient conditions.

K^+ limitation can lead to low chlorophyll and photosynthetic activity, the accumulation of simple carbohydrates, and a decrease in plant growth and yield. Therefore, sufficient K^+ must be present in the soil to facilitate sufficient absorption and transportation, since its distribution to different plant organs depends on its availability and significantly affects the biosynthesis, conversion, and allocation of plant metabolites (Sorger et al. 1965; Evans and Sorger 1966). K^+-deficient conditions can also increase the sensitivity of plants to unfavorable environmental conditions. The plant K level decreases the occurrence of pests and diseases by regulating the production of organic compounds having large molecular weight, such as proteins, cellulose, and starch (Marschner 1995). The availability and level of K^+ also regulate plant metabolism through transcriptional and posttranscriptional regulation of metabolic enzymes. Studies on K^+-deficient *A. thaliana* plants gave insight into the induction of GS/GOGAT pathway, malic enzyme, and the suppression of nitrate uptake (Amtmann and Blatt 2009; Armengaud et al. 2009). K limitation in cultivated land has emerged as a major factor negatively affecting the sustainable growth and production of crop plants worldwide. K-sufficient conditions help sustain the normal metabolism and growth of plants (Sharma et al. 2006). K-limited plants accumulate numerous basic or neutral amino acids, show minor rise in total amino acids and proteins, and contain few acidic amino acids and nitrate. K also plays a key role as an inorganic osmotic solute as well as in the synthesis of other compatible solutes required for stomatal movement and maintenance of water relations. Potassium also governs the loading of photo-assimilates and the functioning of key metabolic enzymes (Ahanger et al. 2014; 2015; Erel et al. 2015). Furthermore, a significant reduction in pyruvic acid was observed in the roots. Most metabolites returned to their prealteration levels within 24 hours of K re-stocking; the exception was sucrose, glucose, and fructose in the shoots. However, in roots, pyruvic acid recovered within minutes, but 2-oxoglutaric acid and malic acid recovered more slowly. K limitation also reversibly changed the activities of many enzymes engaged in sugar metabolism (glucokinase, fructokinase, and invertase), glycolysis (glyceraldehyde 3-phosphate dehydrogenase [GAPDH]), Krebs cycle (malic enzyme), and N assimilation (NR, GS, GDH, and Fd-GOGAT). The noticeable build-up of sugars, the reduction of pyruvic acid in the root cells, and the swift restoration of root sugar and pyruvic acid level suggest that K level in root cytoplasm could be the first one to be recovered upon K re-supply (Leigh and Wyn Jones 1984). Alterations in the shoot metabolite profile (e.g. sugar accumulation and nitrate reduction) are likely derived from the same alterations in the roots. However, following K re-supply, changes in some amino acids were specific to the shoots and took place even before K accumulated, specifying the presence of a root/shoot signal from K itself. Based on the determined metabolite profile, glutamic acid appears to play a vital role in the long-distance signaling of tissue K status (Lacombe et al. 2001; Davenport 2002). K limitation has also been shown to decrease the scopolin content in sunflower (Lehman and Rice 1972), oxylipins in *A. thaliana* (Troufflard et al. 2010), and galanthamine in Narcissus bulbs (Lubbe and Verpoorte 2011). These findings suggest that K availability has crucial role in regulating the production of SMs in plants. Low K also induces oxidative stress owing to overproduction of reactive oxygen species (ROS), which damages cell membrane stability. Proline and ascorbate, which are important nonenzymatic antioxidants, have been found to mount up in plants against low K stress to protect plants from oxidative stress damage (Zeng et al. 2018).

Sulfur (S)

S is another essential mineral element and a central component of plant metabolism. Autotrophic organisms require sulfur for the biosynthesis of various PMs and SMs for their growth and development. The acquisition and assimilation of S in plants have received increased research interest, especially in the context of crop improvement. Any changes in S acquisition and assimilation can lead to drastic metabolic changes that affect sulfur-containing pools of both PMs and SMs.

Sulfur is an integral part of an array of plant metabolites including proteins and amino acids (cysteine, methionine), antioxidants (reduced glutathione), phytohormones, vitamins, sulfolipids, phytochelatins, iron-sulfur (Fe-S) clusters, glucosinolates, and cofactors (biotin and thiamine) having ability to potentially modify the physiological processes in plants (Takahashi et al. 2011; Asgher et al. 2014; Fatma et al. 2016). S application decreases the production of excess ROS and thus protects plants from oxidative damage (Fatma et al. 2014; Khan et al. 2015). Once absorbed by the roots, S becomes assimilated into some key organic metabolites, such as thiol (-SH) groups in cysteine or nonprotein thiol such as glutathione. Glutathione helps to remove excess ROS, which protects the cells from oxidative damage.

S limitation inhibits chlorophyll biosynthesis, Rubisco content, photosynthesis, N assimilation, and protein synthesis and eventually disturbs plant metabolism (Honsel et al. 2012). Lunde et al. (2008) reported that S limitation decreases the concentrations of glucose, fructose, and the glycolytic intermediate PEP but increases the content of starch, flavonoids, and anthocyanins. The consumption of carbohydrates was more strongly inhibited compared with photosynthesis in S-limited plants. The reduction in monosaccharides can be linked to the reduction in photosynthesis, and the increase in starch accumulation is considered a general response to nutrient deficiency. S-limitation has been shown to markedly decrease the concentration of lipids (especially sulfolipids), protein, RNA, S-adenosylmethionine (SAM), and chlorophyll (which requires SAM for biosynthesis), as well as the photosynthetic rate (Nikiforova et al. 2005). S deprivation decreases total glucosinolates and downregulates the genes involved in glucosinolate biosynthesis (Aghajanzadeh 2015). Indole-acetonitrile, a breakdown product of glucosinolate, is produced from Trp and it acts as a precursor in the biosynthesis of indole-3-acetic acid (auxin), which is also upregulated by S-deficiency (Kutz et al. 2002). The increase in the auxin level could explain the increase in lateral root growth in S-deficient plants.

Magnesium (Mg)

Low Mg availability negatively affects morphogenesis, development, and the productivity of plants. Mg is another essential macronutrient element required for plant growth and development. Mg has vital role in carbon fixation being a key constituent of chlorophyll molecule. Mg^{2+} holds central place in the chlorophyll molecule, surrounded by four N atoms from the porphyrin ring. The important regulatory role of Mg^{2+} in photosynthesis is well established. Changes in Mg^{2+} concentration in the chloroplast control the functioning of many important enzymes involved in photosynthesis. Mg^{2+} plays an important role in photosynthesis because it can promote light reactions operating in the stroma. Light triggers the entry of Mg^{2+} into the stroma of the chloroplast in exchange for H^+, thus establishing an optimum level for the carboxylase reaction. Mg^{2+} acts as a cofactor and activates the ribulose bisphosphate carboxylase (RuBP carboxylase,) and fructose-1,6-bisphosphatase. Thus, Mg^{2+} plays a fundamental role in CO_2 assimilation and related metabolic processes, such as sugar and starch production (Marschner 2012). Mg^{2+} also acts as a cofactor in almost all the enzymes activating phosphorylation processes. Mg^{2+} links the pyrophosphate structure of ATP or ADP to the enzyme molecule. Hence, the activation of ATPase by Mg^{2+} is brought about by this linking function (Balke and Hodges 1975). Mg^{2+} is also required for the activation of dehydrogenases and enolase, the aggregation of ribosomes, the modulation of ionic currents across the chloroplast, and vacuolar membranes (i.e. the organelle that regulates ion balance in the cell and stomatal opening (Maathuis and Diatloff 2013).

Mun et al. (2020) examined the effects of excess Mg on *Perilla frutescens* leaves and noted higher chlorophyll, carbohydrates (glucose, xylose, maltose, adonitol, xylitol, glycerol, gluconic acid, glyceric acid, maltose, and fructose), and amino acids (threonine, glycine, serine, and GABA) but low antioxidant metabolites (ABTS, DPPH, and FRAP). In contrast, tryptophan, sucrose, and some organic acids (oxoglutaric acid and lactic acid), flavonoids (apigenin-7-*O*-diglucuronide, apigenin-7-*O*-glucuronide, luteolin-7-*O*-diglucuronoide, apigenin-6,8-diglucoside, luteolin-7-*O*-glucoside, and liquiritigenin), sagerinic acid, cinnamic acid derivatives (dimethoxycinnamic acid, caffeic acid, and salvianolic acid C), and triterpenoids (tormentic acid and corosolic acid) were significantly reduced in Mg-oversupplied plants compared to control plants. In contrast, previous studies have shown that plant SMs (flavonoids and phenolic compounds) also serve the purpose of antioxidants and assist in reducing the oxidative damage induced by abiotic stress (Nakabayashi et al. 2014).

Calcium (Ca)

Ca is an abundant element existing in the lithosphere having a mean value of approximately 3.5% Ca. However, soils composed of chalk or limestone are usually rich in $CaCO_3$ (calcite) and possess 50% Ca. Plant roots absorb Ca in the form of divalent ion (Ca^{2+}) and account for approximately 0.5% of plant dry matter. Inside plant tissues, Ca^{2+} has a high affinity for plant metabolites with a negative charge. Therefore, Ca cannot be remobilized from old plant organs; hence, if Ca availability is low in the soil solution, the Ca concentration can decrease below critical levels and have negative effects on growth and productivity. A sharp decrease in Ca levels, especially in fast-growing tissues, can lead to diseases such as "blossom end rot" in tomatoes, "black heart" in celery, "tip burn" in lettuce, or "bitter-pit" in apples, even when plants are growing in the presence of sufficient amounts of Ca in soil (White 2000; Demidchik and Maathuis 2007; Bárcena et al. 2019).

Ca has a crucial role in cell division, structure and permeability of plasma membrane, cell transduction and elongation, translocation of carbohydrate, and nitrogen metabolism (White 2000; El-Beltagi and Mohamed 2013. Ca plays a role in maintaining the structure of the cell wall and membranes. It readily form complexes with several anionic substances such as the phosphates and carboxyls of phospholipids, sugars and proteins. For example, in plant cell walls, the cross-linking of pectins and glycans occurs through electrostatic coordination of Ca. Ca acts as "glue" that not only contributes to cell wall structure and strength but also decreases the risk of pathogen penetration (Marschner 1995; Maathuis and Diatloff 2013).

This complexation mainly occurs outside the cell membrane and requires a high concentration of apoplastic Ca. Replacement of Ca with other cations or elimination of apoplastic Ca can compromise membrane strength and lead to cellular electrolyte leakage. Since Ca can form insoluble salts with phosphates and sulfates easily, therefore concentration of free Ca in the cytoplasm is normally kept low nearly around ca. 100 nM. This makes Ca an ideal secondary messenger, and various stimuli have been shown to evoke rapid changes in cytosolic free Ca in plants, including responses to biotic and abiotic stress, stomatal regulation, and physical damage (Haynes 1980; Wallace and Mueller 1980). Ca also participates in photosynthetic water oxidation in PSII, which requires a metal cluster that contains a Ca ion (Hochmal et al. 2015). In contrast, optimal amounts of Ca have been shown to significantly improve the production of chlorophylls and carotenoids and maintain high photosynthetic efficiency and stomatal conductance (Elkelish et al. 2019). Ca also plays a role in polyphenolic metabolism. For example, Ca takes part in the biosynthesis of polyphenolic compounds. The application of Ca can increase the phenylalanine ammonia-lyase activity, which leads to the accumulation of polyphenols (Castaneda and Perez 1996; Juric et al. 2020).

In addition, Ca deprivation has been reported to inhibit the photosynthesis and N metabolism of spinach plants by reducing the content of chlorophyll and activities of Rubisco, GDH, GS, and GPT (Chao et al. 2008). Under Ca deficiency, plants face the problem of an enhanced electrolyte leakage rate and decreased AsA content (da Silva et al. 2021). The increase in malondialdehyde might enhance the activity of pectolytic enzymes (Olle and Bender 2009) and lipid peroxidation (Chao et al. 2009).

Boron (B)

B is an important trace mineral nutrient necessary for physiological processes in higher plants. B plays important roles in cell division and elongation, sugar transport, carbohydrate metabolism, N metabolism, cytoskeletal proteins, cell wall synthesis, and lignification, as well as in maintaining the integrity of bio-membranes. Ions (K^+, H^+, Ca^{2+}, KPO_4^{3-}, and Rb^+) are transported across the membranes, and plasmalemma-bound enzymes, polyamines, nucleic acids, ascorbic acid, and phytohormones (indoleacetic acid) are important for mediating ion transport (Shireen et al. 2018). Plant growth and metabolism are adversely affected due to B deficiency. B deficiency or toxicity usually leads to carbohydrate accumulation (hexose sugars (glucose, fructose) and starch contents become high, but the accumulation of sucrose is low. Ruuhola et al. (2011a) reported that B availability affects carbohydrate and phenol metabolism in summer leaves and autumn buds of one-year-old birch seedlings (*Betula pendula* Roth) when treated with three different B levels: B-0 (control), B-30 (low level), and B-100 (high level). B-100 in summer leaves increased the starch accumulation and caused a reduction in hexose sugars (glucose and fructose). In contrast, B-0 seedlings accumulated more hexose sugar (mainly glucose) and less starch content. The autumn buds of B-30 seedlings accumulated fewer sugars (glucose and raffinose) along with total polyols, mainly due to the deposition of other metabolites (i.e. phenols). Moreover, the highest accumulation of condensed tannins was detected in B-30. Similarly, severe boron limitation enhanced the level of hexoses, starch, and sucrose in tobacco (*Nicotiana tabacum* L.) (Camacho-Cristóbal and González-Fontes 1999; Camacho-Cristóbal et al. 2004). The accumulation of sugar in response to B availability might have occurred due to decrease in sink demand because of a reduction in growth (Han et al. 2008, 2009).

Some researchers have examined the B availability effects on polyol accumulation in leaves. Leite and Mather (2008) detected high amount of sorbitol, myo-inositol, mannitol, scyllo-inositol, and arabinose in the leaves of B-supplied *Eucalyptus grandis* seedlings compared with B-deprived seedlings. The phenolic content may constitute about 30–40% part of the total plant biomass (dry weight), which represents a major path to properly channelize the carbon (Ossipov et al. 2001). Phenolic metabolism is also involved in the production of different precursors needed for the biosynthesis of various phenolic metabolites through the shikimate pathway. Extreme B deprivation facilitates phenolic accumulation in different herbaceous species (Camacho-Cristóbal et al. 2002; Hajiboland and Farhanghi 2010). However, moderate B limitation did not affect the accumulation of phenolics in a woody plant (silver birch, Ruuhola et al. 2011b).

Chlorine (Cl)

Chloride (Cl) is an important mineral nutrient highly important for higher plants. It mainly occurs as Cl anion in soil and plants. Glycophytes usually contain 1–20 mg Cl g dw^{-1} (Marschner 2011). Although the demand for Cl for normal growth is low, Cl deprivation occurs when the Cl level is below 20 μmol kg^{-1} dw^{-1}, indicating that Cl is a micronutrient (Clarkson and Hanson 1980). However, the thresholds vary among crop plants. For example, the threshold level of barley (*Hordeum vulgare*) and wheat (*Triticum aestivum*) is below 1.2–4.0 mg g dw^{-1}; for lettuce (*Lactuca sativa*) and spinach, it is 0.14 mg g dw^{-1}; that for rice is 3 mg g shoot dw^{-1}; and that for maize (*Zea mays*) is 0.05–0.11 mg Cl g dw^{-1} (Xu et al. 2000)

Cl has been found in more than 130 organic compounds isolated from higher plants and ferns. It is actively involved in a variety of plant physio-biochemical processes such as osmotic adjustment and stomatal regulation, the oxygen evolution during photosynthetic process, and disease resistance and tolerance (Chen et al. 2010). Cl can also promote the evolution of oxygen and photophosphorylation (Bove et al. 1963). Cl is indirectly involved in stomatal regulation as opening and closing of guard cells are regulated by potassium and anions (Cl). Cl^- is a nonassimilating highly mobile anion, and therefore regulates the cell's osmolarity and the electrical charge balance of cations (Flowers 1988; Marschner 2012). Specifically, it can balance the electric charges of important cations, including calcium

(Ca^{2+}), potassium (K^+), and protons (H^+); Cl^- thus plays a critical role in stabilizing the electric potential of cell membranes and the regulation of pH gradients and electrical excitability. It also affects the absorption and utilization of other important plant nutrients, including nitrogen (N), phosphorus (P), potassium (K),magnesium (Mg), calcium (Ca), manganese (Mn), silicon (Si), zinc (Zn), sulfur (S), copper (Cu), and iron (Fe)in higher plants (Zhong and Ma 1993). Cl reduces NO_3^- uptake by the roots and facilitates the incorporation of N into organic compounds in the leaves of *Catharanthus rhoeus* (Flores et al. 2000). However, Cl can improve N assimilation and increase N use efficiency in tobacco plants (Rosales et al. 2020). Several enzymes are known to be stimulated by Cl (e.g. ATPase, α-Amylase, and asparagine synthetase, and it also promotes starch accumulation in potatoes (Fixen Paul 1993; Geilfus 2018).

Copper (Cu)

Although plants require Cu in low concentrations, it is critically important to plant life. Its concentrations in most plants generally range from 2 to 20 ppm in dry plant material. Two oxidation states Cu^{1+} (unstable) and Cu^{2+} of copper normally prevail that can be substituted with each other; therefore, Cu functions as a redox agent in biochemical reactions. Copper plays essential roles in photosynthesis and mitochondrial respiration, C/N metabolism, lignification, and protection against oxidative stress. Cu acts as cofactor for plastocyanin, cytochrome-c oxidase, copper/zinc superoxide dismutase (Cu/ZnSOD), apoplastic oxidases: ascorbate oxidase, polyphenol oxidase, and diamine oxidase (Yruela 2009). It is also part of the ethylene receptor (Rodriguez et al. 1999) and facilitates the biosynthesis of molybdenum cofactor (Kuper et al. 2004). In contrast, the excessive accumulation of Cu is potentially harmful and can increase malondialdehyde and thus oxidative stress by decreasing the concentration of antioxidant enzymes, including CAT, APX, and GPX, and negatively affecting mineral nutrition (Bankaji et al. 2015). It can also catalyze the synthesis of free radicals and thus damage DNA, proteins, and other biomolecules. Therefore, Cu ions are bound by scavenging proteins such as metallothioneins to prevent their accumulation (Hansch and Ralf 2009). Approximately 70% of the total copper in leaves is bound to various cell organelles/components. The copper bounded by plastocyanin (an electron carrier protein photosystem I) accounts for approximately 50% of plastidic copper (Pilon et al. 2006).

Iron (Fe)

Fe is the second most abundant element in the Earth's crust. Although it is hardly present in living matter (50–100 $\mu g \cdot g^{-1}$ dry matter), it is still an essential micronutrient that is critical for plant life because of its involvement in plant metabolism, including photosynthesis, respiration, and DNA synthesis, hormone biosynthesis (ethylene, gibberellic acid, and jasmonic acid), N assimilation, osmoprotection, the scavenging of reactive oxygen species, and pathogen defense (Hansch and Ralf 2009). Fe is an active cofactor of many enzymes that are important for the biosynthesis of plant hormones, including ethylene, lipoxygenase, and 1-aminocyclopropane acid-1-carboxylic oxidase (Siedow 1991). The activity of amylase reduces when a large amount of iron accumulates because iron forms complexes with carbohydrates such as maltose (Hofner 1970). A high degree of iron accumulation may lead to the iron-induced deprivation of other essential nutrients (e.g. Mn, P, K, Ca, and Mg) in plants (Howeler 1973; Ottow et al. 1982). The assimilation of NH_4^+ into amino acids occurs via the joint action of the GS/GOGAT cycle, which is a primary route of N assimilation in plants. Any reduction in the activities of these enzymes reduces the production of amino acids and thus N assimilation in plants. Amylase invertase, aspartate aminotransferase, and glutamate synthase have been reported to be significantly affected in tea plants under a surplus of Fe (Hemalatha and Venkatesan 2011). In addition, high Fe concentrations decreased the amount of photosynthetic pigments, such as chlorophyll and carotenoids. Monteiro and Winterbourn (1988) reported that high iron levels catalyze the generation of active O_2-free radical species, which may oxidize chlorophyll and subsequently reduce the chlorophyll content.

References

Aghajanzadeh, T. (2015). Sulfur metabolism, glucosinolates and fungal resistance in Brassica. University of Groningen.

Ahanger, M.A. and Agarwal, R.M. (2017a). Salinity stress induced alterations in antioxidant metabolism and nitrogen assimilation in wheat (*Triticum aestivum* L.) as influenced by potassium supplementation. *Plant Physiol. Biochem.* 115: 449–460.

Ahanger, M.A. and Agarwal, R.M. (2017b). Potassium improves antioxidant metabolism and alleviates growth inhibition under water and osmotic stress in wheat (*Triticum aestivum* L.). *Protoplasma* 254 (4): 1471–1486.

Ahanger, M.A., Tyagi, S.R., Wani, M.R., and Ahmad, P. (2014). Drought tolerance: role of organic osmolytes, growth regulators, and mineral nutrients. In: *Physiological Mechanisms and Adaptation Strategies in Plants Under Changing Environment* (ed. P. Ahmad and M. Wani), 25–55. New York: Springer.

Ahanger, M.A., Agarwal, R.M., Tomar, N.S., and Shrivastava, M. (2015). Potassium induces positive changes in nitrogen metabolism and antioxidant system of oat (*Avena sativa* L cultivar Kent). *J. Plant Interact.* 10 (1): 211–223.

Amtmann, A. and Armengaud, P. (2009). Effects of N, P, K and S on metabolism: new knowledge gained from multi-level analysis. *Curr. Opin. Plant Biol.* 12: 275–283.

Amtmann, A. and Blatt, M.R. (2009). Regulation of macronutrient transport. *New Phytol.* 181: 35–52.

Armengaud, P., Sulpice, R., Miller, A.J. et al. (2009). Multi-level analysis of primary metabolism provides new insights into the role of potassium nutrition for glycolysis and nitrogen assimilation in Arabidopsis roots. *Plant Physiol.* 150 (2): 772–785.

Asgher, M., Khan, N.A., Khan, M.I.R. et al. (2014). Ethylene production is associated with alleviation of cadmium- induced oxidative stress by sulfur in mustard types differing in ethylene sensitivity. *Ecotox. Environ. Safety* 106: 54–61.

Ashihara, H. and Stupavska, S. (1984). Comparison of activities and properties of pyrophosphate- and adenosine triphosphate-dependent phosphofructokinase of black gram (*Phaseolus mungo*) seeds. *J. Plant Physiol.* 116: 241–252.

Ashley, M.K., Grant, M., and Grabov, A. (2006). Plant responses to potassium deficiencies: a role for potassium transport proteins. *J. Exp. Bot.* 57 (2): 425–436.

Avigad, G. and Bohrer, P.J. (1984). 6-Phosphofructo-2-kinase and D-fructose-2,6-bisphosphatase activities in mung bean seedlings. *Biochim. Biophys. Acta* 798 (3): 317–324.

Balasundram, N., Sundram, K., and Samman, S. (2006). Phenolic compounds in plants and agri-industrial by-products: antioxidant activity, occurrence, and potential uses. *Food Chem.* 99 (1): 191–203.

Balke, N.E. and Hodges, T.K. (1975). Plasma membrane adenosine triphosphatase of oat roots: activation and inhibition by Mg^{2+} and ATP. *Plant Physiol.* 55 (1): 83–86.

Ballizany, W.L., Hofmann, R.W., Jahufer, M.Z.Z., and Barrett, B.A. (2012). Genotype×environment analysis of flavonoid accumulation and morphology in white clover under contrasting field conditions. *Field Crop Res.* 128: 156–166.

Bankaji, I., Caçador, I., and Sleimi, N. (2015). Physiological and biochemical responses of *Suaeda fruticosa* to cadmium and copper stresses: growth, nutrient uptake, antioxidant enzymes, phytochelatin, and glutathione levels. *Environ. Sci. Pollut. Res.* 22: 13058–13069.

Baranauskiene, R., Venskutonis, P.R., Viskelis, P., and Dambrauskiene, E. (2003). Influence of nitrogen fertilizers on the yield and composition of thyme (*Thymus vulgaris*). *J. Agric. Food Chem.* 51: 7751–7758.

Bárcena, A., Graciano, C., Luca, T. et al. (2019). Shade cloths and polyethylene covers have opposite effects on tip burn development in greenhouse grown lettuce. *Sci. Hortic.* 249: 93–99.

Bove, J.M., Bove, C., Whatley, F.R., and Arnon, D.I. (1963). Chloride requirement for oxygen evolution in photosynthesis. *Zashch Met.* 18b: 683–688.

Camacho-Cristóbal, J.J. and González-Fontes, A. (1999). Boron deficiency causes a drastic decrease in nitrate content and nitrate reductase activity, and increases the content of carbohydrates in leaves from tobacco plants. *Planta* 209 (4): 528–536.

Camacho-Cristóbal, J.J., Anzellotti, D., and González-Fontes, A. (2002). Changes in phenolic metabolism of tobacco plants during short-term boron deficiency. *Plant Physiol. Biochem.* 40 (12): 997–1002.

Camacho-Cristóbal, J.J., Lunar, L., Lafont, F. et al. (2004). Boron deficiency causes accumulation of chlorogenic acid and caffeoyl polyamine conjugates in tobacco leaves. *J. Plant Physiol.* 161 (7): 879–881.

Castaneda, P. and Perez, L.M. (1996). Calcium ions promote the response of citrus lemon against fungal elicitors or wounding. *Phytochemistry* 42: 595–598.

Chao, L., Bofu, P., Weiqian, C. et al. (2008). Influences of calcium deficiency and cerium on growth of spinach plants. *Biol. Trace Elem. Res.* 121 (3): 266–275.

Chao, L., Weiqian, C., Yun, L. et al. (2009). Cerium under calcium deficiency – influence on the antioxidative defense system in spinach plants. *Plant Soil* 323 (1): 285–294.

Chen, W., He, Z.L., Yang, X.E. et al. (2010). Chlorine nutrition of higher plants: progress and perspectives. *J. Plant Nutr.* 33 (7): 943–952.

Clarkson, D.T. and Hanson, J.B. (1980). The mineral nutrition of higher plants. *Annu. Rev. Plant Physiol.* 31: 239–232.

Coruzzi, G. and Bush, D.R. (2001). Nitrogen and carbon nutrient and metabolite signaling in plants. *Plant Physiol.* 125: 61–64.

Davenport, R. (2002). Glutamate receptors in plants. *Ann. Bot.* 90: 549–557.

Demidchik, V. and Maathuis, F.J.M. (2007). Physiological roles of nonselective cation channels in plants: from salt stress to signaling and development. *New Phytol.* 175: 387–404.

El-Beltagi, H.S. and Mohamed, H.I. (2013). Alleviation of cadmium toxicity in *Pisum sativum* L. seedlings by calcium chloride. *NotBot. Horti. Agrobot. Cluj-Napoca* 41: 157–168.

Elkelish, A.A., Alnusaire, T.S., Soliman, M.H. et al. (2019). Calcium availability regulates antioxidant system, physio-biochemical activities and alleviates salinity stress mediated oxidative damage in soybean seedlings. *J. Appl. Bot. Food Qual.* 92: 258–266.

Erel, R., Yermiyahu, U., Ben-Gal, A. et al. (2015). Modification of non-stomatal limitation and photo-protection due to K and Na nutrition of olive trees. *J. Plant Physiol.* 177: 1–10.

Evans, H.J. and Sorger, G.J. (1966). Role of mineral elements with emphasis on the univalent cations. *Annu. Rev. Plant Physiol.* 17: 47–76.

Fatma, M., Asgher, M., Masood, A., and Khan, N.A. (2014). Excess sulfur supplementation improves photosynthesis and growth in mustard under salt stress through increased production of glutathione. *Environ. Exp. Bot.* 107: 55–63.

Fatma, M., Masood, A., Per, T.S., and Khan, N.A. (2016). Nitric oxide alleviates salt stress inhibited photosynthetic performance by interacting with sulfur assimilation in mustard. *Front. Plant Sci.* 7: 521.

Fixen Paul, E. (1993). Crop responses to chloride. *Adv. Agron.* 50: 107–150.

Flores, P., Botella, M.A., Martínez, V., and Cedra, A. (2000). Ionic and osmotic effects on nitrate reductase activity in tomato seedlings. *J. Plant Physiol.* 156: 552–557.

Flowers, T.J. (1988). Chloride as a nutrient and as an osmoticum B. In: *Advances in Plant Nutrition* (ed. P.B. Tinker and A. Läuchli), 55–78. New York, USA: Praeger.

Geilfus, C.M. (2018). Review on the significance of chlorine for crop yield and quality. *Plant Sci.* 270: 114–122.

Hajiboland, R. and Farhanghi, F. (2010). Remobilization of boron, photosynthesis, phenolic metabolism and antioxidant defense capacity in boron-deficient turnip (*Brassica rapa* L.) plants. *Soil Sci. Plant Nutr.* 56: 427–437.

Han, S., Chen, L.S., Jiang, H.X. et al. (2008). Boron deficiency decreases growth and photosynthesis, and increases starch and hexoses in leaves of citrus seedlings. *J. Plant Physiol.* 165 (13): 1331–1341.

Han, S., Tang, N., Jiang, H.X. et al. (2009). CO_2 assimilation, photosystem II photochemistry, carbohydrate metabolism and antioxidant system of citrus leaves in response to boron stress. *Plant Sci.* 176 (1): 143–153.

Hansch, R. and Ralf, R.M. (2009). Physiological functions of mineral micronutrients (Cu, Zn, Mn, Fe, Ni, Mo, B, Cl). *Curr. Opin. Plant Biol.* 12: 259–266.

Haynes, R.J. (1980). Ion exchange properties of roots and ionic interactions within the root apoplasm: their role in ion accumulation by plants. *Bot. Rev.* 46: 75–99.

Hemalatha, K. and Venkatesan, S. (2011). Impact of iron toxicity on certain enzymes and biochemical parameters of tea. *Asian J. Biochem.* 6: 384–394.

Hernandez, G., Valdes-Lopez, O., Ramirez, M. et al. (2009). Global changes in the transcript and metabolic profiles during symbiotic nitrogen fixation in phosphorus stressed common bean plants. *Plant Physiol.* 151: 1221–1238.

Hochmal, A.K., Schulze, S., Trompelt, K., and Hippler, M. (2015). Calcium-dependent regulation of photosynthesis. *Biochim. Biophys. Acta* 1847: 993–1003.

Hofner, W. (1970). Iron and manganese compounds in the Blutungssaft of *Helianthus annuus*. *Physiol. Plant.* 23: 673–677.

Honsel, A., Kojima, M., Haas, R. et al. (2012). Sulfur limitation and early sulfur deficiency responses in poplar: significance of gene expression, metabolites, and plant hormones. *J. Exp. Bot.* 63: 1873–1893.

Howeler, R.H. (1973). Iron induced ringing disease of rice in relation to physicochemical changes in a flooded soil. *Soil Sci. Soc. Am. Proc.* 37: 898–903.

Hurry, V., Strand, A., Furbank, R., and Stitt, M. (2001). The role of inorganic phosphate in the development of freezing tolerance and the acclimatization of photosynthesis to low temperature is revealed by the pho mutants of *Arabidopsis thaliana*. *Plant J.* 24: 383–396.

Jin, S.H., Huang, J.Q., Li, X.Q. et al. (2011). Effects of potassium supply on limitations of photosynthesis by mesophyll diffusion conductance in *Carya cathayensis*. *Tree Physiol.* 31 (10): 1142–1151.

Juric, S., Stracenski, K.S., Król-Kilińska, Ż. et al. (2020). The enhancement of plant secondary metabolites content in *Lactuca sativa* L. by encapsulated bioactive agents. *Sci. Rep.* 10 (1): 1–12.

Khan, M.I.R., Nazir, F., Asgher, M. et al. (2015). Selenium and sulfur influence ethylene formation and alleviate cadmium-induced oxidative stress by improving proline and glutathione production in wheat. *J. Plant Physiol.* 173: 9–18.

Kováčik, J. and Klejdus, B. (2013). Induction of phenolic metabolites and physiological changes in chamomile plants in relation to nitrogen nutrition. *Food Chem.* 1 (142): 334–341.

Kováčik, J., Klejdus, B., Štork, F., and Hedbavny, J. (2011). Nitrate deficiency reduces cadmium and nickel accumulation in chamomile plants. *J. Agric. Food Chem.* 59: 5139–5149.

Kuper, J., Llamas, A., Hecht, H.J. et al. (2004). Structure of the molybdopterin-bound Cnx1G domain links molybdenum and copper metabolism. *Nature* 430: 803–806.

Kutz, A., Muller, A., Hennig, P. et al. (2002). A role for nitrilase 3 in the regulation of root morphology in sulphur-starving *Arabidopsis thaliana*. *Plant J.* 30: 95–106.

Lacombe, B., Becker, D., Hedrich, R. et al. (2001). The identity of plant glutamate receptors. *Science* 292: 1486–1487.

Lehman, R.H. and Rice, E.L. (1972). Effects of deficiencies of nitrogen, potassium, and sulfur on chlorogenic acids and scopolin in sunflower. *Am. Midl. Nat.* 87: 71–80.

Leigh, R.A. and Wyn Jones, R.G. (1984). A hypothesis relating critical potassium concentrations for growth to the distribution and functions of this ion in the plant-cell. *New Phytol.* 1984 (97): 1–13.

Leite, T.S. and Mather, J.A. (2008). A new approach to octopuses' body pattern analysis: a framework for taxonomy and behavioral studies. *Am. Malacol. Bull.* 24 (1): 31–41.

Lubbe, A. and Verpoorte, R. (2011). Cultivation of medicinal and aromatic plants for specialty industrial materials. *Ind. Crop. Prod.* 34 (1): 785–801.

Lunde, C., Zygadlo, A., Simonsen, H.T. et al. (2008). Sulfur starvation in rice: the effect on photosynthesis, carbohydrate metabolism, and oxidative stress protective pathways. *Physiol. Plant.* 134: 508–521.

Maathuis, F.J. (2009). Physiological functions of mineral macronutrients. *Curr. Opin. Plant Biol.* 12 (3): 250–258.

Maathuis, F.J. and Diatloff, E. (2013). Roles and functions of plant mineral nutrients. *J. Plant Nutr.* 953: 1–21.

del Mar Rubio-Wilhelmi, M., Sanchez-Rodriguez, E., Rosales, M.A. et al. (2012a). Ammonium formation and assimilation in PSARK: : IPT tobacco transgenic plants under low N. *J. Plant Physiol.* 169 (2): 157–162.

del Mar Rubio-Wilhelmi, M., Sanchez-Rodriguez, E., Leyva, R. et al. (2012b). Response of carbon and nitrogen-rich metabolites to nitrogen deficiency in PSARK: : IPT tobacco plants. *Plant Physiol. Biochem.* 57: 231–237.

Marschner, H. (1995). *Mineral Nutrition in Higher Plants*. London: Academic Press.

Marschner, H. (2011). *Marschner's Mineral Nutrition of Higher Plants*, 3e. Academic Press.

Marschner, H. (2012). *Mineral Nutrition of Higher Plants*, 3e. London, UK: Academic Press.

Mazid, M., Khan, T.A., and Mohammad, F. (2011). Role of secondary metabolites in defense mechanisms of plants. *Biol. Med.* 3 (2): 232–249.

Monteiro, H.P. and Winterbourn, C.C. (1988). The superoxide dependant transfer of iron from ferritin to transferring and lactoferrin. *Biochem. J.* 256: 923–928.

Morcuende, R., Bari, R., Gibon, Y. et al. (2007). Genome-wide reprogramming of metabolism and regulatory networks of Arabidopsis in response to phosphorus. *Plant Cell Environ.* 30: 85–112.

Mun, H.I., Kim, Y.X., Suh, D.H. et al. (2020). Metabolomic response of *Perilla frutescens* leaves, an edible-medicinal herb, to acclimatize magnesium oversupply. *PLoS One* 15 (7): e0236813.

Nakabayashi, R., Yonekura-Sakakibara, K., Urano, K. et al. (2014). Enhancement of oxidative and drought tolerance in Arabidopsis by over accumulation of antioxidant flavonoids. *Plant J.* 77 (3): 367–379.

Nguyen, P.M., Kwee, E.M., and Niemeyer, E.D. (2010). Potassium rate alters the antioxidant capacity and phenolic concentration of basil (*Ocimum basilicum* L.) leaves. *Food Chem.* 123 (4): 1235–1241.

Nikiforova, V.J., Kopka, J., Tolstikov, V. et al. (2005). Systems rebalancing of metabolism in response to sulfur deprivation, as revealed by metabolome analysis of Arabidopsis plants. *Plant Physiol.* 138: 304–318.

Olle, M. and Bender, I. (2009). Causes and control of calcium deficiency disorders in vegetables: a review. *J. Hortic. Sci. Biotechnol.* 84: 577–584.

Ossipov, V., Haukioja, E., Ossipova, S. et al. (2001). Phenolic and phenolic-related factors as determinants of suitability of mountain birch leaves to an herbivorous insect. *Biochem. Syst. Ecol.* 29: 223–240.

Ottow, J.C.G., Benckiser, G., Santiago, S., and Watanabe, I. (1982). Iron toxicity of wetland rice (*Oryza sativa* L.) as a multiple nutritional stress. *Plant Nutr* 2: 454–460.

Pant, B.D., Pant, P., Erban, A. et al. (2015). Identification of primary and secondary metabolites with phosphorus status-dependent abundance in Arabidopsis, and of the transcription factor PHR 1 as a major regulator of metabolic changes during phosphorus limitation. *Plant Cell Environ.* 38 (1): 172–187.

Paul, M.J. and Driscoll, S.P. (1997). Sugar repression of photosynthesis: the role of carbohydrates in signaling nitrogen defciency through source: sink imbalance. *Plant Cell Environ.* 20: 110–116.

Pilon, M., Abdel-Ghany, S.E., Cohu, C.M. et al. (2006). Copper cofactor delivery in plant cells. *Curr. Opin. Plant Biol.* 9: 256–263.

Rai, A., Saito, K., and Yamazaki, M. (2017). Integrated omics analysis of specialized metabolism in medicinal plants. *Plant J.* 90: 764–787.

Rodriguez, F.I., Esch, J.J., Hall, A.E. et al. (1999). A copper cofactor for the ethylene receptor ETR1 from Arabidopsis. *Science* 283: 996–998.

Rosales, M.A., Franco-Navarro, J.D., Peinado-Torrubia, P. et al. (2020). Chloride improves nitrate utilization and NUE in plants. *Front. Plant Sci.* 11: 442.

Ruuhola, T., Leppänen, T., and Lehto, T. (2011a). Retranslocation of nutrients in relation to boron availability during senescence of Betula pendula Roth. *Plant Soil* 344: 227–240.

Ruuhola, T., Keinänen, M., Keski-Saari, S., and Lehto, T. (2011b). Boron nutrition affects the carbon metabolism of silver birch seedlings. *Tree Physiol.* 31 (11): 1251–1261.

Scheible, W.R., Morcuende, R., Czechowski, T. et al. (2004). Genome-wide reprogramming of primary and secondary metabolism, protein synthesis, cellular growth processes, and the regulatory infrastructure of Arabidopsis in response to nitrogen. *Plant Physiol.* 136 (1): 2483–2499.

Secco, D., Jabnoune, M., Walker, H. et al. (2013). Spatio-temporal transcript profiling of rice roots and shoots in response to phosphate starvation and recovery. *Plant Cell* 25: 4285–4304.

Sharma, G.L., Agarwal, R.M., and Singh, R.P. (2006). Potassium induced changes in certain aspects of nitrogen metabolism in chickpea (*Cicer arietinum* L.). *Physiol. Mol. Biol. Plants* 12: 157–162.

Shireen, F., Nawaz, M.A., Chen, C. et al. (2018). Boron: functions and approaches to enhance its availability in plants for sustainable agriculture. *Int. J. Mol. Sci.* 19 (7): 1856.

Shiwakoti, S., Shannon, D.A., Wood, C.W. et al. (2016). Nitrogen, phosphorus, and potassium effects on biomass yield and flavonoid content of American skullcap (*Scutellaria lateriflora*). *J. Plant Nutr.* 39 (9): 1240–1249.

Siedow, J.N. (1991). Plant lipoxygenase: structure and function. *Ann. Rev. Biol.* 42: 145–188.

da Silva, D.L., de Mello, P.R., Tenesaca, L. et al. (2021). Silicon attenuates calcium deficiency by increasing ascorbic acid content, growth and quality of cabbage leaves. *Sci. Rep.* 11 (1): 1770.

Soetan, K.O., Olaiya, C.O., and Oyewole, O.E. (2010). The importance of mineral elements for humans, domestic animals and plants: a review. *Afr. J. Food Sci.* 4 (5): 200–222.

Sorger, G.J., Ford, R.E., and Evans, H.J. (1965). Effects of univalent cations on the immunoelectrophoretic behavior of pyruvic kinase. *Proc. Natl. Acad. Sci. U. S. A.* 54: 1614–1621.

Strissel, T., Halbwirth, H., Hoyer, U. et al. (2005). Growth-promoting nitrogen nutrition affects flavonoid biosynthesis in young apple (*Malus domestica* Borkh) leaves. *Plant Biol.* 7 (6): 677–685.

Takahashi, H., Kopriva, S., Giordano, M. et al. (2011). Sulfur assimilation in photosynthetic organisms: molecular functions and regulations of transporters and assimilatory enzymes. *Annu. Rev. Plant Biol.* 62: 157–184.

Tiwari, H.S., Agarwal, R.M., and Bhatt, R.K. (1998). Photosynthesis, stomatal resistance and related characteristics, as influenced by potassium under normal water supply and water stress conditions in rice (*Oryza sativa* L.). *Indian J. Plant Physiol.* 3 (4): 314–316.

Troufflard, S., Mullen, W., Larson, T.R. et al. (2010). Potassium deficiency induces the biosynthesis of oxylipins and glucosinolates in *Arabidopsis thaliana*. *BMC Plant Biol.* 10 (1): 1–13.

Wallace, A. and Mueller, R.T. (1980). Calcium uptake and distribution in plants. *J. Plant Nutr.* 2: 247–256.

White, P.J. (2000). Calcium channels in higher plants. Biochi. *Biophys. Acta Biomembr.* 146: 171–189.

Jones, W. and Pollard, A. (1983). Proteins, enzymes and inorganic ions. In: *Inorganic Plant Nutrition* (ed. A. Luchli and R.L. Bieleski), 528–562. New York: Springer.

Xu, G., Magen, H., Tarchitzky, J., and Kafkafi, U. (2000). Advances in chloride nutrition of plants. *Adv. Agron.* 68: 97–150.

Yruela, I. (2009). Copper in plants: acquisition, transport and interactions. *Funct. Plant Biol.* 36 (5): 409–430.

Yuan, H. and Liu, D. (2008). Signaling components involved in plant responses to phosphate starvation. *J. Integr. Plant Biol.* 50: 849–859.

Zeng, J., Quan, X., He, X. et al. (2018). Root and leaf metabolite profiles analysis reveals the adaptive strategies to low potassium stress in barley. *BMC Plant Biol.* 18 (1): 1–13.

Zhong, H. and Ma, G.R. (1993). The influence of chlorine on physiological effects of potato. *J Zhejiang Agric* 5: 83–88.

Zorb, C., Senbayram, M., and Peiter, E. (2014). Potassium in agriculture? Status and perspectives. *J. Plant Physiol.* 171: 656–669.

2

Agricultural Production Relation with Nutrient Applications

Sehrish Sadia[1], *Muhammad Zubair*[2], *Akash Tariq*[3,4,5,6], *Fanjiang Zeng*[3,4,5], *Corina Graciano*[7], *Abd Ullah*[3,4,5,6], *Zeeshan Ahmed*[3,4,5,6], *Zhihao Zhang*[3,4,5,6], *and Khasan Ismoilov*[6,8]

[1]*Department of Biological Sciences, University of Veterinary and Animal Sciences-Lahore, Pattoki, Pakistan*
[2]*Discipline of Zoology, University of Veterinary and Animal Sciences-Lahore, Pakistan*
[3]*Xinjiang Key Laboratory of Desert Plant Roots Ecology and Vegetation Restoration, Xinjiang Institute of Ecology and Geography, Chinese Academy of Sciences, Urumqi, China*
[4]*State Key Laboratory of Desert and Oasis Ecology, Xinjiang Institute of Ecology and Geography, Chinese Academy of Sciences, Urumqi, China*
[5]*Cele National Station of Observation and Research for Desert-Grassland Ecosystems, Cele, China*
[6]*University of Chinese Academy of Sciences, Beijing, China*
[7]*Instituto de Fisiología Vegetal, Consejo Nacional de Investigaciones Científicas y Técnicas, Universidad Nacional de La Plata, Buenos Aires, Argentina*
[8]*CAS Key Laboratory of Biogeography and Bioresource in Arid Land, Chinese Academy of Sciences, Urumqi, China*

Introduction

Agricultural improvements play a significant part in improving food nutrition and security by expanding the variety and amount of food, as an economic change driver, and on the fact that agriculture is the principle kind of revenue for a larger part of individuals who live in high limit scarcity. Acquiring adequate pay from agriculture is essential and vital for the 1.3 billion individuals who work in the area and straightforwardly decides their food security and nutrition. Broad experience across numerous nations over numerous years indicates that all agricultural turn of events and economy-wide development are required to improve food nutrition and security. Food frameworks "comprise of the multiple of components" (climate, individuals, inputs, methods, foundations, organizations, etc.) and processes that connect with the generation, handling, dissemination, processing, arrangement and utilization of food, and the results of these actions and measures (Caron et al. 2018). Agricultural turn of events, since the Second World War, has empowered magnificent advancement in food creation. This was for the most part because of an amalgamation of financial support and development, good progress in innovation and information, and better administration and management along with supply and stock chains. This expanded creation has for the most part happened under strengthening, differentiations, and levels of economies that rely

Plant Ionomics: Sensing, Signaling, and Regulation, First Edition.
Edited by Vijay Pratap Singh and Manzer H. Siddiqui.

progressively upon inputs in particularly creature feed and noninexhaustible fountains of energy. The inquiries of "sustainability, manageability, and agricultural advancement" are especially unpredictable in the view of fact that as the greater part of the subjects tended to up to this point by the HLPE, they demand a drawn out coordinated point of view (Agarwal 2014).

The fundamental difficulties confronting agribusiness organizers and cultivating chiefs in the coming many years' lies in vision world without zero neediness and yearning strengthened by increasing ways of life of rustic areas, where most of the needy people live and their full reliance on farming for living to accomplish their food requirements (Wheeler and Von Braun 2013). The key to remove and wipe out the present sufferings lies in making of a valuable arrangement and scheme that increase prosperous cultivating and empowers peasants and ranchers themselves to accomplish agricultural development, minimize destitution, and also maintain exceptional yields and earnings. Since antiquated occasions, farmers and ranchers have realized that soil and earth fitness and health can be re-established by applying natural fertilizer; appropriately, they employed and utilized to apply barnyard compost consistently and straightforwardly after crop collect and yield (Selim 2018). Subsequently, the custom of utilizing natural composts since yield and harvesting has been combined with rehabilitation and recovery of soil fitness and well-being and improvement of chemical, biological, and physical substance, and natural properties of soil, especially in minimal soils, which are as of now experiencing low natural matter and low local supplement content, low efficiency, and also restrictions and inaccessibility of fundamental supplements (Adeoye et al. 2011). Furthermore, the natural cultivating system gives natural food and food stuff, which is liked by numerous clients in any case of best value. In this specific situation, natural manures, as a result of sluggish delivery, have more prominent impact on resulting crops than inorganic supplement and nutrients, which is immediately lost by water draining and overflow to underground water (Khoshgoftarmanesh et al. 2010). Mineral sustenance alongside accessibility of water, control and management of illnesses, creepy crawlies, grass, weeds, and financial states of the rancher plays and perform a significant contribution in the expanding of crop profitability and productivity. In soil, supplements and nutrients levels and concentrations have been of fascinate interest for a long time as indicator and index of soil richness, fertility, and prolificacy in agriculture. Mineral nourishment alludes and applies to the stock, accessibility, supply, ingestion, movement, and use of inorganically shaped components for development, growth, and improvement of graze and harvest plants (Weil et al. 2003). To accomplish food creation at an ideal level, the utilization of synthetic manures and enhancements in soil prolificacy and richness are basic systems and strategies. It is assessed that 60% of developed and cultivated richness of soils have supplement inadequacy and deficiency or basic poisonousness and toxicity issues and that about half of the total population experiences micronutrient defects and short comings (Cakmak et al. 1995). Consequently, it is assessed and measured that to complete and meet requirements of food and foodstuff in future, and the complete utilization of composts and fertilizations will rise from 133 million tons each year in 1993 to around 200 million tons each year by 2030 (FAO 2000–2022). This situation makes plant victuals, nutrition, and sustenance research, a main concern in farming and agricultural science to overcome and fulfill food quality need in these 1000 years. Public interest and considerations about natural quality and prolonged time productivity and profitability of agro ecosystems has underlined and highlighted the need to create and carry out administration procedures that keep up soil fertility and richness at a sufficient level excluding deteriorate and debasing soil and water reserves and fortunes (Fageria et al. 1997). A large portion of the fundamental plant supplements are crucial and fundamental for human fitness, strength, and well-being and also for livestock and domesticated animal's creation and production.

The goal, intention, and objective of this basic introductory chapter is to give data, facts, information, and relation of significance of minerals and nutrients in expanding crop yields, supplement accessibility and necessities, and graze and harvest classification systems and to review and examine about yield and yield elements for enhancing harvest yields. This data may also support and help in better understanding of relation of nutrients for better agriculture yield.

Soil as a Basic Element in Agriculture

Through photosynthesis plants produce different economically valuable products, i.e. tuber, grain, fiber, vegetables and fruits by changing light energy into biomass. For this purpose, plants require adequate temperature and appropriate light, oxygen, carbon dioxide and various supplements. So, plant produces products that are most crucial and valuable for the endurance of animals as well as human beings, which thus relies intensely upon the accessibility of minerals and different supplements. For proper development and growth, plants also require and demand nutrients and minerals like other organisms. Supplements are essential as plant elements for biochemical responses and for the creation of natural stuff and materials alluded as photosynthates (fats, starches, vitamins, proteins, and nutrients) by the process of photosynthesis. Worldwide food creation and output have to be expanded at the rate of 70% at the ludicrous level by 2050 to fulfill and meet the expansion in food requests and requirements, which quickly enhances because of overcrowding overpopulation (Bruinsma 2009). In horticulture and agriculture, ideal harvest sustenance is a significant and essential necessary conditions for acquiring significant returns and great quality produce. The supplements needed are collected by plants both from soil stores and outside supplement resources (including composts, natural excrements, the environment, etc.). To accomplish these challenging objectives and targets, horticulture and agriculture should develop fundamentally in light of the elements that add and contribute to enhancing the yield creation and production rate (Ali et al. 2010).

Constituents and Ingredients of Soil

A soil takes the form of mineral substances, pore space, and natural and matter substances, which is communal by means of water, air, and living things (Table 2.1). Furthermore, the above components, the soil moreover includes an enormous and differed inhabitants of macro-organisms and micro-organisms that assume and perform a significant part and contribution in plant sustenance and nourishment.

Table 2.1 FAO/UNESCO categorized major soil groups.

Sr No	Variety and character of soil	UNESCO/FAO
1	**Arid region soil formed by vegetation/climate change**	
	Carbonate aggregated soil	Calcisols
	Soluble salts accumulated saline soil	Solonchaks
	Gypsum aggregated soil	Gypsisols
	Na accumulated sodic soil	Solonetz
2	**Sub humid regions formed soil by vegetation/climate change**	
	Ash gray-layered acidic soil	Podzol
	Clay transfer, vile rich brown soil	Luvisols
	Low-lying unwell and peaked soil	Planosols
	Clay transfer, base rich and brown soil	Luvisols
3	**Topographically formed mineral soil**	
	Deep seated and unwell developed soils in elevated sectors	Regosols
	Waterlogged soils	Gleysols
	Mountainous soil	Leptosols
	Lowlands level soil	Fluvisols
4	**Humid tropic regions formed mineral soils**	
	Heavily weathered soil	Nitisols
	Deeply weathered loams	Ferralsols
	Strongly acidic and leached soil	Acrisols
	Strongly weathered and hard soil	Plinthosols
5	**Steppe region formed mineral soils by climate change**	
	High humus filled soil	Greyzem
	Dark and black earths top soil	Chernozems
	Prairie regions soil	Phaeozems
	Brown color chestnut soil	Kastanozems

Source: Shand (2007)/FAO.

Agribusiness and agriculture production is a powerful and dynamic area that is influenced by numerous inconstant and changeable elements, such as diverse agro ecological sectors and kinds of soil surface, trailed by cultivating practices, trainings, advancements, and other related and different factors in commodity and product markets. However, crops have diverse compost requirements and needs and distinctive usage economies and efficiencies, as result of which a crucial and critical focus and preference should be done to check it out at the fertilization schemes and programs when deciding and arranging new soil recovery programs and plans (Fixen et al. 2015).

Essential Nutrients in Agriculture Especially in Plants

In compliance with Stout and Arnon in 1939, an aggregate of just 16 components are fundamental for the growth, development, and full improvement of higher green plants. They laid down a criteria for essential nutrients efficacy and needs in crop production.

- An inadequacy of a fundamental supplement enables and makes it incomprehensible for the plant to complete and carry out the reproductive, vegetative, or regenerative phase of its life span.
- The component is included straight forwardly in the nourishment of the plant very separated from its potential impacts in remedying some negative chemical, microbiological, or compound state of the soil or other growing and culture medium.
- That kind of shortcoming and inadequacy is précised and specified to the component being referred to and can be forestalled or rectified exclusively by providing this component for better growth and development.

The centrality and fundamentality of most micronutrients for larger and higher plants was set up somewhere in the range of 1922 and 1954. Fundamental and essential supplements and nutrients might be outlined as those excluding which plants cannot carry out and finish their life cycle, and also are indispensable by different components, and are straightforwardly associated with plant digestion and metabolism (Table 2.2). Nevertheless, this catalog and rundown may not be deemed and considered as last and ultimate, and it is likely that more components may end up being fundamental for the future (Rice 2007).

Table 2.2 Essential nutrients concentrations and forms taken up by plants.

Nutrients	Symbol	Absorbed form	Concentration	Established by
Macronutrients				
Calcium	Ca	Ca^{2+}	0.2–1.0%	Sprengel (1839)
Potassium	K	K^{2}	1–54%	Sprengel (1839)
Nitrogen	N	NH_4^+	1.5%	de Saussure (1804)
Magnesium	Mg	Mg^{2+}	0.1–0.4%	Sprengel (1839)
Sulphur	S	SO_4^{2-}	0.1–0.4%	Salm-Horstman (1851)
Phosphorus	P	H_2PO^-, HPO_4^{-2}	0.1–0.4%	Sprengel (1839)
Micronutrients				
Chlorine	Cl	Cl^-	0.2–2%	Lipman et al. (1931)
Zinc	Zn	Zn^{2+}	21–150 ppm	Lipman et al. (1931)
Manganese	Mn	Mn^{2+}	20–500 ppm	McHargue (1922)
Boron	Br	H_3BO_3, H_2BO_3	6–60 (ppm^2)	Warrington (1923)
Copper	Cu	Cu^+, Cu^{2+}	5–20 ppm	Lipman et al. (1931)
Iron	Fe	Fe^{2+}	50–250 ppm	Gris (1844)

Based on these 16 components, oxygen and carbon are received from water and carbon gas, i.e. hydrogen from water and carbon from CO_2 gas. These three components are needed and acquired in huge amounts for the production of constituents of plants, for example starch and cellulose. The alternative 13 components are called mineral supplements since they are absorbed in inorganic form (mineral structures).

Beneficial/Valuable Nutrients

For proper growth and development of crops, not essential nutrients but also some beneficial nutrients are most important and crucial as a whole. A portion of these supplements can be of extraordinary valuable and significant and may be vital outside expansion.

- Sodium: Replace and interchange K for beets.
- Aluminum: Valuable and significant for tea plants.
- Silicon: Play vital role in cereals stalk stability.
- Cobalt: Significant for legumes in nitrogen fixation.
- Nickel: An important part of urease enzyme and help in breaking of sol urea.

Some Other Valuable Nutrients

As domestic animals and humans need a few supplements likewise needed by plants, these extra supplements should be deemed in food or on the other hand feed creation, and their inadequacies rectified by suitable means of productions. Furthermore, to plant supplements, the fundamental components for domestic animals and humans are Iodine, cobalt, chromium, and selenium (Shand 2007).

Plant Nutrient Sources

Plants required different nutrients in various forms with different interval of times. Nature and supply of nutrients is directly associated with plant nature and location. Nutrients play a vital and crucial role and impact in proper growth and development of plants.

Plant Nutrients Supply and Nature

For proper growth and development of plant acquired nutrients from water and soil, energy from sun, oxygen, and carbon dioxide from water as well from air and build up and increase from water. For optimum development and growth of plants, nutrients must be present and available in the form of.

- In a form that is easily available and approachable.
- In sufficient, adequate, and well-balanced form.
- The sources of nutrients that are provided by plants include
- Mineral fertilizers
- Groundwater
- Rain or air deposited aerial deposition

- Soil reservoirs
- Organic sources

Some other nutrient sources have been explained in the following paragraphs.

Compost

Act as a soil conditioner and provided as a good nutrient for growth and development. In agriculture, it has a great value. Quality and the utilization process of compost vary from other nutrients.

Biosolids

The wastewater residuals in urban areas is called biosolids. Biosolids are also utilized as nutrients for proper growth and development of crops after their recycle process. Supplements in biosolids differ in amount and structures, depending upon the supply, source, processing, and treatment, stockpiling, and taking care of measures. Their quantity and content in plant supplements and in potential foreign matter have to be routinely investigated.

Manure of Livestock

It is more significant and valuable nutrient for crop proper growth and development. Supplement value of manure differs broadly between utilization, management in farm, and its source. It is generally perceived that low-quality feed for animals brings about manure with low supplement substances (Table 2.3).

Crop Residues

Like roots, leaves furthermore, stems discharge the supplements when they left in/on the soil contents. K is present in high quantity in roots and leaves. Therefore, for many years, it has become the chief source of supplements of good crop production. But if these crop residues are burned then the level and quantity of K reduces at lower level

Table 2.3 General supplement content qualities (g/kg) of crop yield, animals and poultry manures.

Sr No	Supplement	Manure of livestock (g kg^{-1})	Manure of poultry (g kg^{-1})	Remnants of crop (g kg^{-1})
1	S	4–50	5–15	1–2
2	Ca	5–20	40–45	2–5
3	P	4–10	20–25	1–2
4	Mg	3–4	6–8	1–3
5	K	15–20	11–20	10–15
6	N	20–30	25–30	10–15

Source: Adapted from Lam et al. (2000).

due to which it not gives proper results for the growth and development. Crop nutrients and residues supply directly depends upon plant type.

Atmospheric Deposition

For S and N, atmospheric deposition is most valuable and significant in some zones. In developing and emerging countries, S fertilizers are in common practices. S and N fertilizers show rapid, considerable, and valuable results.

Synthetic Fertilizers

Fertilizer industries produces these synthetic fertilizers. These fertilizers are composed of essential and other valuable nutrients which that are more and more valuable for outgrowth and development of crop production (Table 2.4). According to one observation, an average of about 180 million tons of fertilizers are employed by farmers in order to improve the quality, quality goals, and sustainable yield of their crops. Synthetic fertilizers are simple that contain some supplements, while complex fertilizers are composed of several nutrients for good outgrowth. At the industrial scale, these mineral and synthetic fertilizers are manufactured in solid or liquid form usually. Complex or high grade fertilizers are composed of almost 82% of supplements for suitable outgrowth.

Issues Related to Plant Nutrition

It is critical to recognize that reducing nutrient intake when treating all fields uniformly is inefficient, since nutrient surplus and undersupply can coexist. Only balanced fertilization that is tailored to the spatial heterogeneity of soil features has the potential to effectively minimize dispersed nutrient losses (Haneklaus and Schnug 1999).

Table 2.4 Normal supplement substance of some significant fertilizers.

Name	State of matter	K_2O	N	P_2O_5	S
Potassium sulfate	Solid	50	0	0	18
Ammonia	Gas	0	82	0	0
Triple superphosphate	Solid	0	0	46	0
Ammonium sulfate	Solid	0	21	0	24
Ground rock phosphate	Solid	0	0	20–40	0
Calcium ammonium nitrate	Solid	0	20.4–27.0	0	0
Potassium nitrate	Solid	44	13	0	0
Urea	Solid	45–46	0	0	0
Urea ammonium nitrate	Liquid	0	28–32	0	0
Potassium chloride	Solid	60	0	0	0
Diammonium phosphate	Solid	0	18	46	0

When society redefines plant nutrition challenges, new applications for a simple understanding of nutrient usage efficiency in soil-plant processes emerge, such as avoiding waste on abundant soils like those found in the temperate zone and making the most of access-limited environments like those found in the tropics. Increasing the breadth of the domain between the access and excess frontiers, rather than defining a single "economic optimum" limit, could be the key challenge of plant nutrition. To broaden this domain, two methods are discussed: a scientific model of precision farming and an ecological analogue approach focused on filter functions and part complementarity in mixed plant systems. Current plant nutrition knowledge, which is primarily based on monoculture conditions, needs to be supplemented by relationships that exist in more complex processes, such as agroforestry and intercropping, as these can be part of the solution in both abundance and scarcity situations. Simulations using the WaNuLCAS model to investigate the idea of a "safety-net" for mobile nutrients by deep-rooted plants revealed a small but real ability to intercept nutrients on their way out of the system, thus increasing nutrient usage efficiency at the system level. The effects of rhizosphere modification on nutrient mobilization in mixed-species systems were found to be dependent on the degree of root syncopation among plant components, as well as the long-term replenishment of nutrient resources accessed. To summarize, principles and tools to assist farmers in navigating between the Scylla of access and the charibdis of excess problems in plant nutrition do exist, but their application necessitates an understanding of site-specific interactions and different degrees of internal control, rather than relying solely on plant genetic modification aimed at transferring specific processes out of context.

Fertilizers and Fertilization Strategies

Meeting human needs while staying within our planet's ecological boundaries necessitates ongoing reflection and redesign of agricultural technology and practices. Fertilizers are one of these inventions, and their invention and application have been a crucial factor in rising crop production, farm efficiency, and food security. Mineral fertilizers are one of the main drivers of the expanded global food demand needed to feed the world's growing population. About half of today's world food is made up of synthetic nitrogen compounds (Erisman et al. 2008).

Fertilizer use, on the other hand, has an environmental cost, and fertilizers have not proven to be a very cost-effective production factor in lifting many poor farmers out of poverty, particularly in African countries, where the application of unbalanced nutrient compositions in fertilizers on poor soils has had little impact on yield increase. Established mineral fertilizers, mainly comprising N, P, and K, may be applied at the right moment, in the right spot, in the right quantity, and with the right formulation using agronomic practices. However, cumulative success toward reducing harmful side effects remains insufficient for developing countries to achieve the desired transition toward sustainable agriculture. Over the past 5 decades or more, there have been no profound reflections on the nature and function of mineral fertilizers globally, and in comparison to other industries, mineral fertilizer research and development has received pitiful funding (R&D). Along with the application of current fertilizers (Sebilo et al. 2013), more deliberate implementation of knowledge of plant physiological processes, such as the diversity of mineral nutrient absorption pathways, their translocation, and metabolism – as a starting point in understanding the physicochemical "packaging" of nutrients, their structure, number, and timing of application to satisfy plant physiological needs for enhanced instantaneous nutrient availability. In addition to distribution via the root, efforts are redoubled with many other uptake avenues for nutrient delivery to the plant, which are currently at best haphazard, such as above ground sections and seed coating (Bindraban et al. 2015). To improve nutrient absorption, biological mechanisms such as nutrient-specific interactions in plants and soil, plant-microorganism symbiosis, and nanotechnology must be used. To accomplish these objectives, it is hoped that sustained R&D activities will be followed.

References

Adeoye, P., Adebayo, S., and Musa, J. (2011). Growth and yield response of cowpea (*Vigna unguiculata*) to poultry and cattle manure as amendments on sandy loam soil plot. *Agric. J.* 6 (5): 218–221.

Agarwal, B. (2014). Food sovereignty, food security and democratic choice: critical contradictions, difficult conciliations. *J. Peasant Stud.* 41 (6): 1247–1268.

Ali, M.A., Alam, M.R., Molla, M., and Islam, F. (2010). Crop productivity as affected by fertilizer management options in Boro-T. aman cropping pattern at farmers' fields. *Bangladesh. J. Agric. Res.* 35 (2): 287–296.

Bindraban, P.S., Dimkpa, C., Nagarajan, L. et al. (2015). Revisiting fertilisers and fertilisation strategies for improved nutrient uptake by plants. *Biol. Fertil. Soils* 51: 897–911.

Bruinsma, J. (2009). The resource outlook to 2050: by how much do land, water and crop yields need to increase by 2050. In *Expert Meeting on How to Feed the World in*, Vol. 2050 pp. 24-26.

Cakmak, I., Kurz, H., and Marschner, H. (1995). Short-term effects of boron, germanium and high light intensity on membrane permeability in boron deficient leaves of sunflower. *Physiol. Plant.* 95 (1): 11–18.

Caron, P., Loma-Osorio, G.F.y.d., Nabarro, D. et al. (2018). Food systems for sustainable development: proposals for a profound four-part transformation. *Agron. Sustain. Dev.* 38 (4): 1–12.

Erisman, J.W., Sutton, M.A., Galloway, J.N. et al. (2008). How a century of ammonia synthesis changed the world. *Nat. Geosci.* 1: 636–639.

Fageria, N., Baligar, V., and Jones, C. (1997). *Growth and Mineral Nutrition of Field Crop*, 1001. New York: Marcel Dakker. Inc.

FAO (2000–2022). Agricultural production statistics 2000–2020. FAOSTAT Analytical Brief Series No. 41. Rome.

Fixen, P., Brentrup, F., Bruulsema, T. et al. (2015). Nutrient/fertilizer use efficiency: measurement, current situation and trends. In: *Managing Water and Fertilizer for Sustainable Agricultural Intensification*, 270.

Gris, E. (1844). Nouvellesexpériences sur l'action des composésferrugineuxsolubles, appliqués à la vegetation et specialement au traitement de la chlorose et à la debilité des plantes. *C. R. Seances Soc. Biol. Paris* 19: 1118–1119.

Haneklaus, S. and Schnug, E. (1999). Issues of plant nutrition in sustainable agriculture. Proceedings of International Symposium on Plant Nutrition, Quality of Production and Processing, MZLU, Brno: 18-26.

Khoshgoftarmanesh, A.H., Schulin, R., Chaney, R.L. et al. (2010). Micronutrient-efficient genotypes for crop yield and nutritional quality in sustainable agriculture. A review. *Agron. Sustain. Dev.* 30 (1): 83–107.

Lam, J., Takeshita, S., Barker, J.E. et al. (2000). TNF-α induces osteoclastogenesis by direct stimulation of macrophages exposed to permissive levels of RANK ligand. *J. Clin. Invest.* 106 (12): 1481–1488.

Lipman, J.G., Blair, A.W., and Prince, A.L. (1931). The influence of lime on the recovery of total nitrogen in field crops. *Soil Sci.* 32 (3): 217–234.

McHargue, J.S. (1922). The role of manganese in plants. *J. Am. Chem. Soc.* 44 (7): 1592–1598.

Noorkdwijk, M.V. and Cadisch, G. (2002). Access and Excess problems in plant nutrition. *Plant Soil* 247: 25–39.

Rice, R.W. (2007). The physiological role of minerals in the plant. In: *Mineral Nutrition and Plant Disease* (ed. L.E. Datnoff, W.H. Elmer and D.M. Huber), 9–29. St. Paul, MI: APS Press.

Salm-Horstman, F. (1851). *Sand and Water Culture Methods Used in the Study of Plant Nutrition* (ed. E.J. Hewitt). (1966).

de Saussure, T. (1804). *Rechercheschimiques sur la végétation (Chemi-Action Appears to Have Been Most Appropriate.Cal Researches About the Vegetation)*. Paris: Nyon Widow.

Sebilo, M., Mayer, B., Nicolardot, B. et al. (2013). Long-term fate of nitrate fertilizer in agricultural soils. *Proc. Natl. Acad. Sci.* 110 (45): 18185–18189.

Selim, M. (2018). Potential role of cropping system and integrated nutrient management on nutrients uptake and utilization by maize grown in calcareous soil. *Egypt. J. Agron.* 40 (3): 297–312.

Shand, C. (2007). Plant Nutrition for Food Security. A Guide for Integrated Nutrient Management. By RN Roy, A. Finck, GJ Blair and HLS Tandon. Rome: Food and Agriculture Organization of the United Nations (2006), pp. 348, US $70.00. ISBN 92-5-105490-8. Experimental Agriculture. 43(1): 132–132.

Sprengel, C. (1839). *Die Lehrevom Dünger (Fertilizer Science).* Leipzig, Germany: Immanuel Müller.

Warrington, K. (1923). The effect of boric acid and borax on the broad bean and certain other plants. *Ann. Bot.* 37: 629–672.

Weil, R.R., Islam, K.R., Stine, M.A. et al. (2003). Estimating active carbon for soil quality assessment: a simplified method for laboratory and field use. *American J. Altern. Agric.* 3–17.

Wheeler, T. and Von Braun, J. (2013). Climate change impacts on global food security. *Science* 341 (6145): 508–513.

3

Role of Nutrients in the ROS Metabolism in Plants

Muhammad Arslan Ashraf, Rizwan Rasheed, Mudassir Iqbal Shad, Iqbal Hussain, and Muhammad Iqbal

Department of Botany, Government College University Faisalabad, Faisalabad, Pakistan

Introduction

Oxidative stress is a natural biological phenomenon in which the excess of free oxygen radicals surpasses the radical scavenging processes, resulting in an imbalance between antioxidants and reactive oxygen species (ROS) in the biological system (Mahmud et al. 2020). Oxidative stress is a condition characterized by elevated levels of ROS, often known as "oxygen-derived species" or "oxidants." The functionality is physiologically governed by oxygen levels, transduction of signals, and the maintenance of redox balance. Redox management is a rapidly growing field of study that has an impact on practically every aspect, comprising biological systems, which have devised techniques for employing oxidants to their advantage while simultaneously adapting to the coexistence of harmful free radicals (Kapoor et al. 2019). All ROS and reactive nitrogen species (RNS) are produced in a well-controlled manner to aid in the management of homeostasis at the cell level in typical healthy and vigorous tissues, and they play an important role as second messengers that modulate signaling pathways and govern cellular activity (Kapoor et al. 2019).

Aerobic organisms, including plants, rely on oxygen for their energy production processes. Additionally, plants generate O_2 via photosynthesis. It is possible to excite molecular oxygen, resulting in ROS. Hydrogen peroxide (H_2O_2), hydroxyl radical ($OH^{\bullet-}$), singlet oxygen ($^1O_2$), and superoxide radical ($O_2^{\bullet-}$) are among the major ROS (Hasanuzzaman et al. 2020). Under adverse environmental conditions, toxic ROS are produced as a by-product in a variety of cellular locations, including apoplasts, chloroplast region, peroxisomes, and mitochondria, whereas under normal circumstances, the coordinated action of antioxidant defense system components within the plants maintains a balance between the biosynthesis and scavenging of ROS (Hasanuzzaman et al. 2020).

Induction of positive response in antioxidant defense systems and biochemical mechanisms including cellular division, cell differentiation, and stress adaptation responses requires the production of ROS at a lower level during normal growing situations

Plant Ionomics: Sensing, Signaling, and Regulation, First Edition.
Edited by Vijay Pratap Singh and Manzer H. Siddiqui.

(Hasanuzzaman et al. 2020). ROS are highly adaptable signaling molecules due to their numerous features, which comprised varying levels of reactivity, biosynthesis locations, and the ability to traverse cellular membrane (Mittler 2017).

Imbalance between ROS production and quenching produces oxidative damage in cells experiencing stress, impairing typical cellular activity and destroying macromolecules such as DNA, proteins, carbohydrates, and lipids, ultimately leading to cell death (Hasanuzzaman et al. 2020). ROS signaling is controlled by the buildup of ROS in specific cellular compartments (Mittler 2017).

In cells, both free radical and nonradical ROS are generated. The free radicals include alkoxyl radical ($RO^{\bullet-}$), $OH^{\bullet-}$, peroxyl radical ($ROO^{\bullet-}$), and $O_2^{\bullet-}$, while the nonradicals are $^1O_2$ and H_2O_2. Other nonradical ROS discovered in plants include excited carbonyl (RO*) and hypochlorous acid (HOCl). Furthermore, ROS contains some acidic chemicals such as hypobromous acid (HOBr), hypochlorous acid (HOCl), and hypoiodous acid (HOI) as well as radical molecules like carbonate ($CO_3^{\bullet}$) (Hasanuzzaman et al. 2020).

Overproduction of ROS compound is damaging to the cell, whereas it acts as a signal transducer to activate the plant's protection mechanism. Higher ROS generation induces decline in cellular activities, which ultimately results in initiation of defense related processes and stress management (Nadarajah 2020). A further important signaling benefit of ROS is their close relationship with cellular metabolism and homeostasis. Almost every alteration in cellular homeostasis can modify the moderate level of ROS in a specific compartment. Photorespiration-promoting physiological situations also results in increased ROS generation in peroxisomes (Phua et al. 2021).

Despite the abundance of information gained from research on the mitochondrial and/or NOX-associated ROS-induced ROS release (RIRR) mechanisms in animal cells, nothing is known about RIRR in plants. Plants can boost a ROS signal by a variety of ways, the most notable of which is the function of respiratory burst oxidase homolog (RBOH) proteins, which are the plant counterpart of NADPH oxidase (NOXs) (Zandalinas and Mittler 2018).

Oxidative Defense System

When ROS levels are too low or modest, they act as a secondary messenger, triggering a number of events in plant cells, notably gravitropism, stomatal closure, abiotic, and biotic stress tolerance and programmed cell death (PCD) (Xie et al. 2019). Nevertheless, during the past two decades, it has become increasingly clear that large concentrations of all kinds of ROS are extremely damaging to organisms. Plants produce excess ROS as a result of constant exposure to environmental stress factors, which cannot be entirely eliminated by the active oxygen scavenging mechanism. As a result, key physiological functions, such as the PCD pathway, enzyme inhibition, protein denaturation, lipid peroxidation, nucleic acid oxidation, are carried out by plants (Xie et al. 2019).

Over the last few decades, oxidative stress research has primarily concentrated on *Escherichia coli*. However, in the past 10 years, it has shifted from humans and animals to plants such as *Arabidopsis thaliana* and rice. It has significantly improved our understanding of the impact and function of oxidative stress in growth, protection, and ecological responses (Lopes et al. 2016; Guan and Lan 2018). Since elevated levels of ROS results in the loss of different essential intracellular signaling molecules, plants evolved their own antioxidant defense mechanism to manage a dynamic equilibrium of ROS (Xie et al. 2019).

To resist oxidative stress, all organisms have innate cellular defense, which are referred to as antioxidants. Free radicals have a brief life period, measured in nano, milli, or microseconds and they interact spontaneously with proteins, lipids, DNA, and other biomolecules, causing damage and generating toxic substances that includes different lipid adducts and lipid peroxides. Protein damage produces a decrease in enzyme activity, while DNA damage causes carcinogenesis and mutagenesis (Kapoor et al. 2019).

As a result of the imbalance between ROS generation and detoxification, oxidative stress develops, which is a complicated physiological and biochemical phenomenon (Soares et al. 2019). Almost all biotic and abiotic stresses increases ROS production, necessitating in plants an effective and quick process to maintain ROS homeostasis in response to changing environmental conditions (Mittler 2017).

Lipids and proteins in plant cells are the primary targets of oxidative damage caused by ROS. Lipid peroxidation, or the oxidative degradation of polyunsaturated lipids in the cell membrane, exists in every organism and is frequently used as a marker to measure the level of lipid damage under extreme environmental conditions (Xie et al. 2019). Over the past several decades, plenty of investigations have been conducted to determine the mechanism of ROS scavenging and formation. Relying on the plant varieties, their growth phases, types of abiotic stresses, stress severity and duration, reports have showed a rapid threshold value of ROS for being useful or hazardous (Hasanuzzaman et al. 2020).

Plants have adapted methods to deal with the harmful consequences of oxidative stress, which is recognized as a secondary component of many other stressful conditions. Antioxidants (enzymatic and non-enzymatic) are some of the processes that work together to keep ROS equilibrium in the cell (Akyol et al. 2020).

Plants have developed advanced enzymatic ROS scavenging methods, which include ascorbate peroxidase (APX), peroxidase (POX), catalase (CAT), glutathione peroxidase (GPX), and superoxide dismutase (SOD), which are found in many parts of plant cells. Other enzymes, such as glutathione-S-transferase (GST), glutathione reductase (GR), dehydroascorbate reductase (DHAR), and monodehydroascorbate reductase (MDHAR) play role in re-biosynthesis of oxidized non-enzymatic antioxidants like glutathione (glutathione disulfide to glutathione) and ascorbate (MDHA and DHA to ascorbate). Other low-molecular-weight molecules with antioxidant capabilities include tocopherols, carotenoids, and phenolics, in addition to glutathione and ascorbate, which operate as redox buffers in plant cells (Akyol et al. 2020).

Endogenous ascorbate (AsA) levels maintains the antioxidant activity and promote resistance to water shortage. Furthermore, AsA being a natural signalling agent can also affect antioxidant enzyme activities and the expression of resistance genes, therefore reducing the negative impacts of abiotic stressful conditions (Farooq et al. 2020). Additionally, the function of ascorbate in chloroplasts has already been hypothesized during water breakdown in the thylakoid lumen while in the stroma to detoxify hydrogen peroxide produced by Cu/Zn-SOD and Fe-SOD activity (Akram et al. 2017; Niu et al. 2019).

Reactive Oxygen Species (ROS) and Reactive Nitrogen Species (RNS)

The constant generation of RNS and ROS as by-products of metabolism plays an important role in seed physiology. Changes in the level of RNS such as dinitrogen tetroxide, nitrous acid, nitric dioxide, and nitric oxide and ROS such as hydrogen peroxide, superoxide radical, and hydroxyl radical affect longevity, dormancy, and seed germination at all stages of the seed life cycle. Recent research shows that RNS and ROS use nitrosative and oxidative signals to trigger seed germination during the oxidative conditions (Kumar et al. 2021b).

ROS are produced by plant cells and are engaged in physiological and stress responses (Singh et al. 2019). Superoxide anion ($O_2^{\bullet-}$), hydroxyl ion ($OH^{\bullet-}$), and hydrogen peroxide (H_2O_2) are among them, and they are formed in different cellular compartments (Talukdar 2017). Stress causes plants to produce RNS, such as S-nitrosoglutathione (GSNO), peroxynitrite ($ONOO^-$), and nitric oxide (NO), as well as altering their ROS levels, resulting in noticeable imbalance redox homeostasis abnormalities (Espinosa-Vellarino et al. 2020).

Plant growth, development, and agricultural productivity are all harmed by environmental challenges. The metabolism of reactive nitrogen and oxygen species is altered as a result of these unfavorable conditions. High concentrations of reactive species that surpass the capability of antioxidant defense enzymes disrupt redox equilibrium, causing damage to macromolecules including proteins, membrane lipids, and nucleic acids, as well as plant cell death and nitro-oxidative stress (Chaki et al. 2020).

RNS and ROS are well recognized for their functions in numerous signaling pathways, including development and growth, reproduction, nutrition sensing, and defensive responses to biotic and abiotic stimuli, where they are most prominently involved in redox control of thiols on proteins (Sevilla et al. 2015; Waszczak et al. 2015).

The production of ROS and RNS by various biotic agents causes oxidative damage to tissue, which leads to degradation and, in certain cases, impaired clearance of these species, which can have negative consequences for plants. Mitochondria are the sites where oxidative energy is produced. Any sort of mitochondrial failure can induce development of many disorders and the poor cell function, according to distinct biosynthetic reaction in aerobic cells. There is a wealth of knowledge on mitochondrial ROS formation and the reasons of oxidative bursts and function impairment caused by mitochondrial ROS (Kapoor et al. 2019).

ROS has been postulated as a secondary messenger molecule or directly associated in activating a chain of processes that results in germination to address the phenomena. Following imbibition, ROS levels rise, and the accumulation of these molecules, especially H_2O_2 (which is more stable than some other oxidants), may cause protein oxidation (carbonylation, including seed storage protein), which obstructs glycolysis and helps to promote the initiation of the pentose phosphate pathway (PPP) (Kumar et al. 2021b).

Although RNS in plants had been discovered as early as the 1960s (Fewson and Nicholas 1960), they did not expect the same level of consideration as their oxygen counterparts (ROS) until the late 1990s. Unlike ROS, which have been identified as toxic substances and also as signals, RNS have been initially identified as signaling molecules, and the terminology "nitrosative stress" did not appear in the plant biosciences literature until around the 2000s (Turkan 2018). ROS and RNS are well known in plants as critical signals and regulators of a range of activities, including abiotic and biotic stress responses, metabolism, transport, growth and development, PCD, and solute autophagy (Maurya et al. 2018; Demircan et al. 2020). Despite ROS, the processes of RNS synthesis in plants are still largely unknown, and is a significant topic for future research. In addition, research into the signaling processes mediated by RNS has intensified in the past decade, partly in response to the discovery of their associations with ROS (Turkan 2018).

As plants cope with adverse climatic conditions, their cellular levels of potentially damaging RNS and ROS rise. ROS and RNS, on the other hand, can operate as signals for the regulation of growth and development as well as defense against biotic and abiotic stress at low concentrations. They are now known to play a role in almost every aspect of plant metabolism and cell activities (Nieves-Cordones et al. 2019). ROS and RNS are now well recognized as regulators in a variety of signaling procedures, including nutrition sensing, growth and development, reproduction, and defense activity to abiotic and biotic stimuli (Turkan 2017).

RNS and ROS are still the most significant redox molecules. They are the key signaling molecules involved in the regulation of gas exchange, transpiration, germination, biotic and abiotic stress response, cell death, and plant growth and development due to their short half-life, high diffusion capability, and ability to react with different components in the cell (Lindermayr et al. 2021).

ROS Generation and Functions in Plants

Photosynthetic organisms release O_2 freely into the environment (Gupta et al. 2018; Taverne et al. 2018). Although ROS are a natural element of a cell's metabolism, their negative effects manifest when the equilibrium between ROS scavenged and ROS generated is disrupted (Sewelam et al. 2016; Tiwari et al. 2018). ROS is involved in a variety of developmental processes, including hypersensitive responses, PCD, resistance mechanisms, acclimatization, and signaling molecules as a defense response (Di Meo et al. 2016; Dunnill et al. 2017). ROS are formed when molecular oxygen is reduced or activated (Kumar et al. 2021a). ROS are produced as by-products of a variety of metabolic procedures and are found in the peroxisomes, apoplast, cytosol, plastids, and mitochondria on a regular basis (García-Caparrós et al. 2021). Most plants generate ROS in mitochondria, chloroplasts, peroxisomes, and other sections of the cell in an oxidizing environment as a result of photosynthesis and metabolic reactions (Sachdev et al. 2021). Pathogen attack, low and high temperature, nutritional shortage, salt, drought, high or low temperature, and bright light are just few of the environmental conditions that cause plants to produce ROS (Jan et al. 2021).

The chloroplast is one of the most important places in plants for the generation of ROS, with ROS generated both directly and indirectly as a result of the interaction of chlorophyll (chl) with light (Raja et al. 2017; Tominaga et al. 2020). The principal sources of ROS formation are the electron transport chain (ETC) and triplet chl, notably PSI and PSII (Singh et al. 2019; García-Caparrós et al. 2021). In PSI, the Mehler reaction produces $O_2^{\bullet -}$, which is then converted to H_2O_2 by SOD (Bose et al. 2014; Asada 2019).

Abiotic stressors cause the peroxisome to generate more photorespiration. The enzyme glycolate oxidase (GOX) is a key participant in the increased generation of ROS (Kerchev et al. 2016). Xanthine oxidase (XOD) and peroxisomal membrane NADPH oxidase can also create $O_2^{\bullet-}$ at the organelle matrix in the peroxisome (Phua et al. 2021). SODs, which are metaloenzymes, eventually dismutate $O_2^{\bullet-}$ into H_2O_2. SODs of various kinds, such as Mn-SOD and Cu-Zn-SOD, have been found in the peroxisomes of several plant species (Phua et al. 2021). XOD and peroxisomal membrane NADPH oxidase create xanthine at the organelle matrix. (Hasanuzzaman et al. 2020).

The active site for ROS production in plant cell walls is complex structures made up of polysaccharides, phenolics, and proteins (Kärkönen and Kuchitsu 2015). Differential cell wall growth, triggered by ROS and peroxidase, causes the polymerization of glycoproteins and phenolic substances to make cell walls stiff, making stressed plants sensitive to growth suppression (Le Gall et al. 2015; Novaković et al. 2018; Hasanuzzaman et al. 2020).

RNS and ROS Signaling in Plants in Response to Environmental Stresses

The inescapable outcome of aerobic life is the generation of ROS. The ROS are oxidizing species that can cause substantial damage in biological systems (oxidative stress) and are produced in different cell compartments in plants (Del Río 2015).

Together ROS and RNS are significant signaling in plants and crucial regulators of a range of mechanisms such as PCD, metabolism, solute transport, development and growth, response to biotic and abiotic stressors, and autophagy (Turkan 2018).

Plants are subjected to various types of abiotic and biotic stress conditions in the natural environment, which cause rapid modifications in the scavenging and synthesis of ROS. The ROS generation and quenching are compartmentalized, which means they can be produced and removed in distinct cellular compartments such as the endoplasmic reticulum, chloroplasts, peroxisomes, plasma membrane, mitochondria, and apoplast, depending upon the type of stimulus (Czarnocka and Karpiński 2018).

Recent work indicates that ROS and RNS are important players in these signal transduction pathways in plants, despite the fact that they were previously thought to be only harmful (Bor and Turkan 2019). RNS signaling is a hot issue in plant research, and RNS such as S-nitrosothiols (SNOs), peroxynitrite ($ONOO^-$), nitrogen dioxide ($^{\bullet}NO_2$), and nitric oxide (NO) are known to regulate complicated activities in plants not only under normal physiological conditions but also under various stresses (Corpas et al. 2018).

Excessive ROS production is caused by adverse environmental conditions, which leads to oxidative cell damage at large concentrations. Plants have a wide range of antioxidant processes to prevent ROS-dependent cellular damage, and they can employ ROS as a signal in a variety of biological processes at the same time (Del Río 2015).

From the perspective of activating signaling cascades, localized ROS generation in organelles such as mitochondria, peroxisomes, and chloroplasts is critical. Many stimuli produce ROS in plant cells, which drive signal transduction events and evoke specific cellular activities. The impact of these compounds on cellular functions is governed by a balance between their continued generation and detoxification by various antioxidant mechanisms (Del Río 2015).

In plants, NO is a critical mediator in the defense response to pathogen infestations, collaborating with ROS (Cavaco et al. 2021). The fast generation of ROS and NO frequently unchains a specified PCD mechanism. Both ROS and NO play important roles in this process. During reactions to stress conditions and pathogen infections, PCD is a crucial mechanism for regulating several aspects of development and growth, as well as eradicating damaged or infected cells (Čamagajevac et al. 2019). But at the other side, RNS peroxynitrite ($ONOO^-$) is a significant oxidant/nitrating compound produced by the high reactivity of O_2 with NO, and its presence in plant organelles such as peroxisomes has been documented (Del Río 2015; Zhou et al. 2021).

The durations of different ROS and their degrees of reactivity toward molecular components differ. ROS activate plasma membrane-localized Ca^{2+} influx and K^+ efflux channels, which convey stress signals, at an early stage of stress signal perception (Czarnocka and Karpiński 2018). In both plants and mammals, initiation of K^+ efflux by unfavorable conditions has been exhibited to cause PCD (Czarnocka and Karpiński 2018). Lipids, proteins, and nucleic acids are the main cascade targets of ROS during oxidative stress (Czarnocka and Karpiński 2018).

Both glutamate and nitric oxide (GABA, NO) are produced during the nitrogen metabolism and play significant roles in plant growth and stress responses (Majumdar et al. 2016). The RNS peroxynitrite ($ONOO^-$) is a nitrating/oxidant substance that is most typically found in organelles such as peroxisomes and is generated by a strong interaction between NO and O_2 (Del Río 2015; Trapet et al. 2015; Zhou et al. 2021).

Antioxidant Compounds

Protein oxidation via covalent interaction is triggered by ROS or oxidative stress by-products (Krumova and Cosa 2016; Forrester et al. 2018). The majority of protein oxidations are irreversible; however, sulfur-containing amino acids can be reversed (Ahmad et al. 2017). It is widely employed as a marker for ROS or oxidative damage (Pisoschi and Pop 2015). When exposed to environmental conditions, the production of ROS increases, resulting in nucleic acid deterioration, inhibition of enzyme activity, PCD, protein oxidation, and lipid peroxidation, which causes cell damage and death (Czarnocka and Karpiński 2018; Čamagajevac et al. 2019). ROS, on the other hand, have been proposed as secondary messengers in a variety of cellular activities, including environmental stress tolerance mechanisms (Kapoor et al. 2019).

SOD, GR, guaiacol peroxidase (GPX), DHAR, MDHAR, APX, and CAT are examples of enzymatic antioxidants, while phenolics, glutathione (GSH), tocopherols, carotenoids, as well as ascorbate (AsA) are nonenzymatic antioxidants and all play important roles in the reduction of ROS in cells (Kapoor et al. 2019).

The constituents and way of function of the ROS-removing system were found in the past 30 years of the twentieth century (Czarnocka and Karpiński 2018). Earlier, ROS eliminators were thought to be merely a defensive measure against the harmful effects of ROS. Recent studies have shown that ROS-eliminating enzymes and nonenzymatic antioxidants are engaged in ROS-dependent signaling during plant acclamatory response to various environmental challenges in addition to maintaining ROS homeostasis (Noctor et al. 2018). The scavenging systems are further segregated because each ROS has its unique range of chemical characteristics and may mount up in certain cellular compartments (Czarnocka and Karpiński 2018).

By increasing the activities of enzymatic and nonenzymatic antioxidants, genetic engineering technologies have yielded considerable progress in terms of upgrading tolerance to a variety of abiotic stresses (Birnie-Gauvin et al. 2017). The upregulation of SOD is linked to the defense against oxidative stress generated by a variety of abiotic stresses, and it is critical for plant life. Several studies have shown that when plants are stressed, they produce more SOD, which detoxifies lethal hydrogen peroxide and aids in plant survival (Biczak 2016). Glutathione peroxidases besides other isozymes have been reported to degrade hazardous hydrogen peroxide and hydroperoxides to alcohols. Glutathione peroxidases further help to restore lipid peroxidation that occurs as a result of ROS (Kapoor et al. 2019).

In tobacco plants, NO reduces the cytotoxic consequences of salt-induced H_2O_2 (Kolupaev et al. 2015). NO also has antioxidant properties, which are classified based on whether it interacts directly or indirectly with ROS to reduce stress. RNS and its derivatives interact with ROS to produce nonreactive byproducts, reducing oxidative injury. NO may increase the activities of antioxidant enzymes through causing changes that result in effective free radical scavenging. NO may boost the functions of SOD, an enzyme that aids in the conversion of superoxide to H_2O_2, which is then detoxified by APX and CAT (Fatma et al. 2016). If the concentration of superoxide is higher, it can scavenge NO, and vice versa (Saddhe et al. 2019).

Antioxidant-Mediated RNS/ROS Regulation

Oxidative stress is caused by an imbalance between antioxidants and ROS, which results in either an increase in ROS or a decrease in antioxidants (Schieber and Chandel 2014). Cells strive to stabilize the effect of oxidants and restore the redox balance in stressful situations by suppressing or activating genes, transcription factors, and structural proteins (Sosa et al. 2013). One of the most important aspects in defining stress in the body is the relationship between oxidized and reduced glutathione. Changes in DNA structure, triggering of numerous stress-induced transcription factors, protein and lipid alteration, and the synthesis of anti-inflammatory and proinflammatory cytokines are all effects of ROS (Gu et al. 2013).

SODs are members of the metalloenzyme family. They reduce the potential of OH^- production in the Haber–Weiss reaction by catalyzing the elimination of O_2^- by dismutating it into O_2 and H_2O_2. SODs are found in the cytosol (Cu/Zn-SOD), peroxisomes (Cu/Zn-SOD), chloroplasts (iron SOD, Fe-SOD, and Cu/Zn-SOD, copper/zinc SOD), and mitochondria (Mn-SOD, manganese SOD) in plants (Czarnocka and Karpiński 2018). SOD activity and expression have been shown to increase in response to abiotic and biotic stress (Gill et al. 2015; Lu et al. 2017). Furthermore, SOD overexpression shields plants from oxidative stress-mediated damage (Czarnocka and Karpiński 2018).

Plants have a small number of H_2O_2-detoxifying enzymes. Except for CATs, they all need cellular reducing equivalents to detoxify H_2O_2. In animal cells, there is just one CAT isoform, but plant cells have several isoforms (Czarnocka and Karpiński 2018). Catalytic dismutation of two H_2O_2 molecules into O_2 and H_2O is catalyzed by CATs, which are heme-containing enzymes (Czarnocka and Karpiński 2018). Environmental stresses such as bacterial and nematode infection, UV radiation, salinity, heat, cold, and drought, all stimulate the activity and development of CATs (Caverzan et al. 2016).

H_2O_2 seems to be the only ROS that can permeate through membrane aquaporins and transported longer areas within the cell, and it is rather stable compared to the other ROS (Mittler 2017). Even at low concentrations, it is engaged in regulating biological processes and gives tolerance to various abiotic and biotic stress conditions (Kapoor et al. 2019).

Peroxidase primarily oxidizes phenolic compounds (PhOH) to form the phenoxyl radical (PhO), with H_2O_2 acting as an electron acceptor in the reaction and being transformed to $2H_2O$ (Hasanuzzaman et al. 2020). APX (chAPX, chloroplastic APX; mitAPX, mitochondrial APX; cAPX, cytosolic APX; and microbodies including peroxisomal and glyoxysomal APX is a class I heme-peroxidase with various isoforms (Del Río and López-Huertas 2016). All isoforms scavenge H_2O_2; however, without the presence of AsA, the action ceases (Hasanuzzaman et al. 2019). APX eliminates H_2O_2 and oxidizes AsA to create monodehydroascorbate (MDHA) as well as subsequently dehydroascorbate (DHA) in the AsA-GSH cycle (Hasanuzzaman et al. 2020).

Plants have low-molecular, nonenzymatic antioxidants including GSH and AsA, along with flavonoids, proline, carotenoids, and α-tocopherol, in addition to enzymatic ROS scavengers. They play a significant role in retrograde signaling in addition to ROS detoxification (König et al. 2018). AsA is found in vacuoles, chloroplasts, nuclei, mitochondria, peroxisomes, cytosol, and the endoplasmic reticulum (ER). Plants' total AsA content is affected by both abiotic and biotic stresses. Excess light, UV radiation, heavy metals (HM), and high ozone levels have all been demonstrated to raise its content. It participates in redox signaling, gene expression control, and enzymatic activity regulation by donating electrons to a variety of nonenzymatic and enzymatic processes. It either scavenges ROS directly or via the AsA-GSH cycle (Czarnocka and Karpiński 2018). Tocopherols constitute lipophilic antioxidants that are effective at scavenging reactive nitrogen and oxygen species (RNS/ROS) and lipid radicals. Carotenoids are a class of lipid-soluble antioxidants found in plastids of both nonphotosynthetic and photosynthetic plant tissues. Carotenoids safeguard photosynthetic apparatus and so have antioxidative properties. Flavonoids are pigments found in fruits, seeds, and flowers that provide red, blue, and purple coloration. For a long period of time, flavonoids were thought to be primarily used as a UV filter to shield plant tissues (Czarnocka and Karpiński 2018).

Proline, an osmoprotectant, is also a high power oxidative defense molecule and potential cell death suppressor. It can scavenge free radicals while also suppressing lipid peroxidation. Proline transfer takes place between the mitochondria, chloroplasts, and cytosol as a result of its partitioned metabolism (Kaur and Asthir 2015). An increased proline level gives resistance to abiotic stress factors, for example cold temperatures, salt, and drought (Kaur and Asthir 2015).

Role of Nutrients in ROS Metabolism Under Salinity

Nitrogen (N), phosphorus (P), potassium (K), calcium (Ca), magnesium (Mg), and sulfur (S) are considered as main or macronutrients, since they are needed in considerable quantities (1–150 g kg^{-1} dry weight) by the plants. Plants require a balanced supply of all of these critical elements. Metallic micronutrients (Cu, Fe, Mg, Zn, Ni, and Mn) are useful in the plant metabolic activity as inhibitors, activators, or components of enzymes. As a result, Zn plays a critical role in protein synthesis, glucose metabolism, auxin production regulation, cellular membrane integrity, environmental stress resistance, and pollen production (Dubey et al. 2020). Figure 3.1 depicts the schematic mechanism of nutrient-mediated ROS metabolism.

However, in order to meet main challenges in quinoa plants, the researchers looked at stomatal and physiological-based differences in quinoa development under salinity and K, then used a number of analytical methods and a model approach to evaluate the changes in photosynthetic and stomatal aperture. Quinoa had strong maintenance and absorption of Mg, Ca, K, and Cl alongside enhanced stomata density (SD), reduced stomatal aperture, and charge balancing ion to maintain photosynthetic activity, resulting in enhanced growth under salinity (Waqas et al. 2021). Additionally, because ROS are highly reactive and cause damage in plant cells at a variety of levels, including proteins, lipids, nucleic acids, and

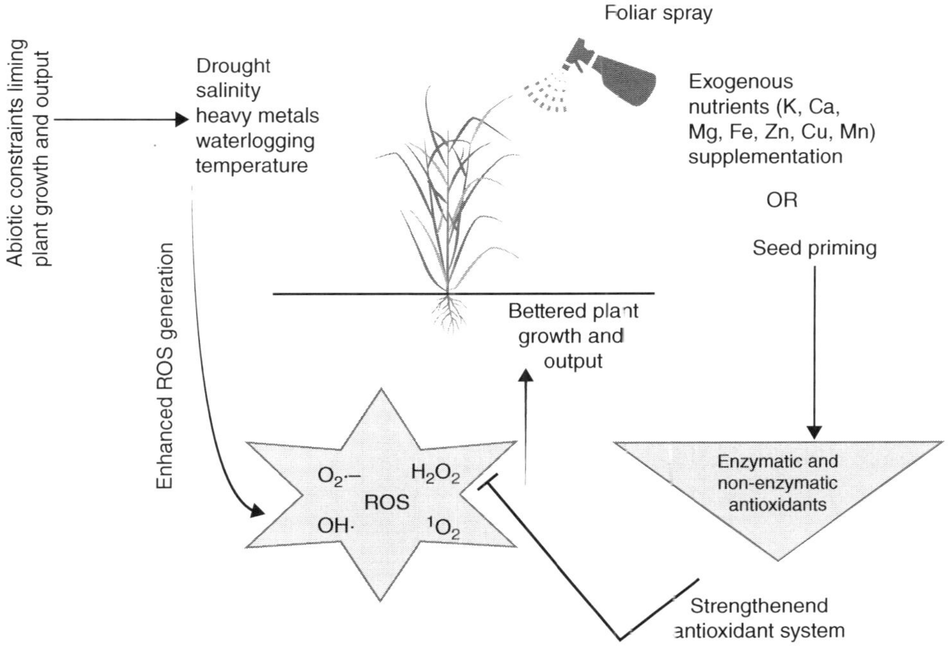

Figure 3.1 Mechanistic representation of nutrient-mediated ROS metabolism in plants under abiotic stress.

photosynthetic components, K deprivation could be a contributing factor for salinity-induced oxidative damages (Waqas et al. 2021).

In another study, two wheat (*Triticum aestivum* L.) genotypes were treated with NaCl stress with and without potassium (K) supplementation. Salt stress generated by NaCl resulted in oxidative stress, which increased lipid peroxidation and affected growth and yield. Introducing K to the nutrition resulted in a considerable increase in growth, as well as good impacts on antioxidant and nitrogen metabolism. Despite the fact that NaCl-induced stress activated the antioxidant defense system, antioxidant enzyme activity and the level of nonenzymatic antioxidants enhanced in K-fed plants. At all developmental stages, an increase in the buildup of osmolytes, which include amino acids, sugars, and free proline, was seen with K supplementation that also increased relative water content and subsequently yield. Potassium improved nitrogen uptake and incorporation while simultaneously lowering Na ions and as a result the Na/K ratio. To some extent, appropriate K can be employed as a strategy to alleviate NaCl stress in wheat (Ahanger and Agarwal 2017b).

The accumulation of suitable osmolytes helps in coping with stressful situations, and it has been suggested that ions such as K may be a crucial predictor of water content in tissue. K ion is involved in various osmotically driven processes at the entire-plant level, including phloem transport, photoassimilate loading, stomatal regulation, and cellular mobility (Zörb et al. 2014; Erel et al. 2015).

Controlling PCD, which is directly linked to posttranslational modulations of K channels, requires maintaining intracellular K homeostasis for regulating adaptive responses to various biotic and abiotic stimuli such as drought, salinity, and oxidative stress. Potassium has been shown to protect agricultural plants from ROS-mediated oxidative stress by increasing the K/Na ratio and modifying antioxidant metabolism (Oyarburo et al. 2015; Ahanger and Agarwal 2017a).

Role of Nutrients in ROS Metabolism Under Drought

Long-term water scarcity (drought) is a primary climatic element that restricts plant development and growth (La et al. 2020). A frequent stress reaction is the buildup of ROS besides proline (Rejeb et al. 2014; Nounjan et al. 2018). Likewise, fast formation of ROS (also known as oxidative burst) has been one of the initial plant responses to a variety of environmental challenges and pathogenic infestations (Lee et al. 2015; Fernández-Crespo et al. 2017; Islam et al. 2017). Multiple investigations have shown that ROS-induced salicylic acid production is mediated by Ca signaling (Seyfferth and Tsuda 2014; Herrera-Vásquez et al. 2015). However, Ca-dependent protein kinases (CPKs) are already recognized to collaborate with hormone signaling to make a significant contribution in innate immunity as a stress signaling (Seyfferth and Tsuda 2014; Aldon et al. 2018; Jing et al. 2020). Moreover, the contradictory roles of ROS and proline in enhancing stress resistance and establishing hypersensitive toxicity in the context of the hormonal signaling system are yet unknown (La et al. 2020).

Plants accumulate osmolytes, such as proline and soluble sugars in water-stressed conditions, that contribute to maintaining cellular water balance and control important ions absorption (such as K, Ca, and Mg) (Sohag et al. 2020). Conversely, drought-induced loss of water might alter the cation–anion ratio, resulting in cell turgor loss, cell division interference, photosynthesis impairment, and carbon metabolism inefficiencies (Sohag et al. 2020).

According to another study, drought is among the key environmental constraints that negatively affect maize (*Zea mays* L.) growth and productivity around the world. Drought resistance in plants has been promoted by foliar sprays of micronutrients, osmoprotectants, or plant growth regulators. The relative efficacy of foliar applications of glycine betaine (GB), salicylic acid (SA), and zinc (Zn) on morphology, antioxidant enzyme activities, chlorophyll contents, gas-exchange attributes, relative water content (RWC), buildup of osmolytes and ROS, and yield traits of maize plants experiencing two soil moisture levels (control: 85%; drought: 50% field capacity) was investigated in a controlled pot experiment. Except for intercellular CO_2 concentration, drought stress effectively decreased morphological parameters, yield and also its components, gas-exchange parameters, RWC, and chlorophyll concentrations, when compared to controlled situations. Under drought, however, foliar treatments significantly improved all of the above characteristics. Drought stress raised the levels of hydrogen peroxide and superoxide anion, as well as the rate of lipid peroxidation as evaluated by malonaldehyde (MDA) content. Under drought stress, meanwhile, foliar treatments considerably lowered MDA and ROS levels. Under both good-watered and drought-stressed circumstances, foliar treatments improved antioxidant enzyme activity, soluble sugar, and proline concentration. Consequently, the treatment of GB was the most beneficial among all substances in improving drought resistance in maize by lowering ROS levels, increasing antioxidant enzyme activity, and increasing osmolytes deposition (Shemi et al. 2021).

Zinc (Zn) is indeed an important micronutrient implicated in many physiological processes and structures, including protein synthesis, pollen formation, DNA and RNA metabolism, fertilization, and glucose and chlorophyll biosynthesis (Shemi et al. 2021). By boosting antioxidant enzymes and detoxifying ROS, Zn plays a crucial role in improving drought stress tolerance (Sofy 2015; Suganya et al. 2020). In addition, Zn combines with plant hormones, enhances the transcription of stress proteins, and promotes antioxidant enzymes, all of which help to fight the impacts of drought (Umair Hassan et al. 2020).

Likewise, Wang and Jin (2007) evaluated the role of exogenous Zn on ROS metabolism, plant growth, and Zn uptake in drought-stressed maize plants. Two independent experiments were carried out. In experiment 1, maize plants were exposed to different Zn levels (0, 3, 9, 27, and 81 mg kg^{-1} soil) in the soil with different soil moisture contents including serious drought (30–35%), mild drought (40–45%), and adequate drought (70–75%). In the plants with adequate drought conditions, exogenous Zn significantly enhanced the shoot dry biomass. The authors did not see any significant difference in different Zn treatments with respect to plant growth. In experiment 2, two Zn levels (0 and 5 mg kg^{-1} soil) and two soil moisture regimes (40–45 and 70–75%) were used to establish the effect of Zn on ROS metabolism that subsequently affected plant growth. The drought stress and Zn deficiency induced a noteworthy accretion in the production of superoxide radical and lipid peroxidation measured as MDA contents. SOD activity was lower in leaves of plants facing Zn deficiency. However, higher SOD activity was recorded in the leaves of plants under drought stress independent of Zn supply. POD activity was higher in plants with Zn supply under well-watered conditions. CAT activity showed a slight modification in response to Zn deficiency and different soil moisture levels. The increase in ROS levels was not alleviated by Zn application; however, Zn supplementation would reduce the water use for biomass production.

Role of Nutrients in ROS Metabolism Under Heavy Metal Stress

With the fast evolution of industry and agriculture, a considerable amount of heavy metal waste and sewerage is currently being introduced into the atmosphere (Soffianian et al. 2014; Sauliutė and Svecevičius 2015). Heavy metal concentration in the environment is further increased by the use of pesticides and inorganic fertilizers in agriculture (Kim et al. 2018; Ahmad et al. 2018a). HM toxicity is a source of agricultural contamination in several regions of the world that may be caused by long-term usage of waste water, improper irrigation, industrial effluent, and phosphatic fertilizers, resulting in crop deterioration. The harmful effects of HM on agricultural crops is mostly due to overly produced ROS, which causes oxidative stress and membrane damage in plants (da Silva et al. 2018; Ahmad et al. 2018b). The findings are in line with those of Basalah et al. (2013), who discovered that applying NO and salicylic acid (SA) singly or in combination might lower Cd content thereby enhances nutritional content in wheat seedlings under Cd stress. Calcium (Ca) supplementation also increased the content of major nutrients while lowering Cd buildup in Cd-stressed mustard plants (Ahmad et al. 2015).

Plants have many defense mechanisms to combat heavy metal stress, including the activation of complex metabolic procedures and an antioxidant defense system that detoxifies ROS production (Vardhini and Anjum 2015). The use of 24-epibrassinolide (24-EBL) increased the uptake of magnesium, calcium, zinc, and iron, and this is attributable to its impact on the function of transport proteins (Song et al. 2016). Under Hg stress, 24-EBL increases the amount of Na and K in *Raphanus sativus*, demonstrating that BRs contribute in regulating plant cell membrane permeability and ions transportation (Kapoor et al. 2016). By boosting the assimilation of Ca, P, K, S, and Mg macronutrients, BRs rescued cadmium-mediated toxicity in *Pisum sativum*, resulting in increased plant physio-biochemical activity (Jan et al. 2018).

According to another investigation, chromium (Cr) interferes with the uptake or deposition of a diverse variety of other metals or minerals, including Mg, P, K, Mn, Ca, and Fe, in both aerial and root sections of plants, resulting in their lower cellular or tissue concentrations (Emamverdian et al. 2015). There existed a strong positive association between Cr and Fe concentration in plant tissues (Emamverdian et al. 2015). There is no Cd-specific transporter, although Cd appears to be carried through various categories of Ca, Fe, and Zn transporters, influencing their absorption and redistribution in plants and thereby causing deficiency of such ions (Gupta et al. 2017).

Metallic micronutrients such as Cu, Mo, Fe, Ni, Mn, and Zn are useful in the plant metabolic activities as component, stimulators, or inhibitors of enzymatic catalyzed reactions (Dubey et al. 2020). As a result, efficient uptake of metallic micronutrients can compete with HM, thereby lowering toxicity and strengthening antioxidant defenses. Kaya et al. (2020) studied the response of exogenous silicon (Si) supplementation on the alleviation of Cd stress in pepper plants. Cd toxicity resulted in a noteworthy depression in plant biomass, ascorbic acid content, K and Ca levels, relative water contents, chlorophyll, glutathione, PSII efficiency, and water potential. Besides, a significant rise in proline, electrolyte leakage, MDA, and H_2O_2 alongside bettered antioxidant enzyme activities was noticed. Furthermore, Cd accumulation in aerial plant parts, H_2S, and NO were also higher in Cd-stressed pepper plants. Exogenous Si notably improved Cd tolerance in pepper plants by lowering oxidative stress, Cd accumulation, and strengthening antioxidant defense system. Likewise, Zaheer et al. (2018) reported the role of exogenous K and Si in the alleviation of Cd stress in *Gladiolus grandiflora* L. Cd toxicity generated larger quantities of H_2O_2 and MDA in plants. However, supplementation of Si and K counteracted Cd effects by lowering H_2O_2 and MDA levels. Further, the antioxidant enzyme activities were also better in plants supplemented with Si and K. The degree of oxidative damage was minimal in plants treated with exogenous K and Si. In another study, Si application protected plants from Cd effects and strengthened oxidative defense system in *Phaseolus vulgaris* plants (Rady et al. 2019). Siddiqui et al. (2012) reported the effects of exogenous K and Ca on nutrients, Cd accumulation, photosynthetic pigments, lipid peroxidation, proline, and antioxidant defense system in *Vicia faba* L. plants exposed to Cd toxicity in the growth medium. The degree of oxidative damage was minimal in plants treated with Ca and K reflected in the form of bettered antioxidant defense system under Cd toxicity. Proline accumulation was also higher in plants supplemented with exogenous Ca and K under Cd stress. In another study, *Gladilous grandiflora* L. plants manifested a remarkable decline in growth and chlorophyll contents due to Cd-induced oxidative injury evident in the form of higher H_2O_2 and MDA. Plants supplemented with exogenous K exhibited increase in the activities of antioxidant enzymes that counteracted Cd-generated oxidative damage by lowering H_2O_2 and MDA levels (Yasin et al. 2018). Liang et al. (2016) reported the ameliorative effects of exogenous sulfur (S) in two *Brassica chinensis* L. cultivars (Pakchoi) under Cd toxicity. Exposure of Pakchoi cultivars to Cd stress resulted in inhibited growth due to enhanced production of superoxide radicals and lipid peroxidation of membranes. Exogenous S improved Cd tolerance by lowering oxidative damage in Pakchoi plants. The decline in oxidative injury due to exogenous S is attributed to S-induced bettered efficiency of ascorbate-glutathione cycle.

Role of Nutrients in ROS Metabolism Under Low- and High-Temperature Stress

Temperature and UV radiation spikes can always have a major effect on plant development, crop production, as well as quality of fruit. Abiotic stresses can activate a variety of metabolic pathways in plant cells, including plastid biosynthesis and pigments or secondary metabolite biosynthesis (Bita and Gerats 2013; Sun et al. 2018). Antioxidant molecules, including carotenoids, ascorbic acid (AscA), and polyphenols, are deposited in responses to heat stress because thermo-resistance involves the modification of metabolic processes involved in ROS elimination (Scarano et al. 2020).

After exposure to cold stress-mediated oxidative stress, and ROS buildup are frequent mechanisms reported in several species (Clemente-Moreno et al. 2020). The generation of ROS in chilling-temperature environments is thought to be caused by an imbalance in photosynthetic primary and secondary processes. Because electron transportation and light capturing are less temperature susceptible than CO_2 assimilating biochemical processes, consequently, PSII excitation pressure is generated (Distelbarth et al. 2013).

Intracellular NO is primarily produced by the enzymes NO synthase as well as nitrate reductase. NO has also been demonstrated to play a role in abiotic stress responses such as heavy metal toxicity, ozone stress, moisture limiting stress, elevated salt stress, extreme temperatures, and low-temperature stress (Fancy et al. 2017).

Magnesium and nitrogen (Mg, N) are structural parts of chlorophyll needed for photosynthetic activity, while phosphorus is needed for storage and energy production. Phosphorus is a structural constituent of nucleic acids. Furthermore, potassium is needed for enzyme activity and osmotic management (Waraich et al. 2011).

Nutrition is significantly important for plant growth and development. It is thought that with the application of ample nourishment, a plant can generate a notable amount of biomass according to per unit of transpired water if compared with plants experiencing nutrient-deficit conditions (Waraich et al. 2012).

Radin and Mathews (1989) discovered that P- and N-deficient plants had a notable decline in root cortical cell hydraulic conductance. Recent studies reported that plant nutrients are important not only for plant development and growth but also for agriculture WUE improvement (Wang et al. 2017; Li et al. 2020). A lot of studies have been published that show the function of nutrients in mitigating different abiotic stress factors, such as Si has positive impacts on enhanced salinity resistance in wheat crop and K for improved salinity resistance in wheat (Waraich et al. 2012). However according Byrnes and Bumb (1998), fertilizer consumption will have to double in the coming 20 years to meet the required increases in food productivity. Plant nutrition research appears to be an extremely important subject in the upcoming years, boosting crop yield and soil fertility maintenance. The ability to establish adaptive responses to escape or manage stress is critical to their existence and yield when they are exposed to environmental challenges. According to growing research, nutrient status of plants has a significant impact on their ability to adapt to harsh climatic situations (Campbell 2017).

It has been found that nitrogen fertilization can help to minimize the negative impacts of abiotic challenges (Noreen et al. 2018). Nitric oxide (NO) is a highly reactive and volatile, freely permeable through membrane with a wide range of regulatory activities in a variety of physiological phenomena, including PCD, seed germination, cell aging, stomata closing, ethylene generation, and leaf elongation as well as a signal molecule, regulating adaptations to biotic and abiotic stress factors like water limitation, salt-induced stress, UV-B radiation, and extreme temperatures (Sami et al. 2018; Bhuyan et al. 2020; Prakash et al. 2021).

Potassium (K) is one of the mineral nutrients that is critical for agricultural plant survival in stressful environments. K is required for several physiological functions, including photosynthetic activity, photosynthate transportation into sink tissues, turgidity regulation, and enzymatic functions under stress (Hasanuzzaman et al. 2018).

Ca is thought to be required for healing from low-temperature stress because it triggers the plasma membrane localized enzyme ATPase, which itself is required to shuttle back nutrients eliminated during cell injury (Qian and Bo 2001; Liu et al. 2014). Ca performs a function in chilling injury resistance because dehydration is indeed the common factor. Calcium has an important part in cellular structure preservation. It stimulates the plasma membrane enzyme ATPase, which restores nutrients loss due to Ca deficit cell membrane degradation and helps the plant to avoid from cold injury. Calcium generally functions as calmodulin that also influences plant metabolic activity as well as promotes plant development in low-temperature environments (Waraich et al. 2012; Sun et al. 2018).

Magnesium promotes root growth and increases root surface area, therefore aids in absorption capability of water and mineral. Because magnesium is a component of chlorophyll, it increases the level of sucrose in the leaves and improves sucrose transportation from the leaves to the roots (Boaretto et al. 2020). Due to temperature-induced stressful conditions (lower or higher), magnesium increases carbohydrate transportation by enhancing phloem export and lowers ROS synthesis as well as photo-oxidative damage to chloroplast (Waraich et al. 2012). In this context, Bradáčová et al. (2016) reported that maize plants treated with Zn and Mn had several times better ROS detoxification due to enhanced antioxidant system. Similarly, Iqbal et al. (2015) also reported better yield, chlorophyll, and strengthened antioxidant system in wheat with Se pretreatent under high temperature stress. Better antioxidant system was also found in sorghum with Se pretreatment under temperature stress (Chu et al. 2010). Wheat plants with exogenous Mo improved ROS metabolism reflected as stimulated antioxidant system under heat stress (Noreen et al. 2018).

The aforementioned literature indicated the notable involvement of different plant nutrients in regulation of ROS metabolism mirrored as bettered antioxidant system. However, the literature is scarce on this topic, and it need further research by the experts. Moreover, the role of nutrients in mitigation of different abiotic stresses by regulation of ascorbate-glutathione cycle is not yet established. Also, the involvement of nutrients in mediating plant physiological responses and their relationship with ROS metabolism are not fully understood and needs to be explored.

References

Ahanger, M.A. and Agarwal, R. (2017a). Potassium up-regulates antioxidant metabolism and alleviates growth inhibition under water and osmotic stress in wheat (*Triticum aestivum* L.). *Protoplasma* 254 (4): 1471–1486.

Ahanger, M.A. and Agarwal, R. (2017b). Salinity stress induced alterations in antioxidant metabolism and nitrogen assimilation in wheat (*Triticum aestivum* L.) as influenced by potassium supplementation. *Plant Physiol. Biochem.* 115: 449–460.

Ahmad, P., Sarwat, M., Bhat, N.A. et al. (2015). Alleviation of cadmium toxicity in *Brassica juncea* L.(Czern. & Coss.) by calcium application involves various physiological and biochemical strategies. *PLos ONE* 10 (1): e0114571.

Ahmad, S., Khan, H., Shahab, U. et al. (2017). Protein oxidation: an overview of metabolism of sulphur containing amino acid, cysteine. *Front. Biosci. (Schol Ed)* 9: 71–87.

Ahmad, I., Akhtar, M.J., Mehmood, S. et al. (2018a). Combined application of compost and Bacillus sp. CIK-512 ameliorated the lead toxicity in radish by regulating the homeostasis of antioxidants and lead. *Ecotoxicol. Environ. Saf.* 148: 805–812.

Ahmad, P., Ahanger, M.A., Alyemeni, M.N. et al. (2018b). Exogenous application of nitric oxide modulates osmolyte metabolism, antioxidants, enzymes of ascorbate-glutathione cycle and promotes growth under cadmium stress in tomato. *Protoplasma* 255 (1): 79–93.

Akram, N.A., Shafiq, F., and Ashraf, M. (2017). Ascorbic acid-a potential oxidant scavenger and its role in plant development and abiotic stress tolerance. *Front. Plant Sci.* 8: 613.

Akyol, T.Y., Yilmaz, O., Uzilday, B. et al. (2020). Plant response to salinity: an analysis of ROS formation, signaling, and antioxidant defense. *Turk. J. Bot.* 44 (1): 1–13.

Aldon, D., Mbengue, M., Mazars, C., and Galaud, J.-P. (2018). Calcium signalling in plant biotic interactions. *Int. J. Mol. Sci.* 19 (3): 665.

Ali, M.M., Shafique, M.W., Gull, S. et al. (2021). Alleviation of heat stress in tomato by exogenous application of sulfur. *Horticulturae* 7 (2): 21.

Asada, K. (2019). Production and action of active oxygen species in photosynthetic tissues. In: *Causes of Photooxidative Stress and Amelioration of Defense Systems in Plants* (ed. C.H. Foyer), 77–104. CRC Press.

Basalah, M.O., Ali, H.M., Al-Whaibi, M.H. et al. (2013). Nitric oxide and salicylic acid mitigate cadmium stress in wheat seedlings. *J. Pure Appl. Microbiol.* 7: 139–148.

Bhuyan, M.B., Hasanuzzaman, M., Parvin, K. et al. (2020). Nitric oxide and hydrogen sulfide: two intimate collaborators regulating plant defense against abiotic stress. *Plant Growth Regul.* 90 (3): 409–424.

Biczak, R. (2016). Quaternary ammonium salts with tetrafluoroborate anion: phytotoxicity and oxidative stress in terrestrial plants. *J. Hazard. Mater.* 304: 173–185.

Birnie-Gauvin, K., Costantini, D., Cooke, S.J., and Willmore, W.G. (2017). A comparative and evolutionary approach to oxidative stress in fish: a review. *Fish Fish.* 18 (5): 928–942.

Bita, C. and Gerats, T. (2013). Plant tolerance to high temperature in a changing environment: scientific fundamentals and production of heat stress-tolerant crops. *Front. Plant Sci.* 4: 273.

Boaretto, R.M., Hippler, F.W., Ferreira, G.A. et al. (2020). The possible role of extra magnesium and nitrogen supply to alleviate stress caused by high irradiation and temperature in lemon trees. *Plant Soil* 457 (1): 57–70.

Bor, M. and Turkan, I. (2019). Is there a room for GABA in ROS and RNS signalling? *Environ. Exp. Bot.* 161: 67–73.

Bose, J., Rodrigo-Moreno, A., and Shabala, S. (2014). ROS homeostasis in halophytes in the context of salinity stress tolerance. *J. Exp. Bot.* 65 (5): 1241–1257.

Bradáčová, K., Weber, N.F., Morad-Talab, N. et al. (2016). Micronutrients (Zn/Mn), seaweed extracts, and plant growth-promoting bacteria as cold-stress protectants in maize. *Chem. Biol. Technol. Agric.* 3 (1): 1–10.

Byrnes, B.H. and Bumb, B.L. (1998). Population growth, food production and nutrient requirements. In: *Mineral Nutrition of Crops: Mechanisms and Implications* (ed. Z. Rengel), 1–27. New York: The Haworth Press.

Čamagajevac, I.Š., Maronić, D.Š., Pfeiffer, T.Ž. et al. (2019). Nitric oxide and hydrogen peroxide in plant response to biotic stress. In: *Nitric Oxide and Hydrogen Peroxide Signaling in Higher Plants* (ed. D.K. Gupta, J.M. Palma, and F.J. Corpas), 221–243. Springer.

Campbell, L.C. (2017). Managing soil fertility decline. In: *Nutrient use in Crop Production* (ed. Z. Rengel), 29–52. CRC Press.

Cavaco, A.R., Matos, A.R., and Figueiredo, A. (2021). Speaking the language of lipids: the cross-talk between plants and pathogens in defence and disease. *Cell. Mol. Life Sci.* 1–17.

Caverzan, A., Casassola, A., and Brammer, S.P. (2016). Antioxidant responses of wheat plants under stress. *Genet. Mol. Biol.* 39 (1): 1–6.

Chaki, M., Begara-Morales, J.C., and Barroso, J.B. (2020). *Oxidative Stress in Plants.* Multidisciplinary Digital Publishing Institute.

Chu, J., Yao, X., and Zhang, Z. (2010). Responses of wheat seedlings to exogenous selenium supply under cold stress. *Biol. Trace Elem. Res.* 136: 355–363.

Clemente-Moreno, M.J., Omranian, N., Sáez, P.L. et al. (2020). Low-temperature tolerance of the Antarctic species Deschampsia antarctica: a complex metabolic response associated with nutrient remobilization. *Plant Cell Environ.* 43 (6): 1376–1393.

Corpas, F.J., Freschi, L., Rodríguez-Ruiz, M. et al. (2018). Nitro-oxidative metabolism during fruit ripening. *J. Exp. Bot.* 69 (14): 3449–3463.

Czarnocka, W. and Karpiński, S. (2018). Friend or foe? Reactive oxygen species production, scavenging and signaling in plant response to environmental stresses. *Free Radic. Biol. Med.* 122: 4–20.

Del Río, L.A. (2015). ROS and RNS in plant physiology: an overview. *J. Exp. Bot.* 66 (10): 2827–2837.

Del Río, L.A. and López-Huertas, E. (2016). ROS generation in peroxisomes and its role in cell signaling. *Plant Cell Physiol.* 57 (7): 1364–1376.

Demircan, N., Cucun, G., and Uzilday, B. (2020). Mitochondrial alternative oxidase (AOX1a) is required for the mitigation of arsenic-induced oxidative stress in *Arabidopsis thaliana*. *Plant Biotechnol. Rep.* 14: 235–245.

Di Meo, S., Reed, T.T., Venditti, P., and Victor, V.M. (2016). Role of ROS and RNS sources in physiological and pathological conditions. *Oxidative Med. Cell. Longev.* 2016: 1245049.

Distelbarth, H., Nägele, T., and Heyer, A. (2013). Responses of antioxidant enzymes to cold and high light are not correlated to freezing tolerance in natural accessions of *Arabidopsis thaliana*. *Plant Biol.* 15 (6): 982–990.

Dubey, R., Gupta, D.K., and Sharma, G.K. (2020). Chemical stress on plants. In: *New Frontiers in Stress Management for Durable Agriculture* (ed. A. Rakshit, H.B. Singh, A.K. Singh et al.), 101–128. Springer.

Dunnill, C., Patton, T., Brennan, J. et al. (2017). Reactive oxygen species (ROS) and wound healing: the functional role of ROS and emerging ROS-modulating technologies for augmentation of the healing process. *Int. Wound J.* 14 (1): 89–96.

Emamverdian, A., Ding, Y., Mokhberdoran, F., and Xie, Y. (2015). Heavy metal stress and some mechanisms of plant defense response. *Sci. World J.* 2015: 756120.

Erel, R., Yermiyahu, U., Ben-Gal, A. et al. (2015). Modification of non-stomatal limitation and photoprotection due to K and Na nutrition of olive trees. *J. Plant Physiol.* 177: 1–10.

Espinosa-Vellarino, F.L., Garrido, I., Ortega, A. et al. (2020). Effects of Antimony on reactive oxygen and nitrogen species (ROS and RNS) and antioxidant mechanisms in tomato plants. *Front. Plant Sci.* 11: 674.

Fancy, N.N., Bahlmann, A.K., and Loake, G.J. (2017). Nitric oxide function in plant abiotic stress. *Plant Cell Environ.* 40 (4): 462–472.

Farooq, A., Bukhari, S.A., Akram, N.A. et al. (2020). Exogenously applied ascorbic acid-mediated changes in osmoprotection and oxidative defense system enhanced water stress tolerance in different cultivars of safflower (*Carthamus tinctorious* L.). *Plan. Theory* 9 (1): 104.

Fatma, M., Masood, A., Per, T.S. et al. (2016). Interplay between nitric oxide and sulfur assimilation in salt tolerance in plants. *Crop J.* 4 (3): 153–161.

Fernández-Crespo, E., Navarro, J.A., Serra-Soriano, M. et al. (2017). Hexanoic acid treatment prevents systemic MNSV movement in *Cucumis melo* plants by priming callose deposition correlating SA and OPDA accumulation. *Front. Plant Sci.* 8: 1793.

Fewson, C. and Nicholas, D. (1960). Utilization of nitric oxide by micro-organisms and higher plants. *Nature* 188: 794–796. https://doi.org/10.1038/188794a0.

Forrester, S.J., Kikuchi, D.S., Hernandes, M.S. et al. (2018). Reactive oxygen species in metabolic and inflammatory signaling. *Circ. Res.* 122 (6): 877–902.

García-Caparrós, P., De Filippis, L., Gul, A. et al. (2021). Oxidative stress and antioxidant metabolism under adverse environmental conditions: a review. *Bot. Rev.* 87 (4): 421–466.

Gill, S.S., Anjum, N.A., Gill, R. et al. (2015). Superoxide dismutase – mentor of abiotic stress tolerance in crop plants. *Environ. Sci. Pollut. Res.* 22 (14): 10375–10394.

Gu, X., Sun, J., Li, S. et al. (2013). Oxidative stress induces DNA demethylation and histone acetylation in SH-SY5Y cells: potential epigenetic mechanisms in gene transcription in Aβ production. *Neurobiol. Aging* 34 (4): 1069–1079.

Guan, G. and Lan, S. (2018). Implications of antioxidant systems in inflammatory bowel disease. *Biomed. Res. Int.* 2018: 1290179.

Gupta, D., Pena, L.B., Romero-Puertas, M.C. et al. (2017). NADPH oxidases differentially regulate ROS metabolism and nutrient uptake under cadmium toxicity. *Plant Cell Environ.* 40 (4): 509–526.

Gupta, S.K., Sharma, M., Deeba, F., and Pandey, V. (2018). Role of reactive oxygen species in photophosphorylation and damage to D1 protein: past and present. In: *Reactive Oxygen Species Plants* (ed. V.P. Singh, S. Singh, D.K. Tripathi et al.), 165. Wiley.

Hasanuzzaman, M., Bhuyan, M., Nahar, K. et al. (2018). Potassium: a vital regulator of plant responses and tolerance to abiotic stresses. *Agronomy* 8 (3): 31.

Hasanuzzaman, M., Bhuyan, M., Anee, T.I. et al. (2019). Regulation of ascorbate-glutathione pathway in mitigating oxidative damage in plants under abiotic stress. *Antioxidants* 8 (9): 384.

Hasanuzzaman, M., Bhuyan, M., Parvin, K. et al. (2020). Regulation of ROS metabolism in plants under environmental stress: a review of recent experimental evidence. *Int. J. Mol. Sci.* 21 (22): 8695.

Herrera-Vásquez, A., Carvallo, L., Blanco, F. et al. (2015). Transcriptional control of glutaredoxin GRXC9 expression by a salicylic acid-dependent and NPR1-independent pathway in Arabidopsis. *Plant Mol. Biol. Report.* 33 (3): 624–637.

Iqbal, M., Hussain, I., Liaqat, H. et al. (2015). Exogenously applied selenium reduces oxidative stress and induces heat tolerance in spring wheat. *Plant Physiol. Biochem.* 94: 95–103.

Islam, M., Lee, B.-R., Park, S.-H. et al. (2017). Cultivar variation in hormonal balance is a significant determinant of disease susceptibility to *Xanthomonas campestris* pv. campestris in *Brassica napus*. *Front. Plant Sci.* 8: 2121.

Jan, S., Alyemeni, M.N., Wijaya, L. et al. (2018). Interactive effect of 24-epibrassinolide and silicon alleviates cadmium stress via the modulation of antioxidant defense and glyoxalase systems and macronutrient content in *Pisum sativum* L. seedlings. *BMC Plant Biol.* 18 (1): 1–18.

Jan, R., Asaf, S., Numan, M., and Kim, K.-M. (2021). Plant secondary metabolite biosynthesis and transcriptional regulation in response to biotic and abiotic stress conditions. *Agronomy* 11 (5): 968.

Jing, T., Ardiansyah, R., Xu, Q. et al. (2020). Reprogramming of cell fate during root regeneration by transcriptional and epigenetic networks. *Front. Plant Sci.* 11: 317.

Kapoor, D., Rattan, A., Gautam, V., and Bhardwaj, R. (2016). Mercury-induced changes in growth, metal & ions uptake, photosynthetic pigments, osmoprotectants and antioxidant defence system in *Raphanus sativus* L. seedlings and role of steroid hormone in stress amelioration. *J. Pharmacogn. Phytochem.* 5 (4): 259.

Kapoor, D., Singh, S., Kumar, V. et al. (2019). Antioxidant enzymes regulation in plants in reference to reactive oxygen species (ROS) and reactive nitrogen species (RNS). *Plant Gene* 19: 100182.

Kärkönen, A. and Kuchitsu, K. (2015). Reactive oxygen species in cell wall metabolism and development in plants. *Phytochemistry* 112: 22–32.

Kaur, G. and Asthir, B. (2015). Proline: a key player in plant abiotic stress tolerance. *Biol. Plant.* 59 (4): 609–619.

Kaya, C., Akram, N.A., Ashraf, M. et al. (2020). Exogenously supplied silicon (Si) improves cadmium tolerance in pepper (*Capsicum annuum* L.) by up-regulating the synthesis of nitric oxide and hydrogen sulfide. *J. Biotechnol.* 316: 35–45.

Kerchev, P., Waszczak, C., Lewandowska, A. et al. (2016). Lack of GLYCOLATE OXIDASE1, but not GLYCOLATE OXIDASE2, attenuates the photorespiratory phenotype of CATALASE2-deficient Arabidopsis. *Plant Physiol.* 171 (3): 1704–1719.

Kim, H.S., Seo, B.-H., Kuppusamy, S. et al. (2018). A DOC coagulant, gypsum treatment can simultaneously reduce As, Cd and Pb uptake by medicinal plants grown in contaminated soil. *Ecotoxicol. Environ. Saf.* 148: 615–619.

Kolupaev, Y.E., Karpets, Y.V., and Dmitriev, A. (2015). Signal mediators in plants in response to abiotic stress: calcium, reactive oxygen and nitrogen species. *Cytol. Genet.* 49 (5): 338–348.

König, K., Vaseghi, M.J., Dreyer, A., and Dietz, K.J. (2018). The significance of glutathione and ascorbate in modulating the retrograde high light response in *Arabidopsis thaliana* leaves. *Physiol. Plant.* 162 (3): 262–273.

Krumova, K. and Cosa, G. (2016). *Singlet Oxygen: Applications in Biosciences and Nanosciences* (ed. S. Nonell and C. Flors), 1e, vol. 1, 1–21. Royal Society of Chemistry.

Kumar, A., Guleria, S., Ghosh, D. et al. (2021a). Managing reactive oxygen species – some learnings from high altitude extremophytes. *Environ. Exp. Bot.* 189: 104525.

Kumar, S.J., Chintagunta, A.D., Reddy, Y.M. et al. (2021b). Implications of reactive oxygen and nitrogen species in seed physiology for sustainable crop productivity under changing climate conditions. *Curr. Plant Biol.* 26: 100197.

La, V.H., Lee, B.-R., Islam, M. et al. (2020). Characterization of glutamate-mediated hormonal regulatory pathway of the drought responses in relation to proline metabolism in *Brassica napus* L. *Plan. Theory* 9 (4): 512.

Le Gall, H., Philippe, F., Domon, J.-M. et al. (2015). Cell wall metabolism in response to abiotic stress. *Plan. Theory* 4 (1): 112–166.

Lee, B.-R., Jin, Y.-L., Park, S.-H. et al. (2015). Genotypic variation in N uptake and assimilation estimated by ^{15}N tracing in water deficit-stressed *Brassica napus*. *Environ. Exp. Bot.* 109: 73–79.

Li, L., Pan, S., Melzer, R., and Fricke, W. (2020). Apoplastic barriers, aquaporin gene expression and root and cell hydraulic conductivity in phosphate-limited sheepgrass plants. *Physiol. Plant.* 168 (1): 118–132.

Liang, T., Ding, H., Wang, G. et al. (2016). Sulfur decreases cadmium translocation and enhances cadmium tolerance by promoting sulfur assimilation and glutathione metabolism in *Brassica chinensis* L. *Ecotoxicol. Environ. Saf.* 124: 129–137.

Lindermayr, C., Oracz, K., Cuypers, A. et al. (2021). Highlights of POG 2019-Plant Oxygen Group Conference. *Front. Plant Sci.* 12: 72.

Liu, A., Chen, S., Chang, R. et al. (2014). Arbuscular mycorrhizae improve low temperature tolerance in cucumber via alterations in H_2O_2 accumulation and ATPase activity. *J. Plant Res.* 127 (6): 775–785.

Lopes, A.C.B.A., Peixe, T.S., Mesas, A.E., and Paoliello, M.M. (2016). Lead exposure and oxidative stress: a systematic review. *Rev. Environ. Contam. Toxicol.* 236: 193–238.

Lu, F., Liang, X., Lu, H. et al. (2017). Overproduction of superoxide dismutase and catalase confers cassava resistance to *Tetranychus cinnabarinus*. *Sci. Rep.* 7 (1): 1–13.

Mahmud, J.A., Hasanuzzaman, M., Khan, M.I.R. et al. (2020). β-Aminobutyric acid pretreatment confers salt stress tolerance in *Brassica napus* L. by modulating reactive oxygen species metabolism and methylglyoxal detoxification. *Plan. Theory* 9 (2): 241.

Majumdar, R., Barchi, B., Turlapati, S.A. et al. (2016). Glutamate, ornithine, arginine, proline, and polyamine metabolic interactions: the pathway is regulated at the post-transcriptional level. *Front. Plant Sci.* 7: 78.

Maurya, V.K., Kumar, D., Pathak, C., and Tiwari, B.S. (2018). Involvement of reactive species of oxygen and nitrogen in triggering programmed cell death in plants. In: *Biotic and Abiotic Stress Tolerance in Plants* (ed. S. Vats), 257–278. Springer.

Mittler, R. (2017). ROS are good. *Trends Plant Sci.* 22 (1): 11–19.

Nadarajah, K.K. (2020). ROS homeostasis in abiotic stress tolerance in plants. *Int. J. Mol. Sci.* 21 (15): 5208.

Nieves-Cordones, M., López-Delacalle, M., Ródenas, R. et al. (2019). Critical responses to nutrient deprivation: a comprehensive review on the role of ROS and RNS. *Environ. Exp. Bot.* 161: 74–85.

Niu, L., Yu, J., Liao, W., and Yu, J. (2019). Hydrogen peroxide and nitric oxide signaling network. In: *Nitric Oxide and Hydrogen Peroxide Signaling in Higher Plants*, 17–36. Springer.

Noctor, G., Reichheld, J.P., and Foyer, C.H. (2018). ROS-related redox regulation and signaling in plants. *Semin. Cell Dev. Biol.* 80: 3–12, 1084–9521. https://doi.org/10.1016/j.semcdb.2017.07.013.

Noreen, S., Fatima, Z., Ahmad, S., and Ashraf, M. (2018). Foliar application of micronutrients in mitigating abiotic stress in crop plants. In: *Plant Nutrients and Abiotic Stress Tolerance* (ed. M. Hasanuzzaman, M. Fujita, H. Oku et al.), 95–117. Springer.

Nounjan, N., Chansongkrow, P., Charoensawan, V. et al. (2018). High performance of photosynthesis and osmotic adjustment are associated with salt tolerance ability in rice carrying drought tolerance QTL: physiological and co-expression network analysis. *Front. Plant Sci.* 9: 1135.

Novaković, L., Guo, T., Bacic, A. et al. (2018). Hitting the wall – sensing and signaling pathways involved in plant cell wall remodeling in response to abiotic stress. *Plan. Theory* 7 (4): 89.

Oyarburo, N.S., Machinandiarena, M.F., Feldman, M.L. et al. (2015). Potassium phosphite increases tolerance to UV-B in potato. *Plant Physiol. Biochem.* 88: 1–8.

Phua, S.Y., De Smet, B., Remacle, C. et al. (2021). Reactive oxygen species and organellar signaling. *J. Exp. Bot.* https://doi.org/10.1093/jxb/erab218.

Pisoschi, A.M. and Pop, A. (2015). The role of antioxidants in the chemistry of oxidative stress: a review. *Eur. J. Med. Chem.* 97: 55–74.

Prakash, V., Singh, V.P., Tripathi, D.K. et al. (2021). Nitric oxide (NO) and salicylic acid (SA): a framework for their relationship in plant development under abiotic stress. *Plant Biol.* 23: 39–49.

Qian, Z.Z.S.L.R. and Bo, W.J. (2001). Effect of Ca~(2+) pretreatment on the Ca~(2+)-ATPase activity in the mesophyll cells of pepper seedling under heat stress. *Acta Photophysiol. Sinica* 27: 451–454.

Radin J.W. and Mathews M.A. (1989). Water transport properties of cortical cells in roots of nitrogen- and phosphorus-deficient cotton seedlings. *Plant Physiol.* 89: 264–268.

Rady, M.M., Elrys, A.S., El-Maati, M.F.A., and Desoky, E.S.M. (2019). Interplaying roles of silicon and proline effectively improve salt and cadmium stress tolerance in *Phaseolus vulgaris* plant. *Plant Physiol. Biochem.* 139: 558–568.

Raja, V., Majeed, U., Kang, H. et al. (2017). Abiotic stress: Interplay between ROS, hormones and MAPKs. *Environ. Exp. Bot.* 137: 142–157.

Rejeb, K.B., Abdelly, C., and Savouré, A. (2014). How reactive oxygen species and proline face stress together. *Plant Physiol. Biochem.* 80: 278–284.

Sachdev, S., Ansari, S.A., Ansari, M.I. et al. (2021). Abiotic stress and reactive oxygen species: generation, signaling, and defense mechanisms. *Antioxidants* 10 (2): 277.

Saddhe, A.A., Malvankar, M.R., Karle, S.B., and Kumar, K. (2019). Reactive nitrogen species: paradigms of cellular signaling and regulation of salt stress in plants. *Environ. Exp. Bot.* 161: 86–97.

Sami, F., Faizan, M., Faraz, A. et al. (2018). Nitric oxide-mediated integrative alterations in plant metabolism to confer abiotic stress tolerance, NO crosstalk with phytohormones and NO-mediated post translational modifications in modulating diverse plant stress. *Nitric Oxide* 73: 22–38.

Sauliutė, G. and Svecevičius, G. (2015). Heavy metal interactions during accumulation via direct route in fish: a review. *Zool Ecol.* 25 (1): 77–86.

Scarano, A., Olivieri, F., Gerardi, C. et al. (2020). Selection of tomato landraces with high fruit yield and nutritional quality under elevated temperatures. *J. Sci. Food Agric.* 100 (6): 2791–2799.

Schieber, M. and Chandel, N.S. (2014). ROS function in redox signaling and oxidative stress. *Curr. Biol.* 24 (10): R453–R462.

Sevilla, F., Camejo, D., Ortiz-Espín, A. et al. (2015). The thioredoxin/peroxiredoxin/sulfiredoxin system: current overview on its redox function in plants and regulation by reactive oxygen and nitrogen species. *J. Exp. Bot.* 66 (10): 2945–2955.

Sewelam, N., Kazan, K., and Schenk, P.M. (2016). Global plant stress signaling: reactive oxygen species at the cross-road. *Front. Plant Sci.* 7: 187.

Seyfferth, C. and Tsuda, K. (2014). Salicylic acid signal transduction: the initiation of biosynthesis, perception and transcriptional reprogramming. *Front. Plant Sci.* 5: 697.

Shemi, R., Wang, R., Gheith, E.-S.M. et al. (2021). Effects of salicylic acid, zinc and glycine betaine on morpho-physiological growth and yield of maize under drought stress. *Sci. Rep.* 11 (1): 1–14.

Siddiqui, M.H., Al-Whaibi, M.H., Sakran, A.M. et al. (2012). Effect of calcium and potassium on antioxidant system of *Vicia faba* L. under cadmium stress. *Int. J. Mol. Sci.* 13 (6): 6604–6619.

da Silva, C.J., Canatto, R.A., Cardoso, A.A. et al. (2018). Oxidative stress triggered by arsenic in a tropical macrophyte is alleviated by endogenous and exogenous nitric oxide. *Braz. J. Bot.* 41 (1): 21–28.

Singh, A., Kumar, A., Yadav, S., and Singh, I.K. (2019). Reactive oxygen species-mediated signaling during abiotic stress. *Plant Gene* 18: 100173.

Soares, C., Carvalho, M.E., Azevedo, R.A., and Fidalgo, F. (2019). Plants facing oxidative challenges – a little help from the antioxidant networks. *Environ. Exp. Bot.* 161: 4–25.

Soffianian, A., Madani, E.S., and Arabi, M. (2014). Risk assessment of heavy metal soil pollution through principal components analysis and false color composition in Hamadan Province, Iran. *Environ. Sys. Res.* 3 (1): 1–14.

Sofy, M. (2015). Application of salicylic acid and zinc improves wheat yield through physiological processes under different levels of irrigation intervals. *Int. J. Plant Res.* 5: 136–156.

Sohag, A.A.M., Tahjib-Ul-Arif, M., Polash, M.A.S. et al. (2020). Exogenous glutathione-mediated drought stress tolerance in rice (*Oryza sativa* L.) is associated with lower oxidative damage and favorable ionic homeostasis. *Iran J. Sci. Technol. Trans. A Sci.* 44 (4): 955–971.

Song, Y., Dong, Y., Tian, X. et al. (2016). Role of foliar application of 24-epibrassinolide in response of peanut seedlings to iron deficiency. *Biol. Plant.* 60 (2): 329–342.

Sosa, V., Moliné, T., Somoza, R. et al. (2013). Oxidative stress and cancer: an overview. *Ageing Res. Rev.* 12 (1): 376–390.

Suganya, A., Saravanan, A., and Manivannan, N. (2020). Role of zinc nutrition for increasing zinc availability, uptake, yield, and quality of maize (*Zea mays* L.) grains: an overview. *Commun. Soil Sci. Plant Anal.* 51: 2001–2021.

Sun, T., Yuan, H., Cao, H. et al. (2018). Carotenoid metabolism in plants: the role of plastids. *Mol. Plant* 11 (1): 58–74.

Talukdar, D. (2017). Balancing roles of reactive oxygen species in plants' response to metalloid exposure. In: *Reactive Oxygen Species in Plants: Boon Or Bane-Revisiting the Role of ROS* (ed. V.P. Singh, S. Singh, D.K. Tripathi et al.), 51. Wiley.

Taverne, Y.J., Merkus, D., Bogers, A.J. et al. (2018). Reactive oxygen species: radical factors in the evolution of animal life: a molecular timescale from Earth's earliest history to the rise of complex life. *BioEssays* 40 (3): 1700158.

Tiwari, S., Tiwari, S., Singh, M. et al. (2018). Generation mechanisms of reactive oxygen species in the plant cell: an Overview. In: *Reactive Oxygen Species in Plants: Boon or Bane-revisiting the Role of ROS* (ed. V.P. Singh, S. Singh, D.K. Tripathi et al.), 1–22. Wiley.

Tominaga, J., Takahashi, S., Sakamoto, A., and Shimada, H. (2020). Arabidopsis BSD2 reveals a novel redox regulation of Rubisco physiology in vivo. *Plant Signal Behav.* 15 (4): 1740873.

Trapet, P., Kulik, A., Lamotte, O. et al. (2015). NO signaling in plant immunity: a tale of messengers. *Phytochemistry* 112: 72–79.

Turkan, I. (2017). *Emerging Roles for ROS and RNS–Versatile Molecules in Plants*. UK: Oxford University Press.

Turkan, I. (2018). ROS and RNS: key signalling molecules in plants. *J. Exp. Bot.* 69 (14): 3313.

Umair Hassan, M., Aamer, M., Umer Chattha, M. et al. (2020). The critical role of zinc in plants facing the drought stress. *Agriculture* 10 (9): 396.

Vardhini, B.V. and Anjum, N.A. (2015). Brassinosteroids make plant life easier under abiotic stresses mainly by modulating major components of antioxidant defense system. *Front. Environ. Sci.* 2: 67.

Wang, H. and Jin, J.Y. (2007). Effects of zinc deficiency and drought on plant growth and metabolism of reactive oxygen species in maize (*Zea mays* L.). *Agric. Sci. China* 6 (8): 988–995.

Wang, Z., Sadras, V.O., Hoogmoed, M. et al. (2017). Shifts in nitrogen and phosphorus uptake and allocation in response to selection for yield in Chinese winter wheat. *Crop Pasture Sci.* 68 (9): 807–816.

Waqas, M., Yaning, C., Iqbal, H. et al. (2021). Synergistic consequences of salinity and potassium deficiency in quinoa: linking with stomatal patterning, ionic relations and oxidative metabolism. *Plant Physiol. Biochem.* 159: 17–27.

Waraich, E.A., Ahmad, R., Ashraf, M.Y. et al. (2011). Improving agricultural water use efficiency by nutrient management in crop plants. *Acta Agric. Scand B-Soil Plant Sci.* 61 (4): 291–304.

Waraich, E., Ahmad, R., Halim, A., and Aziz, T. (2012). Alleviation of temperature stress by nutrient management in crop plants: a review. *J. Soil Sci. Plant Nutr.* 12 (2): 221–244.

Waszczak, C., Akter, S., Jacques, S. et al. (2015). Oxidative post-translational modifications of cysteine residues in plant signal transduction. *J. Exp. Bot.* 66 (10): 2923–2934.

Xie, X., He, Z., Chen, N. et al. (2019). The roles of environmental factors in regulation of oxidative stress in plant. *Biomed. Res. Int.* 2019: 9732325.

Yasin, N.A., Zaheer, M.M., Khan, W.U. et al. (2018). The beneficial role of potassium in Cd-induced stress alleviation and growth improvement in *Gladiolus grandiflora* L. *Int. J. Phytoremediation* 20 (3): 274–283.

Zaheer, M.M., Yasin, N.A., Ahmad, S.R. et al. (2018). Amelioration of cadmium stress in gladiolus (*Gladiolus grandiflora* L.) by application of potassium and silicon. *J. Plant Nutr.* 41 (4): 461–476.

Zandalinas, S.I. and Mittler, R. (2018). ROS-induced ROS release in plant and animal cells. *Free Radic. Biol. Med.* 122: 21–27.

Zhou, X., Joshi, S., Khare, T. et al. (2021). Nitric oxide, crosstalk with stress regulators and plant abiotic stress tolerance. *Plant Cell Rep.* 40: 1395–1414.

Zörb, C., Senbayram, M., and Peiter, E. (2014). Potassium in agriculture–status and perspectives. *J. Plant Physiol.* 171 (9): 656–669.

4

Polyamines Metabolism and their Regulatory Mechanism in Plant Development and in Abiotic Stress Tolerance

Savita Bhardwaj[1], Tunisha Verma[1], Monika Thakur[2], Rajeev Kumar[1], and Dhriti Kapoor[1]

[1] *Department of Botany, School of Bioengineering and Biosciences, Lovely Professional University, Phagwara, Punjab, India*
[2] *Division Botany, Department of Bio-Sciences, Career Point University, Hamirpur, Himachal Pradesh, India*

Introduction

As sessile organisms, plants are naturally confronting several environmental constraints such as biotic and abiotic stresses like drought, salinity, heat stress, flooding, heavy metal stress, which ultimately impacts plant metabolic processes and yield. Metabolic shift takes place due to the alteration in the number of cellular metabolites in response to abiotic stresses. These changes against abiotic stresses are beneficial in improving the plant potential to tolerate stress environments. Amino acids, soluble sugars, organic acids, polyamines (PAs), and lipids are the metabolites that are significantly recognized to increase the ability of plants to tolerate stress (Fariduddin et al. 2013). In modern agronomic systems, proper maintenance of plant productivity and quality against poor ecological systems is one of the major problems. In this context, plants hold several protective strategies to mitigate abiotic stress-mediated adverse effects such as signaling molecules and plant hormones. Among plant hormones, PAs are widely recognized as a new class of growth compound that regulate various plant growth and metabolic aspects and also perform as stress messengers in plants against abiotic stresses (Gill and Tuteja 2010).

PAs are one of the important plant growth regulators which is small polycationic molecule with small molecular mass. There are various types of PAs to which diamine putrescine (Put), triamine spermidine (Spd), and tetraamine spermine (Spm) are of maximum importance in plants and among these Spm occurs mainly in eukaryotes and some bacteria while Put and Spd are widespread in entirely all living cells (Tiburcio et al. 2014). PAs shows various beneficial physiological roles in plants to regulate plant growth and development such as plant growth, photosynthetic activity, organ and embryo formation, flowering, fruit ripening, stimulation of antioxidant defense system etc. under normal as well as under biotic and abiotic stress conditions. Physiological impacts of PAs were primarily

Plant Ionomics: Sensing, Signaling, and Regulation, First Edition.
Edited by Vijay Pratap Singh and Manzer H. Siddiqui.

considered to be structural but their polycationic behavior allows the anionic biomolecules to be in equilibrium in the cell. Moreover, PAs are also known to regulate plant cellular progressions for instance, cell division, cell expansion, and molecular regulation, i.e. at transcriptional and translational level (Sequera-Mutiozabal et al. 2017). Amounts of PAs in plants may vary with plant species, particular developmental phase, nature of nitrogen nutrition also depends on biotic or abiotic stresses. However, plants possess various mechanisms to regulate PAs homeostasis to maintain optimal metabolic functions and also to stabilize the equilibrium of C:N proportion in plant tissues (Wuddineh et al. 2018).

Distribution, Biosynthesis, and Catabolism of Polyamines

Distribution

PAs are class of natural compounds with aliphatic nitrogen structure present in both eukaryotic and prokaryotic cells (Liu et al. 2016), as well as in plant RNA viruses and tumours. They have a wide range of biological processes that can be used in a variety of ways. PAs such as Put, spermidine (Spd), and spermine (Spm) are most common low molecular aliphatic polycations predominately found in free form. Insoluble bound PAs are covalently bounded to macromolecules such as nucleic acids and proteins, while soluble PAs are covalently conjugated to small molecules such as phenolic compounds. In addition to this, rare PAs observed in numerous biological processes such as homospermidine, 1,3-diaminopropane, cadaverine, and canavalmine are present in certain plants, animals, algae, and bacteria under special conditions.

In plants, PAs have tissue- and organ-specific transport patterns. Put, for example, was found to be the most abundant PA in leaves, with levels three times greater than that of Spd and Spm, despite Spd being the most abundant PA in other organs. The concentration of various PAs differs significantly, depending on plant type, organs, tissues, and developmental level. In general, higher PA biosynthesis and PA content are linked to more robust plant growth and metabolism (Zhao and Qin 2004; Cai et al. 2006). Within cells, various forms of PAs have different localization patterns. Put was observed to accumulate in the cytoplasm and Spm in the cell wall of carrot cells (Cai et al. 2006).

Polyamine Biosynthesis

Polyamine biosynthetic pathways are conserved with certain modifications from bacteria to animals and plants (Tabor and Tabor 1984; Minguet et al. 2008; Pegg 2009; Fuell et al. 2010) Three PAs (Put, Spd, and Spm) are synthesized from Larginine (via Lornithine) and Lmethionine in eukaryotic cells by a series of six interdependent enzyme reactions. The central product of the traditional PA biosynthetic pathway is Put. It contains two amino groups and is a synthetic precursor of Spd and Spm. There are three different pathways of Put biosynthesis in plants. In the first pathway, arginine decarboxylase (ADC) removes the carbon atom present at position 8 from (Arg) resulting in the formation of agmatine (Agm) and CO_2, then Agm loses the nitrogen atom at carbon 2 position to form carbamoyl, and then finally NCPA is hydrolyzed by N-carbamoylputreseine amidohydrolase (NCPAH) and removes carboamyl group to form Put, CO_2, and NH_3. In plants, this is the primary Put synthesis pathway (Docimo et al. 2012). Agmatine iminohydrolase (AIH) and N-carbamoylputrescine amidohydrolase (CPA), as well as ornithine decarboxylase (ODC), produce the diamine Put (ODC). Then, by adding aminopropyl groups to Put and Spd, respectively, spermidine synthase (SPDS) and spermine synthase (SPMS) synthesize Spd and Spm from decarboxylated S-adenosylmethionine (dcSAM), which is formed from methionine by SAM synthase and SAM decarboxylase (SAMDC). ACC synthase and ACC oxidase use S adenosylmethionine for ethylene biosynthesis. In the second pathway, arginase produces ornithine (Orn) from Arg, and then ODC eliminates the carboxyl group of Orn's no.1 carbon atom to produce Put and CO_2 (Docimo et al. 2012). Additionally, copper amine oxidase (CuAO) and polyamine oxidase (PAO) degrade Put, Spd, and Spm, forming 4aminobutanal, 4 aminobutanal, and N(3aminoproyl) 4aminobutanal, respectively, as well as H_2O. Many studies have discovered that Arabidopsis lacks ODC and has a defect in the corresponding enzyme function, indicating that Put biosynthesis by ODC an alternative pathway in certain plant species, and Put may be synthesized indirectly from arginine by ADC in these plants (Hussain et al. 2011). The third pathway involves conversion of Arg to citrulline (Cit), which is then decarboxylated into citrulline decarboxylase (CDC) to form Put (Han 2016; Ouyang et al. 2017; de Oliveira et al. 2018). Since the Cit pathway has only been discovered in sesame, the first two pathways in plants are more natural. The permanent competitive inhibitors difluoromethylarginine (DFMA) and difluoromethylornithine (DFMO) inhibit the activities of ADC and ODC, respectively (Grossi et al. 2016; Yamamoto et al. 2016). Put and aminopropyl residues, which are increasingly given by methionine, are used to make spermidine and Spm (Vuosku et al. 2012). Each PA has been found to be catabolized by a specific oxidase (Kaur-Sawhney et al. 2003). Figure 4.1 shows the diagrammatic representation of biosynthetic pathway of PAs in plants.

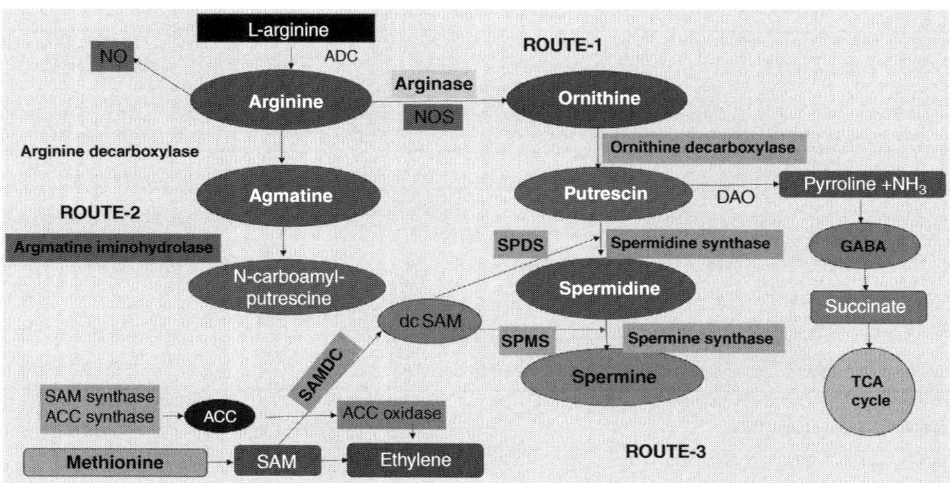

Figure 4.1 Diagrammatic representation of polyamines biosynthetic pathway in plants.

Catabolism

The catabolic pathways are maintained by PA concentration in the cells (Bagni and Tassoni 2001; Cona et al. 2006). Copper amine oxidases (CuAOs) catalyzes PA turnover, while the Flavin-dependent PA oxidases (PAOs) catalyze PAs oxidative deamination; the former is homodimeric enzymes that catalyzes breakdown of Put or Cad to (GABA) with the help of intermediates like 4-aminobutanal and Δ1-pyrroline. PAO is a FAD-dependent enzyme that is abundant in monocots and oxidizes higher PAs like Spd and Spm (Šebela et al. 2001). This suggest that plant PAOs are interested in the catabolism at the end of their life cycles (Kusano et al. 2008). These enzymes are found in tissues cell wall where lignification, suberization, and wall stiffening take place (Slocum 1991). In mammalian cells, Spm oxidase (SMO), a FAD-dependent amine oxidase that guides the back conversion of Spm to Spd while also producing 3-aminopropanol and H_2O_2 (Wang et al. 2001; Vujcic et al. 2002). In a reaction catalyzed by proline dehydrogenase, diaminopropane can be converted to b-alanine, while pyroline can be catabolized to g-aminobutyric acid (GABA) by pyrroline dehydrogenase (PDH). Following that, the g-aminobutyric acid is transaminated and oxidized to succinic acid, which is then inserted into Krebs cycle. As a result, the pathway means that carbon and nitrogen are recycled from Put.

Apart from terminal degradation of PAs to form C and N, thereby bridging PA metabolism with the tricarboxylic acid (TCA) cycle, it has now been established that PA catabolism is critical for plant growth and production. Furthermore, the metabolism of PAs is linked to NO production (Pál et al. 2015), which is important for plant growth signaling (Agurla et al. 2017). As a result, the relationship between PA metabolism and PA hormone effect on plant metabolism on plant signaling compounds can be studied to learn more about plant growth and development and the processes underlying their action.

Role of Polyamines in Plant Development

Ciamician and Ravenna demonstrated the existence of Put in *Datura stramonium* in 1911, which is possibly the first preference to PAs in plants (Bagni and Tassoni 2001). The major PAs in plant cells are the diamine Put, triamine spermidine (Spd), and tetramine spermine (Spm). Because of their cationic existence, they have contributed to numerous biological processes in plant morphogenesis, growth, embryogenesis, organ development, and leaf senescence (Kumar et al. 1997; Walden et al. 1997; Malmberg et al. 1998; Bouchereau et al. 1999; Alcázar et al. 2006b; Liu et al. 2010), in addition to stimulating DNA replication, translation, transcription, etc. Crop germination and seedling growth are the two most essential and fragile stages in the plant life cycle. Crop germination is adversely affected by PA content.

PAs are involved in a variety of fundamental activities, including plant growth and floral initiation, in addition to the germination period. PAs are abundant in meristematic tissues and highly active in developing tissues in higher plants but decrease as the cell-elongation region approaches (Flores and Galston 1984). Tassoni et al. (2010) investigated the roles of each PA in plant cells and discovered that Spd is the most important PA for plant development. Pollen formation is supported by spermidine conjugates. Thermospermine, made from Spd in a similar way, is essential for stem elongation (Tassoni et al. 2010). Exogenous Spd improved photosynthetic rate and plant growth by increasing activity of Rubisco and decreasing carbohydrate accumulation in leaves, resulting in less photosynthesis feedback inhibition (Cheng et al. 2009).

PAs are family of growth regulators found in plants (Xue et al. 2009). Exogenous PAs primarily occur in roots and then transferred and accumulated in shoot apices upon floral initiation (Martin-Tanguy 2010). Similar kind of results were noticed when PAs were applied exogenously that affected flower bud differentiation in chrysanthemum for initiation and maintenance of flower bud differentiation (Xu 2015). PAs were found to be in abundance in Arabidopsis flowers and greatly encourage flowering response (Applewhite et al. 2010). Promotion of flowering by exogenous PAs has been shown to promote flowering in some medicinal plants. Exogenous Spd, for example increased floral bud growth by 20% in tobacco tissue cultures, while when cultures lacked Spd, all buds were vegetative. In cut rose flowers, the use of Spm (10 ppm) improved flower quality and expanded shelf life to three days (Tatte et al. 2015) Whereas plants with higher levels of Put and Spd in their leaves had more flower buds, berries, and a greater mean floral diameter in comparison to lower ones (Li et al. 2014). Mahgoub et al. (2006) stated that the application of Put externally increased the number of flowers/plants, fresh, and dry weight of *Dianthus caryophyllus* (L.) flowers. Youssef et al. (2004) published similar findings on *Datura innoxia* (Mill.). PAs are thought to be regulators in the embryogenesis process in both angiosperms and gymnosperms, From the multicell proembryo, globular, heart-shaped, and torpedo stages to the cotyledon stage, the forms and abundance of PAs vary (Krasuska et al. 2013). PAs applied exogenously and synthesis of PA can be used to control nucleic acid synthesis and protein translation in both directions. This may affect the formation of organelles including the plastids, ER, and mitochondria, as well as the microtubular structures (Vondráková et al. 2015). Many findings have shown that PAs to be important in inducing division of cell and fostering regeneration in tissues and cell cultures (Yadav and Rajam 1997; Vondráková et al. 2015).

PAs effect the formation of secondary plant metabolites, in response to their effects on germination, growth, and developmental stage. PAs effect the formation of secondary metabolites in 2 forms. One by direct effect of PAs on the formation of metabolites and other by indirect effect. For example, Bais et al. (2000) discovered that combining of Put and Spd increased betalain and thiophene content by biomass accumulation. Put signaling was shown to be responsible for ginsenoside synthesis and aggregation in *Panax quinquefolius* (L.) by *Alternaria panax* and *Cylindrocarpon destructans* (Yu et al. 2016). PAs can be metabolized into a wide variety of plant secondary compounds through pathways that are still poorly understood.

Throughout the stages of plant growth, the functions of metabolic PA enzymes and the contents of PAs alter. A drop in PA levels seems to be a major precursor to senescence signals, or it may be the senescence signal itself (Dutra et al. 2012). Exogenous Spd and Spm treatments boost the number of PAs in cut flowers, by extending their life and improving their consistency (Yang and He 2001; Cao 2010). Simões et al. (2018) found that spraying GA3 + Spm on *Anthurium andraeanum* slowed the senescence of cut flowers deposited at 20 °C and increased the consistency of the inflorescences. PAs tended to slow aging by inhibiting the biosynthesis of ethylene (Woo et al. 2013; Anwar et al. 2015). A PA synthesis improved the lifetime and deaccelerated flower senescence (Han 2016).

Cross talk of PA and NO has suggested and that these two can play a vital role in a variety of other enzymatic activities, such as nitrate reductase. However, in reality, PAs, nitrates, and NO all can play a vital role in how plants integrate "N signals." Recently, numerous findings have stated that copper amine oxidase1 (CuAO1) of *A. thaliana* play crucial role in ABA- and PA-induced NO biosynthesis and ABA signal transduction (Yamasaki and Cohen 2006; Wimalasekara et al. 2011; Hancock 2012; Rosales et al. 2012). The commonly used nitric oxide donor sodium nitroprusside (SNP), according to Filippou et al. (2013), controls proline and PA metabolism in the leaves of *Medicago truncatula* plants. In a recent research, Moschou et al. (2012) proposed that the changes in PA levels influence PCD and/or lignification during xylem differentiation through a "N signaling" components.

Polyamines as Biochemical Markers for Abiotic Stress Tolerance

Drought Stress

Drought stress is a major abiotic stress that cause high crop yield loss around the world and alter plant hormonal, metabolic, and molecular responses. PAs play a significant role in increasing plant tolerance to drought stress conditions (Hussain et al. 2011). PAs are known to maintain membrane or DNA structure, protein conformation, and to control the function of several enzymes such as antioxidant enzyme and H^+-ATPase, under stress conditions (Liu et al. 2005). PAs application improved the plant tolerance to drought stress by increasing plant growth, accumulation of osmoprotectants, plant water status, chlorophyll amount, protein content, and reduced the oxidative damage by declining the level of MDA in *Vigna radiata* plants (Sadeghipour 2019). Nayyar et al. (2005) observed that exogenous application of Put and Spd significantly improved chickpea and soybean tolerance to drought stress. Supplementation of Spd and Spm remarkably improved the zeatin (Z) + zeatin riboside (ZR) and ABA amounts and reduced the ethylene evolution rate in wheat grains, which ultimately caused increased grain filling in wheat under drought stress (Liu et al. 2016). PAs increased the drought stress tolerance in rice plants by increasing the plant biomass, net photosynthetic activity, formation of free proline, anthocyanins, and soluble phenolics content. Increase in biomass and net photosynthetic activity was attributed to the PA-induced elevation in the leaf water status and water use efficiency. Moreover, PAs also alleviated the drought stress by reducing the level of H_2O_2 to diminish oxidative damage on cellular membranes (Farooq et al. 2009).

ATP amount, chlorophyll content index (CCI), chlorophyll stability index (CSI), rubisco activity, endogenous PA level, membrane stability index (MSI), activities of antioxidative enzymes, N_2, K^+, Ca^{2+}, Mg^{2+}, and grain yield were improved by the application of PAs, whereas ROS level and MDA content were reduced Ghassemi et al. 2018). Seed germination was promoted through seed soaking in Spd and Spm under drought stress through enhancing the endogenous indole-3-acetic acid (IAA), Z+ ZR, ABA, and gibberellins (GA) amount in wheat seeds. PAs application also increased the deprivation of seed starch and elevated the content of soluble sugars in seeds, which was helpful in promoting seed germination in wheat under drought stress conditions (Yang et al. 2016). Activities of antioxidative enzymes, photosynthetic pigments level, and content of proline were stimulated by the Spd and Spm application in valerian to decrease the drought-induced membrane damages. This increment in the photosynthetic pigments was associated with the high energy supply for better plant development and productivity (Mustafavi et al. 2016). Exogenous application of Spd significantly alleviated the drought-induced adverse impacts by reducing the level of O_2^-, H_2O_2, and MDA. Spd also increased the relative water content (RWC), chlorophyll amount, and activities of antioxidative enzyme, i.e. superoxide dismutase (SOD), peroxidase (POX), catalase (CAT), and ascorbate peroxidase (APX) and reduction in IAA and GA3 in creeping bentgrass (Li et al. 2015). Exogenous application of Spd under drought stress improved the plant growth, photosynthetic pigment amount, photosynthesis rate, photochemical quenching parameters, endogenous PA, IAA, ZR, and GA3 level while decreased the salicylic acid (SA) and jasmonate (JA) amount in *Zea mays* (Li et al. 2018).

Salinity Stress

Spd and Spm application in rice remarkably inhibited the salinity-induced electrolytes leakage and also improved the chlorophyll content, photosynthetic photochemical reactions, and expression of chloroplast-encoded genes like psbA, psbB, psbE, and rbcL (Chattopadhayay et al. 2002). The PA-treated citrus plants exhibited increased expression of PA biosynthesis and catabolism-related genes and also mitigated the salinity-induced oxidative stress by reducing the level of ROS through controlling the transcript expression and functioning of different antioxidant enzymes. Moreover, PAs stimulated the protein S-nitrosylation, while protein carbonylation and tyrosine nitration were inhibited (Tanou et al. 2014). Salinity stress triggered the oxidative stress by excessive generation of O_2^- and H_2O_2 in rice plants. But, exogenous application of Put declined this oxidative stress and cellular disintegration via increasing the nonenzymatic (anthocyanin and flavonoid content) and enzymatic antioxidants (CAT, GPX, and GR) (Ghosh et al. 2012).

Application of spermidine significantly alleviated the salinity triggered adverse effects in *Cucumis sativus* by improving plant growth, photosynthetic activity, PA amount, proline level, and activities of antioxidative enzymes (Duan et al. 2008). Spd application increased the salinity tolerance in *Poa pratensis* by enhancing the proline levels and by maintaining the ion and PA metabolism. Spd also increased the chlorophyll amount and the level of K^+, Ca^{2+}, Mg^{2+}, and K^+/Na^+ ratio under salinity stress (Puyang et al. 2016). Chlorophyll, phenolic compounds, and protein amount were decreased by the salinity stress, but the application of PA reversed this salinity-induced effects by increasing their content in wheat seedlings (Rahdari and Hoseini 2013).

PAs application was observed to increase the growth characteristics, enhanced the activities of antioxidative enzymes, i.e. superoxide dismutase (SOD) and catalase (CAT), and reduced the H_2O_2 level. Moreover, PAs reduced the Na^+/K^+ ratio and Cl^- to balance the ion exchange and better Na^+/K^+ ratio in pistachio seedlings under salinity stress conditions (Kamiab et al. 2014). Growth characteristics and yield components, leaf pigments level, and carbohydrates, protein, and amylase activity were declined by the salinity stress in *Vigna sinensis* plants. However, spm application remarkably alleviated the adverse impacts of salinity stress by improving the growth and yield components, and protective impact of Spm was primarily attributed to the enhanced chlorophyll and protein amount and endogenous PAs level (Alsokari 2011). Exogenous application of Spd and Spm mitigated the harmful effects of salinity stress in rice seedlings by increasing plant growth, inhibition of cellular injuries, maintaining K^+ /Na^+ ratio, and escalated the amounts of compatible solutes and functioning of antioxidative enzymes to improve plant tolerance to salinity stress (Roychoudhury et al. 2011).

Heavy Metal Stress

There are three main forms of PAs present in plants: tetra-amine spermine (Spm), tri-amine spermidine (Spd), and their di-amine precursor Put, although many PAs including cadaverine also have been known (Gupta et al. 2013; Liu et al. 2015). PAs are being mentioned as a group of plant growth regulators that also operate as second messengers, affecting several growth and development processes in plants (Hussain et al. 2011). PAs are necessary for normal growth and development due to transcriptional and translational stimulation (Kusano et al. 2008), but their physiological utility is exhibited by their presence in abiotic stress tolerance, such as metal stress in plants (Pál et al. 2017). They can remove free radicals as well as reactive species and thus act as potent antioxidants across different forms of stress (Rangan et al. 2014).

Cadmium is highly soluble in water or widely absorbed by plants, causing toxic symptoms like inhibition of growth, photosynthesis process system effect, chlorophyll depletion, stomatal closure suppression, $NH4^+$ deposition, oxidative stress, as well as DNA damage (Moussa and El-Gamal 2010; Lehotai et al. 2011; Kao 2014). Cadmium also affects a variety of physiological mechanisms, including respiration, photosynthetic activity, cell expansion, the plant–water interaction, nitrogen fixation, or nutrient availability, leading to lower productivity (Perfus-Barbeoch et al. 2002; Hsu and Kao 2005). The noticeable sign of cadmium toxicity includes leaf folding, necrosis, as well as stomatal closure (DalCorso et al. 2010). Many other studies have been done to assess the role of PAs in plant adaptation to cadmium stress (Groppa et al. 2007, 2008; Kumar et al. 2010; Yang et al. 2010, 2013; Serrano-Martínez and Casas 2011). Put has been deemed a physiological indicator of abiotic environmental stresses, among three types of PAs (Put, Spd, and Spm) (Groppa et al. 2007; Yang et al. 2013). Cadmium stress affected PAs homeostasis in *Hydrocharis dubia* leaves via increasing Put content as well as decreasing Spm and Spd levels, and also increasing the concentration of ADC, DAO, and PAO enzymes (Yang et al. 2013). Serrano-Martínez and Casas (2011) investigated the effect of various cadmium doses (0.05, 0.1, and 0.2 mM) on plant development, antioxidant capacity, as well as endogenous PA status in carnation (*Dianthus caryophyllus* L.) plants, or even the efficacy of a recovery process in cadmium-free medium.

Chromium occurs in many oxidative states, but trivalent Cr (III) and hexavalent Cr (VI) species are more common in plants; nevertheless, Cr (VI) is more toxic than Cr (III) (III). Excessive Cr deposition in plants induces oxidative stress in plants, as shown by decreased carbon dioxide fixation or photosynthetic activity owing to chloroplast disorganization, electron transport suppression, as well as Calvin cycle enzyme inactivation (Shanker et al. 2005). Chromium exposure produces free radicals that induce oxidative damage to DNA, RNA, enzymes, or pigments (Choudhary et al. 2011). Increased levels of chromium effectively diminished fresh weight or photosynthetic pigment while increasing lipid peroxidation, hydrogen peroxide, and proline levels, as well as antioxidant capacity.

Copper is a significant transition metal that plays a role in a variety of physiological functions, although doses of copper that are greater than their optimum levels could be harmful to plant tissues (Fargasova 2004; Emamverdian et al. 2015). According to many studies, copper is primarily accrued in the root, with little upward movement in the shoot. Copper stress, which is deemed the initial effect of copper sensitivity, has a direct impact on root growth and development. Increased copper might cause over accumulation of free radicals oxidative, resulting in the production of reactive oxygen species (ROC) like O•, H_2O_2, and OH•. Such reactive species damage nucleic acids such as DNA, RNA, protein, or lipids, resulting in lipid peroxidation, membrane injury, as well as enzyme disruption. The phytotoxicity of mercury is evident in its mechanisms of sulfhydryl reactivity by increased level of ROC and cellular absorption of Hg^{2+} (mercuric ion) and MeHg (methylmercury), which allow covalent bonds to form with cysteine residues of proteins, reducing cellular antioxidants, which are the main line of defense toward Hg substances (Valko et al. 2006).

Cellular injury, loss of membrane fluidity, biochemical dysfunction of the plant, increased DNA oxidation, lipid peroxidation, and decreased nutrient acquisition capability, as well as photosynthetic ability, are indeed effects. Cell division and elongation disturbances cause a major reduction in yield or productivity (Elbaz et al. 2010). Mercury is a nonessential metal for plants, so they use protective chemicals like proline, glycine betaine, sugars, or other antioxidative enzymes to prevent it out of their environment (Cho and Park 2000). This analysis is supported by accounts of increased endogenous PAs levels in plants caused by excessive metal toxicity (Groppa et al. 2007; Zhao et al. 2008; Ding et al. 2010). Thus, the involvement of regulation of the antioxidative mechanism, which includes all nonenzymatic antioxidants such as glutathione (GSH), non-protein thiols (NPSH), and enzymatic antioxidants such as superoxide dismutase (SOD), ascorbate peroxidase (APX), and glutathione reductase (GR), is responsible for the adverse effects of mercury on plant physiology (Ortega-Villasante et al. 2005).

Temperature Stress

Low- and high-temperature stresses are the two main types of temperature stress. Cold stress and icing stress are the two types of low-temperature stress. Only a few studies have examined the physiological roles of PAs in plants that are exposed to high temperatures. The effects of high thermal stress on PA synthesis in Chinese kale leaves were significant; total PAs and put contents raised after six days of high thermal treatment, but the rises were not maintained across significantly for a longer period (Yang and Yang 2002). PAs might stimulate photosynthesis as well as ensure maximum antioxidant potential or osmotic response efficiency in high-temperature stress (Tian 2012). Antioxidant enzymes may excavate ROS to resist membrane lipid peroxidation as well as maintain cell membrane intact (Ouyang et al. 2017). Plants use PAs for a variety of applications, or the physiological mechanisms that enable them to tolerate high temperatures vary by organisms. Different PAs exhibit different patterns of change in different plant species when exposed to high temperatures (Shao et al. 2015). PAs have the potential to adhere to the phospholipid region of the cell membrane, preventing cytolysis and improving cold tolerance

(Li and He 2012). There are, furthermore, many different perspectives on the interaction between Put and plant chilling stress (Wu and Yuan 2008).

The Put content of sweet pepper and zucchini fruits, which are grown massively while preserved at chilling temperatures, was induced through chilling effect. Processing in a CO_2-modified environment minimized cold harm as well as prevented Put absorption, ensuring that Put accrued as a result of chilling stress (Serrano et al. 1997, 1998). According to Roy et al., Put condition caused chilling harm, as well as enhanced Spm, which might have been a defensive reaction to colder stress. They discovered that in loquat fruit preserved at low temperatures, the Put, Spm, and Spd content gradually increased. Exogenous Spm prevented Put absorption as well as reduced chilling harm by improving the availability of endogenous Spm and Spd (Yong-Hua et al. 2000; Roy and Wu 2001). Another view was that Put deposition was positively correlated with plant cold tolerance, implying that Put absorption was a defense response of plants to chilling impact (Yong et al. 2003b). Sun et al. investigated the biochemical and physiological measures of Anthurium as well as raeanum in chilling stress at 6°C in winter using Put as well as D-Arg at various concentrations (0.5, 1.0, 1.5, and 2.0 mmol l^{-1}). Put application increased activities of antioxidant enzymes, root function, nitrogen metabolism, chlorophyll content, and proline content, while declining malondialdehyde content, according to the investigators. The strongest effect can be achieved with 1.0 mmol l^{-1} Put, while D-Arg treatment minimized chilling damage (Sun et al. 2018). PA supplementation improved the resistance of stevia plants to cold temperatures, with similar outcomes (Peynevandi et al. 2018). The function of PAs in signal transduction, instead of their absorption, induces abiotic stress tolerance (Pál et al. 2015).

Crosstalk of Polyamines with Other Signaling Molecules

Nitric Oxide

Application of Put and NO in combination increased the plant tolerance to cadmium toxicity in *Vigna radiata* through the induction of several plant defense mechanisms such as stimulation of plant antioxidative enzymes and nonenzymatic antioxidant and also enhanced the phytochelatin synthesis (Nahar et al. 2016).Transient NO production was observed through the application of PAs under limited water availability conditions in *Cucumis sativus* seedlings, and it was concluded that NO functions as a downstream signal of PAs (Arasimowicz-Jelonek et al. 2009) (Figure 4.2). NO production was also induced by treating wheat roots with Spm which primarily impacted root growth; however, Spm triggered response was moderately inhibited by cPTIO, a NO scavenger (Groppa et al. 2008). NO markedly improved the salinity stress tolerance in sunflower seedlings by positively controlling PA homeostasis via decreasing the rate of PA breakdown, elevated levels of PA biosynthetic enzymes, and maintenance of PA dispersal between free, conjugated, and bound forms (Tailor et al. 2019). RWC, electrolyte leakage, and lipid peroxidation were significantly increased by the salinity stress in the leaves of chickpea. However, NO and PA application remarkably alleviated salinity stress by reducing RWC, electrolyte leakage, and lipid peroxidation, and both PA and NO exhibited positive influence on the activity of plant antioxidative enzymes, i.e. SOD, CAT, POX, and APX under salt stress (Sheokand et al. 2008).

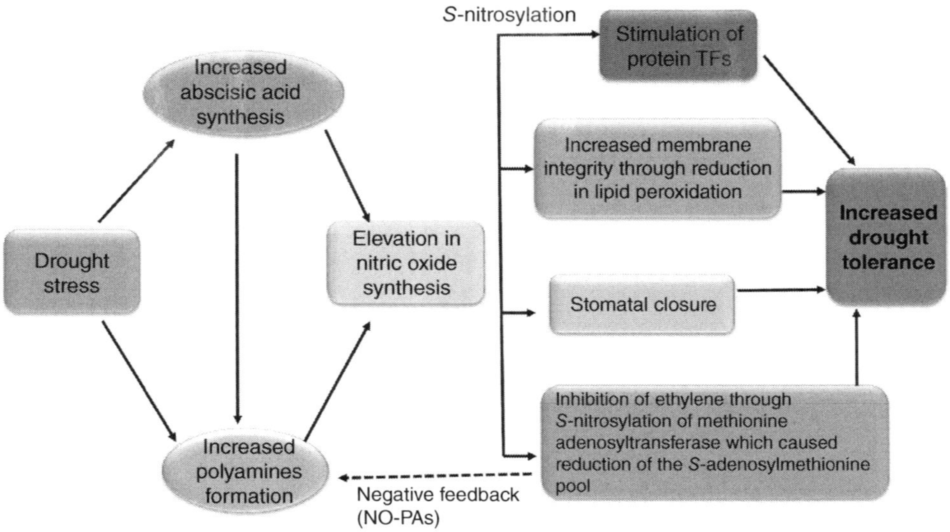

Figure 4.2 Diagrammatic representation of crosstalk between PAs and nitric oxide to improve plant drought tolerance. *Source:* Adapted from Arasimowicz-Jelonek et al. (2009).

Spd appeared as an important moiety in water deficit stimulated pathways, connected with NO production, which stimulated plant antioxidant defence system to improve drought tolerance in *Trifolium repens* (Peng et al. 2016).

Plant Growth Regulators

Chromium-induced phytotoxicity was reduced by the supplementation of epibrassinolide (EBL) and Spd in *R. sativus* (Choudhary et al. 2012a). Combined application of EBL and Spd alleviated the adverse impacts of Zn and salinity stress by improving the plant growth and biochemical aspects in *V. radiata* (Mir et al. 2015). EBL was observed to mitigate salinity stress in *truncatula-Sinorhizobium meliloti* symbiosis through increasing the level of the Spm. EBL application increased the PAs level in the stem and also decreased the number of nodules, indicating that interaction between PAs and BRs was significant for the alleviation of salinity stress and for nodule inhibition in *Medicago truncatula-Sinorhizobium* meliloti symbiosis (López-Gómez et al. 2016). Supplementation of EBR and Spd in combination increased the Cu tolerance in radish by stimulating the expression of genes that influence the metabolism of indole-3-acetic acid (IAA) and abscisic acid (ABA) and PA enzyme-related genes. Mitigation of Cu stress was attributed to the EBR and Spd mediated altered expression of genes for Cu homeostasis (Choudhary et al. 2012b). Spm regulated the production of ethylene by hindering the accumulation of *ACC synthase* transcript in tomato fruit (Alexander and Grierson 2002). Activity of PA oxidase (PAO) was increased by the elevated transcript levels of *AtPAO2* and *AtPAO4* genes, which subsequently resulted in higher H_2O_2 generation in *Arabidopsis* guard cells, suggesting the significant association between ethylene-triggered H_2O_2 generation and

PAO activity (Hou et al. 2013). Both ABA-dependent and ABA-independent routes are stimulated by the water deficit conditions, and PA catabolism was regulated by the ABA by triggering the PAO activity (Guo et al. 2014), whereas ABA production was induced by the PAs (Marco et al. 2011). Endogenous ABA amount was regulated by the Put in *Arabidopsis* plants against low-temperature circumstances (Cuevas et al. 2008). Moreover, PA synthesis was observed to be upregulated by ABA in *Arabidopsis thaliana* under drought and salinity stress (Alcázar et al. 2006a).

Conclusion

PAs function in plant developmental processes, from flowering to senescence, as well as their effects on plant growth and development. Plant stress tolerance could be attributed to the inhibitory action of PAs, which could be accomplished by exogenous application or modulation of PA biosynthetic as well as metabolic functions. The main enzyme in PA biosynthesis, ADC, is included in this PA-related response to stress. The expression of ADC, ODC, and SAMDC has been used to regulate the concentrations of intracellular PAs. Transgenic mechanisms for manipulating PA synthesis have proven to be an effective method for studying PA physiological processes in higher plants. Besides that, a precise metabolite profiling between allelic variants would provide a mechanism for unravelling genetic determinants with genomic supported assessments to know the linkages of PAs biosynthetic pathway, catabolism, as well as conjugation, and also PAs signaling. Genomic methods to control environmental stresses throughout plant systems, a prime example of changing climate, are with a specific focus on plants. As a result, the presence of natural variation would serve as an opportunity to analyze the current genetic variation to classify genotypes of appropriate allelic variants for efficiently imparting stress resistance.

References

Agurla, S., Gayatri, G., and Raghavendra, A.S. (2017). Polyamines increase nitric oxide and reactive oxygen species in gurard cells of *Arabidopsis thaliana* during stomatal closure. *Protoplasma* 225: 153–162.

Alcázar, R., Cuevas, J.C., Patrón, M. et al. (2006a). Abscisic acid modulates polyamine metabolism under water stress in *Arabidopsis thaliana*. *Plant Physiol.* 128: 448–455.

Alcázar, R., Marco, F., Cuevas, J.C. et al. (2006b). Involvement of polyamines in plant response to abiotic stress. *Biotechnol. Lett.* 28 (23): 1867–1876.

Alexander, L. and Grierson, D. (2002). Ethylene biosynthesis and action in tomato: a model for climacteric fruit ripening. *J. Exp. Bot.* 53 (377): 2039–2055.

Alsokari, S.S. (2011). Synergistic effect of kinetin and spermine on some physiological aspects of seawater stressed Vigna sinensis plants. *Saudi J. Biol. Sci.* 18 (1): 37–44.

Anwar, R., Mattoo, A.K., and Handa, A.K. (2015). Polyamine interactions with plant hormones: crosstalk at several levels. *Int. J. Hydrog. Energy* 38: 1039–1051.

Applewhite, P.B., Kaur-Sawhney, R., and Galston, A.W. (2010). A role for spermidine in the bolting and flowering of Arabidopsis. *Physiol. Plant.* 108: 314–320.

Arasimowicz-Jelonek, M., Floryszak-Wieczorek, J., and Kubiś, J. (2009). Interaction between polyamine and nitric oxide signaling in adaptive responses to drought in cucumber. *J. Plant Growth Regul.* 28 (2): 177–186.

Bagni, N. and Tassoni, A. (2001). Biosynthesis, oxidation and conjugation of aliphatic polyamines in higher plants. *Amino Acids* 20: 301–317.

Bouchereau, A., Aziz, A., Larher, F., and Martin-Tanguy, J. (1999). Polyamines and environmental challenges: recent development. *Plant Sci.* 140: 103–125.

Cai, Q., Zhang, J., Guo, C., and Al, E. (2006). Reviews of the physiological roles and molecular biology of polyamines in higher plants. *J. Fujian Educ. Coll.* 7: 118–124.

Cao, D. (2010). *Effects of Polyamines on Seed Quality and Germination of Super Sweet Corn Seeds during Development*. Hangzhou: Zhejiang University.

Chattopadhayay, M.K., Tiwari, B.S., Chattopadhyay, G. et al. (2002). Protective role of exogenous polyamines on salinity-stressed rice (*Oryza sativa* L.) plants. *Physiol. Plant.* 116 (2): 192–199.

Cheng, L., Zou, Y., Ding, S. et al. (2009). Polyamine accumulation in transgenic tomato enhances the tolerance to high temperature stress. *J. Integr. Plant Biol.* 51: 489–499.

Cho, U.H. and Park, J.O. (2000). Mercury-induced oxidative stress in tomato seedlings. *Plant Sci.* 156 (1): 1–9.

Choudhary, S.P., Kanwar, M., Bhardwaj, R. et al. (2011). Epibrassinolide ameliorates Cr (VI) stress via influencing the levels of indole-3-acetic acid, abscisic acid, polyamines and antioxidant system of radish seedlings. *Chemosphere* 84 (5): 592–600.

Choudhary, S.P., Kanwar, M., Bhardwaj, R. et al. (2012a). Chromium stress mitigation by polyamine-brassinosteroid application involves phytohormonal and physiological strategies in *Raphanus sativus* L. *PLoS One* 7 (3): e33210.

Choudhary, S.P., Oral, H.V., Bhardwaj, R. et al. (2012b). Interaction of brassinosteroids and polyamines enhances copper stress tolerance in *Raphanus sativus*. *J. Exp. Bot.* 63 (15): 5659–5675.

Cona, A., Rea, G., Angelini, R. et al. (2006). Functions of amine oxidases in plant development and defence. *Trends Plant Sci.* 11: 80–88.

Cuevas, J.C., López-Cobollo, R., Alcázar, R. et al. (2008). Putrescine is involved in Arabidopsis freezing tolerance and cold acclimation by regulating abscisic acid levels in response to low temperature. *Plant Physiol.* 148 (2): 1094–1105.

DalCorso, G., Farinati, S., and Furini, A. (2010). Regulatory networks of cadmium stress in plants. *Plant Signal. Behav.* 5 (6): 663–667.

Ding, C., Shi, G., Xu, X. et al. (2010). Effect of exogenous spermidine on polyamine metabolism in water hyacinth leaves under mercury stress. *Plant Growth Regul.* 60 (1): 61–67.

Docimo, T., Reichelt, M., Schneider, B., and Al, E. (2012). The first step in the biosynthesis of cocaine in *Erythroxylum coca*: the characterization of arginine and ornithine decarboxylases. *Plant Mol. Biol.* 78: 599–615.

Duan, J., Li, J., Guo, S., and Kang, Y. (2008). Exogenous spermidine affects polyamine metabolism in salinity-stressed *Cucumis sativus* roots and enhances short-term salinity tolerance. *J. Plant Physiol.* 165 (15): 1620–1635.

Dutra, N.T., Silveira, V., de Azevedo, I.G. et al. (2012). Polyamines affect the cellular growth and structure of proembryogenic masses in *Araucaria angustifolia* embryogenic cultures through the modulation of proton pump activities and endogenous levels of polyamines. *Physiol. Plant.* 148: 121–132.

Elbaz, A., Wei, Y.Y., Meng, Q. et al. (2010). Mercury-induced oxidative stress and impact on antioxidant enzymes in *Chlamydomonas reinhardtii*. *Ecotoxicology* 19 (7): 1285–1293.

Emamverdian, A., Ding, Y., Mokhberdoran, F., and Xie, Y. (2015). Heavy metal stress and some mechanisms of plant defense response. *Sci. World J.* 2015: 756120.

Fargasova, A. (2004). Toxicity comparison of some possible toxic metals (Cd, Cu, Pb, Se, Zn) on young seedlings of *Sinapis alba* L. *Plant Soil Environ.* 50 (1): 33–38.

Fariduddin, Q., Varshney, P., Yusuf, M., and Ahmad, A. (2013). Polyamines: potent modulators of plant responses to stress. *J. Plant Interact.* 8 (1): 1–16. https://doi.org/10.1080/17429145.2012.716455.

Farooq, M., Wahid, A., and Lee, D.J. (2009). Exogenously applied polyamines increase drought tolerance of rice by improving leaf water status, photosynthesis and membrane properties. *Acta Physiol. Plant.* 31 (5): 937–945.

Filippou, P., Antoniou, C., and Fotopoulos, V. (2013). The nitric oxide donor sodium nitroprusside regulates polyamine and proline metabolism in leaves of *Medicago truncatula* plants. *Free Radic. Biol. Med.* 56: 172–183.

Flores, H.E. and Galston, A.W. (1984). Osmotic stress-induced polyamine accumulation in cereal leaves. I. Physiological parameters of the response. *Plant Physiol.* 75: 102–109.

Fuell, C., Elliott, K.A., Hanfrey, C.C. et al. (2010). Polyamine biosynthetic diversity in plants and algae. *Plant Physiol. Biochem.* 48 (7): 513–520.

Ghassemi, S., Farhangi-Abriz, S., Faegi-Analou, R. et al. (2018). Monitoring cell energy, physiological functions and grain yield in field-grown mung bean exposed to exogenously applied polyamines under drought stress. *J. Soil Sci. Plant Nutr.* 18 (4): 1108–1125.

Ghosh, N., Das, S.P., Mandal, C. et al. (2012). Variations of antioxidative responses in two rice cultivars with polyamine treatment under salinity stress. *Physiol. Mol. Biol. Plants* 18 (4): 301–313.

Gill, S.S. and Tuteja, N. (2010). Polyamines and abiotic stress tolerance in plants. *Plant Signal. Behav.* 5 (1): 26–33. https://doi.org/10.4161/psb.5.1.10291.

Groppa, M.D., Ianuzzo, M.P., Tomaro, M.L., and Benavides, M.P. (2007). Polyamine metabolism in sunflower plants under long-term cadmium or copper stress. *Amino Acids* 32 (2): 265–275.

Groppa, M.D., Rosales, E.P., Iannone, M.F., and Benavides, M.P. (2008). Nitric oxide, polyamines and Cd-induced phytotoxicity in wheat roots. *Phytochemistry* 69 (14): 2609–2615.

Grossi, M., Phanstiel, O., Rippe, C., and Al, E. (2016). Inhibition of polyamine uptake potentiates the anti-proliferative effect of polyamine synthesis inhibition and preserves the contractile phenotype of vascular smooth muscle cells. *J. Cell. Physiol.* 231: 1334–1342.

Guo, Z., Tan, J., Zhuo, C. et al. (2014). Abscisic acid, H_2O_2 and nitric oxide interactions mediated cold-induced S-adenosylmethionine synthetase in *Medicago sativa* subsp. falcata that confers cold tolerance through up-regulating polyamine oxidation. *Plant Biotechnol. J.* 12 (5): 601–612.

Gupta, K., Dey, A., and Gupta, B. (2013). Plant polyamines in abiotic stress responses. *Acta Physiologiae Plantarum* 35 (7): 2015–2036.

Han, L. (2016). *Studies on Mechanism of Low Temperature Storage and Polyamine Impact in Cut Flowers of Herbaceous Peony Postharvest Senescence*. Shandong: Shandong Agricultural University.

Hancock, J.T. (2012). NO synthase? Generation of nitric oxide in plants. *Period. Biol.* 114 (1): 19–24.

Hou, Z.H., Liu, G.H., Hou, L.X. et al. (2013). Regulatory function of polyamine oxidase-generated hydrogen peroxide in ethylene-induced stomatal closure in *Arabidopsis thaliana*. *J. Integr. Agric.* 12 (2): 251–262.

Hsu, Y.T. and Kao, C.H. (2005). Abscisic acid accumulation and cadmium tolerance in rice seedlings. *Physiol. Plant.* 124 (1): 71–80.

Hussain, S.S., Ali, M., Ahmad, M., and Siddique, K.H. (2011). Polyamines: natural and engineered abiotic and biotic stress tolerance in plants. *Biotechnol. Adv.* 29 (3): 300–311.

Kamiab, F., Talaie, A., Khezri, M., and Javanshah, A. (2014). Exogenous application of free polyamines enhance salt tolerance of pistachio (*Pistacia vera* L.) seedlings. *Plant Growth Regul.* 72 (3): 257–268.

Kao, C.H. (2014). Role of abscisic acid in abiotic stress tolerance in rice. *Crop Environ. Bioinform.* 11: 57–64.

Kaur-Sawhney, R., Tiburcio, A.F., Atabella, T., and Galston, A.W. (2003). Polyamines in plants: an overview. *J. Cell Mol. Biol.* 2: 1–12.

Krasuska, U., Ciacka, K., Bogatek, R., and Al, E. (2013). Polyamines and nitric oxide link in regulation of dormancy removal and germination of Apple (*Malus domestica* Borkh.) embryos. *J. Plant Growth Regul.* 33: 590–601.

Kumar, A., Altabella, T., Taylor, M., and Tiburcio, A.F. (1997). Recent advances in polyamine research. *Trends Plant Sci.* 2: 124–130.

Kumar, M., Kumari, P., Gupta, V. et al. (2010). Differential responses to cadmium induced oxidative stress in marine macroalga *Ulva lactuca* (Ulvales, Chlorophyta). *Biometals* 23 (2): 315–325.

Kusano, T., Berberich, T., Tateda, C., and Takahashi, Y. (2008). Polyamines: essential factors for growth and survival. *Planta* 228 (3): 367–381.

Lehotai, N., Pető, A., Bajkán, S. et al. (2011). in vivo and in situ visualization of early physiological events induced by heavy metals in pea root meristem. *Acta Physiol. Plant.* 33 (6): 2199–2207.

Li, Y.D. and He, J.G. (2012). Advance in metabolism and response to stress of polyamines in plant. *Acta Agric. Boreali Sinica* 27: 240–245.

Li, C., Pei, Z., and Gan, L. (2014). Effects of photoperiod on flowering and polyamine contents of nobile-type Dendrobium. *Plant Physiol. J.* 50: 1167–1170.

Li, Z., Zhou, H., Peng, Y. et al. (2015). Exogenously applied spermidine improves drought tolerance in creeping bentgrass associated with changes in antioxidant defense, endogenous polyamines and phytohormones. *Plant Growth Regul.* 76 (1): 71–82.

Li, L., Gu, W., Li, J. et al. (2018). Exogenously applied spermidine alleviates photosynthetic inhibition under drought stress in maize (*Zea mays* L.) seedlings associated with changes in endogenous polyamines and phytohormones. *Plant Physiol. Biochem.* 129: 35–55.

Liu, H.P., Yu, B.J., Zhang, W.H., and Liu, Y.L. (2005). Effect of osmotic stress on the activity of H^+-ATPase and the levels of covalently and noncovalently conjugated polyamines in plasma membrane preparation from wheat seedling roots. *Plant Sci.* 168 (6): 1599–1607.

Liu, H.P., Dong, B.H., Zhang, Y.Y. et al. (2010). Relationship between osmotic stress and the levels of free, conjugated and bound polyamines in leaves of wheat seedlings. *Plant Sci.* 166: 1261–1267.

Liu, J.H., Wang, W., Wu, H. et al. (2015). Polyamines function in stress tolerance: from synthesis to regulation. *Front. Plant Sci.* 6: 827.

Liu, Y., Liang, H., Lv, X. et al. (2016). Effect of polyamines on the grain filling of wheat under drought stress. *Plant Physiol. Biochem.* 100: 113–129.

López-Gómez, M., Hidalgo-Castellanos, J., Lluch, C., and Herrera-Cervera, J.A. (2016). 24-Epibrassinolide ameliorates salt stress effects in the symbiosis *Medicago truncatula-Sinorhizobium meliloti* and regulates the nodulation in cross-talk with polyamines. *Plant Physiol. Biochem.* 108: 212–221.

Mahgoub, M.H., El-Ghorab, A.H., and Bekheta, H.M. (2006). Effect of some bioregulators on the endogenous phytohormones, chemical composition, essential oil and its antioxidant activity of carnation (*Dianthus caryophyllus* L.). *Mansoura Univ. J. Agri. Sci.* 31: 4229–4245.

Malmberg, R.L., Watson, M.B., Galloway, G.L., and Yu, W. (1998). Molecular genetic analyses of plant polyamines. *Crit. Rev. Plant Sci.* 17: 199–224.

Marco, F., Alcázar, R., Tiburcio, A.F., and Carrasco, P. (2011). Interactions between polyamines and abiotic stress pathway responses unraveled by transcriptome analysis of polyamine overproducers. *Omics J. Integr. Biol.* 15 (11): 775–781.

Martin-Tanguy, J. (2010). Conjugated polyamines and reproductive development: biochemical, molecular and physiological approaches. *Physiol. Plant.* 100: 675–688.

Minguet, E.G., Vera-Sirera, F., Marina, A. et al. (2008). Evolutionary diversification in polyamine biosynthesis. *Mol. Biol. Evol.* 25 (10): 2119–2128.

Mir, B.A., Khan, T.A., and Fariduddin, Q. (2015). 24-epibrassinolide and spermidine modulate photosynthesis and antioxidant systems in *Vigna radiata* under salt and zinc stress. *Int. J. Adv. Res.* 3: 592–608.

Moschou, P.N., Wu, J., Cona, A. et al. (2012). The polyamines and their catabolic products are significant players in the turnover of nitrogenous molecules in plants. *J. Exp. Bot.* 63 (14): 5003–5015.

Moussa, H.R. and El-Gamal, S.M. (2010). Effect of salicylic acid pretreatment on cadmium toxicity in wheat. *Biol. Plant.* 54 (2): 315–320.

Mustafavi, S.H., Shekari, F., and Maleki, H.H. (2016). Influence of exogenous polyamines on antioxidant defence and essential oil production in valerian (*Valeriana officinalis* L.) plants under drought stress. *Acta Agriculturae Slovenica* 107 (1): 81–91.

Nahar, K., Hasanuzzaman, M., Alam, M.M. et al. (2016). Polyamine and nitric oxide crosstalk: antagonistic effects on cadmium toxicity in mung bean plants through upregulating the metal detoxification, antioxidant defense and methylglyoxal detoxification systems. *Ecotoxicol. Environ. Saf.* 126: 245–255.

Nayyar, H., Satwinder, K.A.U.R., Kumar, S. et al. (2005). Involvement of polyamines in the contrasting sensitivity of chickpea (*Cicer arietinum* L.) and soybean (*Glycine max* (L.) Merrill.) to water deficit stress. *Botanical Bulletin of Academia Sinica* 46: 333–338.

de Oliveira, L.F., Navarro, B.V., Cerruti, G. et al. (2018). Polyamines and amino acid related metabolism: the roles of arginine and ornithine are associated with the embryogenic potential. *Plant Cell Physiol.* 59: 1084–1098.

Ortega-Villasante, C., Rellán-Alvarez, R., Del Campo, F.F. et al. (2005). Cellular damage induced by cadmium and mercury in *Medicago sativa*. *J. Exp. Bot.* 56 (418): 2239–2251.

Ouyang, J., Song, C., and Chen, D. (2017). Research progress on heat-tolerance mechanism and transports of polyamfines in plant. *Mol. Plant Breed* 15: 3286–3294.

Pal Bais, H. & Ravishankar G. (2002). Role of Polyamines in the ontogeny of plants and their biotechnological applications. Plant cell, tissue and organ culture, 69(1), 1–34.

Pál, M., Szalai, G., and Janda, T. (2015). Speculation: polyamines are important in abiotic stress signaling. *Plant Sci.* 237: 16–23.

Pál, M., Csávás, G., Szalai, G. et al. (2017). Polyamines may influence phytochelatin synthesis during Cd stress in rice. *J. Hazard. Mater.* 340: 272–280.

Pegg, A.E. (2009). Mammalian polyamine metabolism and function. *IUBMB Life* 61 (9): 880–894.

Peng, D., Wang, X., Li, Z. et al. (2016). NO is involved in spermidine-induced drought tolerance in white clover via activation of antioxidant enzymes and genes. *Protoplasma* 253 (5): 1243–1254.

Perfus-Barbeoch, L., Leonhardt, N., Vavasseur, A., and Forestier, C. (2002). Heavy metal toxicity: cadmium permeates through calcium channels and disturbs the plant water status. *Plant J.* 32 (4): 539–548.

Peynevandi, K.M., Razavi, S.M., and Zahri, S. (2018). The ameliorating effects of polyamine supplement on physiological and biochemical parameters of *Stevia rebaudiana* Bertoni under cold stress. *Plant Prod. Sci.* 21: 123–131. https://doi.org/10.1080/1343943X.2018.1437756.

Puyang, X., An, M., Xu, L. et al. (2016). Protective effect of exogenous spermidine on ion and polyamine metabolism in Kentucky bluegrass under salinity stress. *Hortic. Environ. Biotechnol.* 57 (1): 11–19.

Rahdari, P. and Hoseini, S.M. (2013). Role of poly amines (Spermidine and Putrescine) on protein, chlorophyll and phenolic compounds in wheat (*Triticum aestivum* L.) under salinity stress. *J Sci Res Rep* 1: 19–24.

Rangan, P., Subramani, R., Kumar, R. et al. (2014). Recent advances in polyamine metabolism and abiotic stress tolerance. *Biomed. Res. Int.* 2014: 239621.

Rosales, E.P., Iannone, M.F., Groppa, M.D., and Benavides, M.P. (2012). Polyamines modulate nitrate reductase activity in wheat leaves: involvement of nitric oxide. *Amino Acids* 42 (2): 857–865.

Roy, M. and Wu, R. (2001). Arginine decarboxylase transgene expression and analysis of environmental stress tolerance in transgenic rice. *Plant Sci.* 160 (5): 869–875.

Roychoudhury, A., Basu, S., and Sengupta, D.N. (2011). Amelioration of salinity stress by exogenously applied spermidine or spermine in three varieties of indica rice differing in their level of salt tolerance. *J. Plant Physiol.* 168 (4): 317–328.

Sadeghipour, O. (2019). Polyamines protect mung bean [*Vigna radiata* (L.) Wilczek] plants against drought stress. *Biologia Futura* 70 (1): 71–78.

Šebela, M., Luhová, L., Brauner, F. et al. (2001). Light microscopic localisation of aminoaldehyde dehydrogenase activity in plant tissues using nitroblue tetrazolium-based staining method. *Plant Physiol. Biochem.* 39: 831–839.

Sequera-Mutiozabal, M., Antoniou, C., Tiburcio, A.F. et al. (2017). Polyamines: emerging hubs promoting drought and salt stress tolerance in plants. *Curr. Mol. Biol. Rep.* 3 (1): 28–36. https://doi.org/10.1007/s40610-017-0052-z.

Serrano, M., Martínez-Madrid, M.C., Pretel, M.T. et al. (1997). Modified atmosphere packaging minimizes increases in putrescine and abscisic acid levels caused by chilling injury in pepper fruit. *J. Agric. Food Chem.* 45 (5): 1668–1672.

Serrano, M., Pretel, M.T., Martínez-Madrid, M.C. et al. (1998). CO_2 treatment of zucchini squash reduces chilling-induced physiological changes. *J. Agric. Food Chem.* 46 (7): 2465–2468.

Serrano-Martínez, F. and Casas, J.L. (2011). Effects of extended exposure to cadmium and subsequent recovery period on growth, antioxidant status and polyamine pattern in in vitro cultured carnation. *Physiol. Mol. Biol. Plants* 17 (4): 327–338.

Shanker, A.K., Cervantes, C., Loza-Tavera, H., and Avudainayagam, S. (2005). Chromium toxicity in plants. *Environ. Int.* 31 (5): 739–753.

Shao, C.G., Wang, H., and Yu-Fen, B.I. (2015). Relationship between endogenous polyamines and tolerance in *Medicago sativa* L. under heat stress. *Acta Agrestia Sinica* 23: 1214–1219.

Sheokand, S., Kumari, A., and Sawhney, V. (2008). Effect of nitric oxide and putrescine on antioxidative responses under NaCl stress in chickpea plants. *Physiol. Mol. Biol. Plants* 14 (4): 355–362.

Simões, A.D.N., Diniz, N.B., Vieira, M.R.D.S., and Al, E. (2018). Impact of GA3 and spermine on postharvest quality of anthurium cut flowers (*Anthurium andraeanum*) cv. *Arizona. Sci. Horticult.* 241: 178–186.

Slocum, R.D. (1991). Tissue and subcellular localisation of polyamines and enzymes of polyamine metabolism. In: *Biochemistry and Physiology of Polyamines in Plants* (ed. R.D. Slocum and H.E. Flores), 93–105. Boca Raton: CRC Press.

Sun, X., Wang, Y., and Tan, J. (2018). Effects of exogenous putrescine and D-Arg on physiological and biochemical indices of anthurium under chilling stress. *Jiangsu J. Agric. Sci* 34: 152–157.

Tabor, C.W. and Tabor, H. (1984). Polyamines. *Annu. Rev. Biochem.* 53: 749–790.

Tailor, A., Tandon, R., and Bhatla, S.C. (2019). Nitric oxide modulates polyamine homeostasis in sunflower seedling cotyledons under salt stress. *Plant Signal Behav.* 14 (11): 1667730.

Tanou, G., Ziogas, V., Belghazi, M. et al. (2014). Polyamines reprogram oxidative and nitrosative status and the proteome of citrus plants exposed to salinity stress. *Plant Cell Environ.* 37 (4): 864–885.

Tassoni, A., Franceschetti, M., and Bagni, N. (2010). Polyamines and salt stress response and tolerance in *Arabidopsis thaliana* flowers. *Plant Physiol. Biochem.* 46: 607–613.

Tatte, S., Alka, S., and Ahlawat, T.R. (2015). Effect of PAs on postharvest quality and vaselife of rose var. Samurai. *The Bioscan* 10: 675–678.

Tian, J. (2012). Physiological regulation function and proteomics research of exogenous spermidine on alleviating high temperature stress the of cucumber seedlings. PhD dissertation. Nanjing Agric. Unive, 7–21.

Tiburcio, A.F., Altabella, T., Bitrián, M., and Alcázar, R. (2014). The roles of polyamines during the lifespan of plants: From development to stress. *Planta* 240 (1): 1–18. https://doi.org/10.1007/s00425-014-2055-9.

Valko, M., Rhodes, C., Moncol, J. et al. (2006). Free radicals, metals and antioxidants in oxidative stress-induced cancer. *Chem. Biol. Interact.* 160 (1): 1–40.

Vondráková, Z., Eliášová, K., Vágner, M., and Al, E. (2015). Exogenous putrescine affects endogenous polyamine levels and the development of Picea abies somatic embryos. *Plant Growth Regul.* 75: 405–414.

Vujcic, S., Diegelmann, P., Bacchi, C.J. et al. (2002). Identifcation and characterization of a novel Xavin-containing spermine oxidase of mammalian cell origin. *Biochem. J.* 367: 665–675.

Vuosku, J., Suorsa, M., Ruottinen, M. et al. (2012). Polyamine metabolism during exponential growth transition in Scot's pine embryogenic cell culture. *Tree Physiol.* 32: 1274–1287.

Walden, R., Cordeiro, A., and Tiburcio, A.F. (1997). Polyamines: small molecules triggering pathways in growth and development. *Plant Physiol.* 113: 1009–1013.

Wang, Y., Devereux, W., Woster, P.M. et al. (2001). Cloning and characterization of a human polyamine oxidase that is inducible by polyamine analogue exposure. *Cancer Res.* 61: 5370–5537.

Wimalasekara, R., Tebart, F., and Scherer, G.F.E. (2011). Polyamines, polyamine oxidases and nitric oxide in development, abiotic and biotic stresses. *Plant Sci.* 181: 593–603.

Woo, H.R., Kim, H.J., Nam, H.G., and Lim, P.O. (2013). Plant leaf senescence and death–regulation by multiple layers of control and implications for aging in general. *J. Cell Sci.* 126: 4823–4833.

Wu, M. and Yuan, L. (2008). Research progress in the relationship between polyamine and plant resistance. *J. Anhui Agric. Sci.* 36: 3516–3518.

Wuddineh, W., Minocha, R., and Minocha, S.C. (2018). Polyamines in the context of metabolic networks. *Methods Mol. Biol.* 1694: 1–23. https://doi.org/10.1007/978-1-4939-7398-9_1.

Xu, L. (2015). The effect of polyamineon flower bud differentiation and bud germination of chrysanthemum. *Shandong Agric. Univ.* 31–36.

Xue, B., Zhang, A., and Jiang, M. (2009). Involvement of polyamine oxidase in abscisic acid-induced cytosolic antioxidant defense in leaves of maize. *J. Integr. Plant Biol.* 51: 225–234.

Yadav, J.S. and Rajam, M.V. (1997). Spatial distribution of free and conjugated polyamines in leaves of *Solanum melongena* L. associated with differential morphogenetic capacity: efficient somatic embryogenesis with putrescine. *J. Exp. Bot.* 48: 1537–1545.

Yamamoto, A., Shim, I.S., and Fujihara, S. (2016). Inhibition of putrescine biosynthesis enhanced salt stress sensitivity and decreased spermidine content in rice seedlings. *Biol. Plant.* 61: 385–388.

Yamasaki, H. and Cohen, M.F. (2006). NO signal at the crossroads: polyamine induced nitric oxide synthesis in plants. *Trends Plant Sci.* 11: 522–524.

Yang, C. and He, S. (2001). The relationship between polyamine and membrane lipid peroxidase during the senescence of cut rose flowers. *Acta Botanica* 21: 1157–1161.

Yang, Y. and Yang, X. (2002). Effect of temperature on endogenous polyamine content in leaves of Chinese Kale Seedlings. *J. South China Agric. Univ* 23: 9–12.

Yang, H., Shi, G., Wang, H., and Xu, Q. (2010). Involvement of polyamines in adaptation of *Potamogeton crispus* L. to cadmium stress. *Aquat. Toxicol.* 100 (3): 282–288.

Yang, H.Y., Shi, G.X., Li, W.L., and Wu, W.L. (2013). Exogenous spermidine enhances *Hydrocharis dubia* cadmium tolerance. *Russ. J. Plant Physiol.* 60 (6): 770–775.

Yang, L., Hong, X., Wen, X.X., and Liao, Y.C. (2016). Effect of polyamine on seed germination of wheat under drought stress is related to changes in hormones and carbohydrates. *J. Integr. Agric.* 15 (12): 2759–2774.

Yong, W., Wangjin, L., Zhaoqi, Z., and Hui, X. (2003b). ABA and putrescine treatents alleviate chilling injury in banana fruits during storage at 8 °C. *Zhi wu Sheng li yu fen zi Sheng wu xue xue bao= Journal of Plant Physiology and Molecular Biology* 29 (6): 549–554.

Yong-Hua, Z.H.E.N.G., San-Yu, L.I., Yu-Fang, X.I. et al. (2000). Polyamine changes and chilling injury in cold-stored loquat fruits. *J. Integr. Plant Biol.* 42 (8): 824.

Youssef, A.A., Mahgoub, M.H., and Talaat, I.M. (2004). Physiological and biochemical aspects of *Matthiola incana* L. plants under the effect of putrescine and Kinetin treatments. *Egypt J App Sci* 19: 492–510.

Yu, Y., Zhang, W.B., Li, X.Y. et al. (2016). Pathogenic fungal elicitors enhance ginsenoside biosynthesis ofadventitious roots in Panax quinquefolius during bioreactor culture. *Ind. Crop. Prod.* 94: 729–735.

Zhao, F.G. and Qin, P. (2004). Protective effect of exogenous polyamines on root tonoplast functions against salt stress in barley seedlings. *Plant Growth Regul.* 42: 97–103.

Zhao, J., Shi, G., and Yuan, Q. (2008). Polyamines content and physiological and biochemical responses to ladder concentration of nickel stress in *Hydrocharis dubia* (Bl.) Backer leaves. *Biometals* 21 (6): 665–674.

5

Mycorrhizal Symbiosis and Nutrients Uptake in Plants

Kashif Tanwir[1], Saghir Abbas[1], Muhammad Shahid[2], Hassan Javed Chaudhary[3], and Muhammad Tariq Javed[1]

[1] *Department of Botany, Faculty of Life Sciences, Government College University, Faisalabad, Pakistan*
[2] *Department of Bioinformatics and Biotechnology, Government College University, Faisalabad, Pakistan*
[3] *Department of Plant Sciences, Faculty of Biological Sciences, Quaid-i-Azam University, Islamabad, Pakistan*

Introduction

In worldwide terrestrial environments, an ancient symbiotic association exists between soil fungi and host plant roots, cyanobacteria, and algae (Sosa-Hernández et al. 2019). Fungi also acts as mediators of organic matter present in soil (Frey 2019). This association between semi–aquatic green algae and aquatic fungi is known to be the ancestor of terrestrial plants. Terrestrial plants were thought to be originated in response to the continuous symbiotic association of algae and fungus (Zhang et al. 2020). In this association between fungi and plant roots, the fungus obtains carbon from their photosynthetic partner, whereas in response it provides protection and essential nutrients including zinc (Zn), nitrogen (N), iron (Fe), and phosphorus (P). It was assessed that about 90% of alive plants form mycorrhizal association with fungus (Van der Heijden et al. 2015; Pressel et al. 2016; Prasad et al. 2017; Brundrett and Tedersoo 2018). Plants are capable of adapting themselves according to the changing environment due to external or internal stresses, mycorrhizal associations with fungi and bacteria, and availability of mineral nutrients (Nath et al. 2016). The phenotypic modifications in plant roots were mostly affected by availability of nutrients and mycorrhizal and bacterial colonization. The mostly structural changes in root are influenced by the nutrients accessibility in soil (Diagne et al. 2020). The essential plant nutrients, i.e. N and P present in soil act as a signal to start various molecular mechanisms of plants to uptake these nutrients like modifications of differentiation and cell division patterns in roots of plants. The root structure was also altered by the availability of these plant nutrients in soil. The primary growth of root, formation of root, root hair density, and lateral growth of root were also modified by the availability of internal and external plant nutrients (Martin et al. 2016). Also, the transportation and synthesis of plant growth

Plant Ionomics: Sensing, Signaling, and Regulation, First Edition.
Edited by Vijay Pratap Singh and Manzer H. Siddiqui.

hormones (ethylene, cytokinins, and auxin) were regulated which enhanced the root growth (Lopez-Bucio et al. 2003; Montesinos-Navarro et al. 2019).

Mycorrhizal Association and Its Types

Mycorrhiza are ancient, common, taxonomically diverse, widespread and have risen numerous times by convergence in fungi and plants. These associations also provide chance to inspect the evolution history of fungus and plant interactions (Martin et al. 2017; Montesinos-Navarro et al. 2019). In rhizospheric soil, roots of plants interact with a numerous supportive microorganisms, arbuscular mycorrhizal fungus among those that are known as one of the highly important group of soil biotic life important for sustainability of ecosystem (Jeffries and Barea 2012; Barea et al. 2013; Sosa-Hernández et al. 2019). These fungus belonging to Glomeromycota phylum formed symbiotic interactions with various plants of Gymnosperms, Pteridophytes, and Angiosperms (Prasad et al. 2017). It is primitive kind of interface that is evolved about 460 million years (Smith and Read 2008). These associations are also termed as "ecosystem engineers" because they enhanced the accumulation of essential minerals from soil (Bücking and Kafle 2015). They alter the supply and transport of essential plant nutrients inside soil ecosystem by forming a wide and interlinked soil network of extraradical fungal mycelium (ERM) (Barea et al. 2014; Diagne et al. 2020). These are mainly controlled by the transfer of carbon (C) from roots of plant to the fungal and followed by the reciprocal transport of ammonium (NH_4^+) and phosphate from fungal mycelium to plant tissues (Barea et al. 2014; Begum et al. 2019). These fungal hyphae formed specialized subcellular, tree-like root corticular structure termed as "arbuscules," and it is a derivative of Latin word "arbusculum," which is used for little tree or bush (Gutjahr and Parniske 2013; Lopez-Raez and Pozo 2013; Montesinos-Navarro et al. 2019). Though all these structures exist inside the plant rhizospheric soil rather than leaf or any other structure, therefore the early developmental stages and exchange of signals for the formation of symbiotic association between fungus and plants are still unknown (Genre 2012). These mycorrhizal associations have different forms and varieties on the basis of various fungal clades and types of plants (Van der Heijden et al. 2015).

Endomycorrhiza

The most commonly existing type of mycorrhizal associations is endomycorrhiza. This type of association, also known as arbuscular mycorrhizas (AM), is composed of a specific cluster of fungi known as Glomeromycotina (Figure 5.1). These fungi usually inhabit plant roots and occasionally in rhizomes of vascular bundles containing plant species (Pressel et al. 2016). These fungi are also sometimes form associations with thalli of some ancient plant groups like hornworts (Anthocerotophyta) and liverworts (Marchantiophyta) (Desiro et al. 2013; Pressel et al. 2016). This type of symbiosis is phylogenetically common in terrestrial plants. The fungal hyphae grows in apoplastic empty spaces between corticular plant cells and also penetrate inside cells to form specialized structures termed as arbuscules. These arbuscules are the structures formed by fungus penetration inside the plant cells and meant for ensuring the nutrient supply between fungal and plant (Bonfante and Genre 2010).

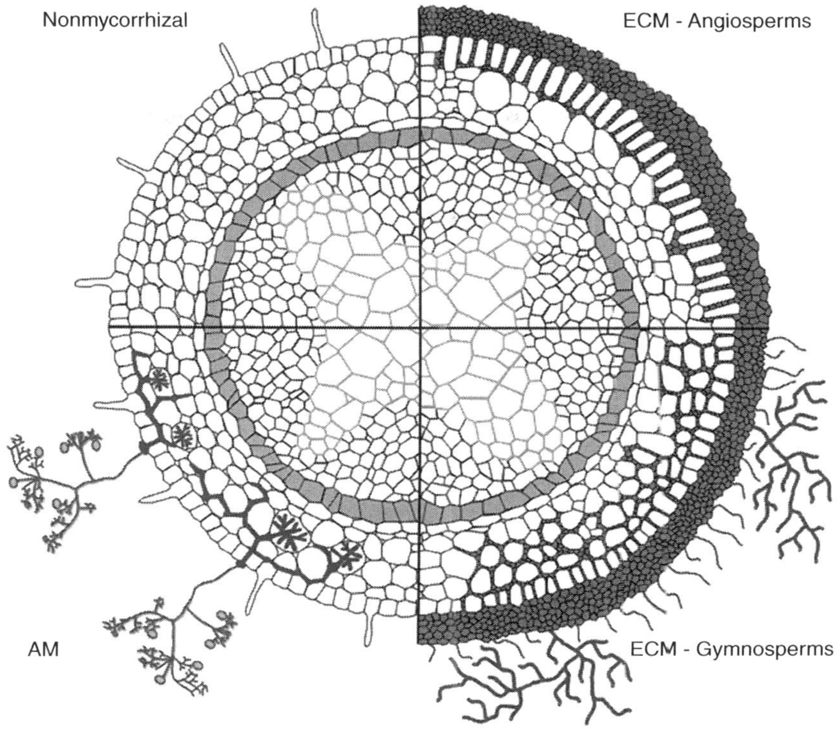

Figure 5.1 The anatomy of root and types of fungal interactions in angiosperms and gymnosperms. *Source:* Bücking et al. (2012)/IntechOpen/Public domain CC BY 3.0.

Sometimes, they also form several vesicles that are inflated intraradical hyphae, which is meant for the storage of reserve food (Smith and Read 2008; Berbee et al. 2017). Some fungi also form coil endomycorrhizas belonging to group Mucoromycotina (CMm), which is alike to other symbiotic fungi. They formed hyphal coils inside plant cells, lumps, and various other intercelluar fungal structures with varying thickness of cell walls (Strullu-Derrien et al. 2014; Field et al. 2016).

Glomeromycotina and Mucoromycotina are two sister clades inside the fungal Mucoromycota phylum (Tang et al. 2016). Similar to AM, Mucoromycotina also penetrates in the thallus of ancient land plants and also in the roots of ferns and lycopods (Desiro et al. 2013; Strullu-Derrien et al. 2014; Rimington et al. 2015; Field et al. 2016).

Ectomycorrhiza (ECM)

The second type of symbiotic association is ectomycorrhiza (ECM), in this association a layer or cover of fungus hyphae developed around the root and an intercellular infiltration pattern developed by fungal hyphae in between cortical cells that are named as Hartig net (Smith and Read 2008; Liu et al. 2020) (Figure 5.1). The fungus belonging to genus Basidiomycota, Ascomycota, and some species of Endogone (Mucoromycotina) also developed ECM (Yamamoto et al. 1995). This type of mycorrhizal association exists in most species of families (such as Myrtaceae, Dipterocarpaceae, Fagaceae, and Caesalpinioideae). Plants belongings to these families are mostly trees and some are shrubs present in temperate, boreal, and mediterranean regions of the World (Smith and Read 2008; Tedersoo and Brundrett 2017). In ECM, the fungal sheath is not formed, and hyphal coils are colonized around root cells (Smith and Read 2008; Liu et al. 2020). Almost 10 000 species of fungi and around 8000 different plant species have known to have ecto-mycorrhizal associations (He et al. 2019). Extraradical mycorrhizal fungi missing their capability to destroy polysaccharides (pectins, pectates, and cellulose) present in cell wall of plants, and this limits their infiltration of mycelium into the intercellular spaces of plant roots (Shao et al. 2018).

Establishment of Arbuscular Mycorrhiza in Soil

Arbuscular mycorrhizal association is an enormously complex and energetic interaction between two partners (plants and fungi), which have highly coordinated and centrally controlled at molecular levels (Aroca et al. 2013; Gutjahr and Parniske 2013; Barea et al. 2014; Bonfante and Desiro 2015; Mohanta and Bae 2015; Pozo et al. 2015; Lopez-Raez 2016). The formation of this fungal interaction comprised three major stages: (i) the first step is a growth of asymbiotic hyphae, it is a place for germination of spores and also grow autonomous hyphae for a small period; (ii) the second step is a presymbiotic growth phase, it is the point where hyphal growth was enhanced due to the signal of host; and (iii) the third one is symbiotic stage, where fungal hyphae infiltrates into plant root cells and form intraradical mycelium (IRM) and extraradical hyphae for nutrients absorption (López-Ráez 2016). Mostly, it was confirmed that arbuscular mycorrhizal fungus inhabits plant roots from three major kinds of soil-based propagules (extraradical hyphae, fragments, and spores of mycorrhizal roots), and all these structures are intricate in the establishment of well-developed soil mycelial network (Barea et al. 2014).

Growth of Asymbiotic Hyphae

In first step, AM colonization starts with the establishment of fungal hyphae that is formed from propagules developed in soil such as mycorrhizal root fragments or resting spores or from native plants (Koltai and Kapulnik 2009). Subsequent to germination, the fungi utilized reserve glycogen and triacylglyceride (TAG) present in spores to maintain growth of young mycelium because it is incapable of using C from organic matter of soil as a source of energy (Leigh et al. 2009). But, if the host is not present (asymbiotic phase), these hyphae can survive more than few days. They have obligate biotrophic nature and only limited life span due to which the growth of these asymbiotic mycelium stops earlier to the depletion of spore reserves. This may lead toward the germination of new event and mycelium withdraw their cytoplasm from spore and becomes dormant (Genre 2012). Such tentative hyphal growth and development patterns were changed intensely when the hyphae spread around the neighborhood of a plant root (Nasim 2013). However, the development of mycelium germ tube rises considerably, and fungal hyphae subdivides quickly in the soil and penetrates into the roots of host plant (Lopez-Raez et al. 2012), which indicates that they used something as a food which is released from the host roots (Harrison 2005).

Presymbiotic Stage

The root of plants released various compounds in the soil, and "strigolactones" are one of those chemical compounds, which are known as "rhizospheric signals of plant," which are taking part in the stimulation of growth at presymbiotic, spore germination, branching, and hyphal growth stages of arbuscular mycorrhiza fungi (Lopez-Raez et al. 2012). These compounds also enhanced the availability of host and root colonization of symbiotic fungus (Kumar et al. 2015). Many studies have confirmed the absence or availability of low concentration of strigolactones or impaired biosynthesis manifolds reduced the growth and establishment of mycorrhizal fungus and plant roots association (Vogel et al. 2010; Lopez-Raez 2016). Commonly, the strigolactones are existing in low amount in rhizospheric exudates (Akiyama and Hayashi 2006), but under limited supply of nutrients and their concentration increased which leads plant toward mycorrhizal colonization (Koltai and Kapulnik 2009). Phosphate starvation also enhanced the exudation and biosynthesis of strigolactones, which ultimately promotes the colonization of mycorrhizal fungi (Yoneyama et al. 2012).

Several studies have validated that some chemical compounds released by host plant are perceived through a specific receptors by mycorrhizal fungus, which encourage hyphal branching of fungus, specifying that discrimination of nonhost and host plant occurs during that growth stage (Harrison 2005; Nasim 2013; Wang et al. 2017). These outcomes responsively specify that mycorrhizal fungus have various developed mechanisms which are switch on by the release of various chemicals from host plant root. These chemicals activate various transcriptional paths that bring several morpho-physiological modifications in fungal hyphae to enhance its growth (Nasim 2013). Therefore, the strigolactones have a central role in establishing symbiotic connection between plant host and fungus through widespread hyphal branching (Aroca et al. 2013; Lopez-Raez 2016; Huang et al. 2020). Plants receive diffusible fungal signals through lysine motif receptor kinases at plasma membrane that are called as Myc factor after the initial growth of arbuscular mycorrhizal fungus around host plant root cells starts (Oldroyd 2013). This will induce the formation of symbiotic association by inducing physiological changes in intracellular environment of the host plant root cells without any physical connection (Genre and Bonfante 2010). These Myc factors are chemically recognized as a combination of various nonsulfur-containing and sulfur-containing lipochito-oligosaccharides, and these compounds also have some structural resemblances with rhizobial nod factors (Maillet et al. 2011). These Myc factors also activate various other metabolic and regulation processes of plants including regulation of signal transduction genes, starch accumulation, lateral root growth stimulation, epidermal cell calcium recurrent oscillations, and resemblance with rhizobium-legume symbiosis (Maillet et al. 2011; Gutjahr and Parniske 2013; Bonfante and Desiro 2015).

Different Symbiotic Stages of Fungal Mycelium Growth

The presymbiotic stage is lagged by straight plant–fungi interaction such as symbiotic phase, in which different cellular and developmental alterations take place in both plant and fungus. The activation of transcriptional symbiosis-related genes takes place via Myc factor-based induction of nuclear Ca^{2+}-spiking, which is formed by decodation of CCaMK (nuclear contained Ca-calmodulin kinase) and phosphorylation of CYCLOPS a transcriptional factor (Carbonnel and Gutjahr 2014; Singh et al. 2014). The fungal mycelium of glomeromycetes formed a flattened, branched, and swollen structure of fungal cells termed as aspersorium (hyphopodium) through penetration in atrichoblasts of root epidermis (Smith and Read 2008; Genre 2012). The formation of aspersorium indicates the start of third growth phase (symbiotic phase). The epidermal cells of root also facilitate the formation of aspersorium through relocating their cell nucleus and repositioning their cytoplasm, which provides empty space for the penetration of fungal hyphae (Genre 2012). All the processes in which activation of epidermal cells earlier and after the production of fungal hyphae prepenetration gadget, late mycorrhizal development, and even cells containing for arbuscules in late stages are controlled through the activation of a proline containing protein (ENOD11) (Mohanta and Bae 2015). Appressorium was converted into a fungal hypha that approaches and penetrates into root cortex by using intracellular path through epidermal cells (Barea et al. 2014; Huang et al. 2020). The integrity of host cell was also sustained through the invagination of root cortex plasma membrane that multiplies and surrounds the penetrating fungal hypha, which physically separates the cell cytoplasm from fungus hyphae (Bonfante and Desiro 2015). The establishment of prepenetration gadget in fungus (*M. truncatula*) was regulated by three essential ion channel regulating genes DMI-1, DMI-2, and DMI-3 (Genre and Bonfante 2010; Mohanta and Bae 2015). The mycelium of fungi after penetration in internal cortical cells continuously divides and branched to form a specific tree-like structure called as arbuscules, which is a place for the transfer of minerals between fungus and symbiotic host plant (Smith and Read 2008). The formation of PPA and subsequent fungal growth were also controlled by another set of fungus protein encoded SYM genes (CYCLOPS and CCaMK) (Bonfante and Genre 2010; Genre 2012). The reorganization of structural infected cells was also regulated by another gene VAPYRIN (Gutjahr and Parniske 2013). In circumstance of vapyrin mutants of Petunia and Medicago, the penetration of fungal hyphae in rhizodermal is often terminated, and in some situations, the arbuscules are not developed due to cortical colonization and only minor corticular hyphal protrusions form (Pumplin et al. 2010; Gutjahr and Parniske 2013). Formation of small arbuscule inner side of the cortical cells lumen assures the induction of mycorrhizal association between plant and its fungal companion (Genre 2012; Wang et al. 2017).

The two major types of arbuscular mycorrhizal associations are formed on the base of morphological characteristics. The first type of colonization is called Arum type, while the other one is Paris type (Garg and Chandel 2010). The Arum type of colonization mostly exists in legumes and generally in the roots of those plants that contain large number of apoplastic channels in its root anatomy (Genre 2012). Arbuscular mycorrhizal fungi required more intercellular air spaces for the growth of intercellular hyphae and enter root cells by developing small-sized lateral branches which form terminal arbuscules (Shah 2014). On the other side, Paris type of colonization develops intracellular coil hyphae, which contains intercalary arbuscules and spreads nonstop from one cell to other in roots (Shah 2014). Additionally, some storage structures formed from lipids in root apoplast are also present in some fungi, which are called as vesicles and they are different from arbuscules (Rouphael et al. 2015). A highly condensed network of extra-radical mycelium (ERM) is formed in response to the growth of IRM in soil (Malbreil et al. 2014), which facilitate the fungus to uptake minerals from soil, mainly those minerals that have less mobility and availability in soil including NH_3 and phosphate (Barea et al. 2014). Additionally, ERM also spreads its colonization to other neighboring plants of same or different species and also interacts with other micro-organisms of the soil. Therefore, in arbuscular mycorrhiza, both fungi and plants connected with each other every time with a web of hyphae and plant roots (Giovannetti et al. 2014), where interchange of signals, minerals, and water takes place (Song et al. 2010; Rouphael et al. 2015). The ERM develops various new chlamydospores and enhanced the fungal propagation, which complete its lifecycle. Consequently, formation of new functional arbuscular mycorrhizal symbiosis includes a high degree of fungal and plant coordination (Genre 2012).

Root Modifications for Accumulation of Nutrients

Plants are capable of adapting themselves according to the changing environment due to external or internal stresses, mycorrhizal associations with fungi and bacteria, and accessibility of mineral nutrients (Nath et al. 2016). The phenotypic modifications in plant roots were mostly affected by the availability of nutrients and mycorrhizal and bacterial colonization (Figure 5.1). The most structural changes in root are influenced by the nutrients accessibility in soil. The essential plant nutrients, i.e. N and P present in soil act as a signal to start various molecular mechanisms of plants to uptake these nutrients like modifications of diffrentiation and patterns of cell division in roots of plants. The root structure was also alters by the presence of these plant minerals in soil. The preliminary root growth, root hair formation, and lateral growth of root were also modified by the availability of internal and external plant nutrients. Also, the transportation and synthesis of plant growth hormones (ethylene, cytokinins, and auxin) were regulated that enhanced the root growth (Lopez-Bucio et al. 2003). The plants roots residing arbuscular mycorrhiza have higher root hairs, lateral root branching, and developed root apex as compared to roots without fungal symbiosis (Berta et al. 1995). The roots residing mycorrhiza containing root hairs similarly alike nonmycorrhizal roots, though their length and density may be different (Orfanoudakis et al. 2010). The plants containing mycorrhizal roots use mycorrhizal pathway (MP) for nutrients accumulation while roots without fungal symbiosis use direct pathway (DP) for nutrients uptake. In MP, nutrients are uptake by ERM and transported to IRM through which these nutrients move for several centimeters and reached into the root cells using apoplast interfacial. The roots with inoculated arbuscular mycorrhizal fungi contain transporters for the transport of NH_4^+ and orthophosphate (Pi) (Bucher 2007; Guether et al. 2009b; Smith and Smith 2011). In the meantime, the perifungal membrane present around intracellular coils and arbuscules also contains H^+-ATPase for active transport (Rosewarne et al. 2007). The roots without fungal symbiosis uptake nutrients from root surrounding rhizosphere zone of soil through root hairs, while roots with fungal symbiosis uptake essential nutrients via mycorrhizal hyphae, which was prolonged many centimeters away from rhizosphere of roots and can transport plant nutrients to root corticular cells through MP (Figure 5.2). This MP for the transport of essential nutrients is highly controlled and provide speedy transport system to plants. Furthermore, mycorrhizal symbiosis also provides rapid transport of nutrients to plants facing various kinds of environmental

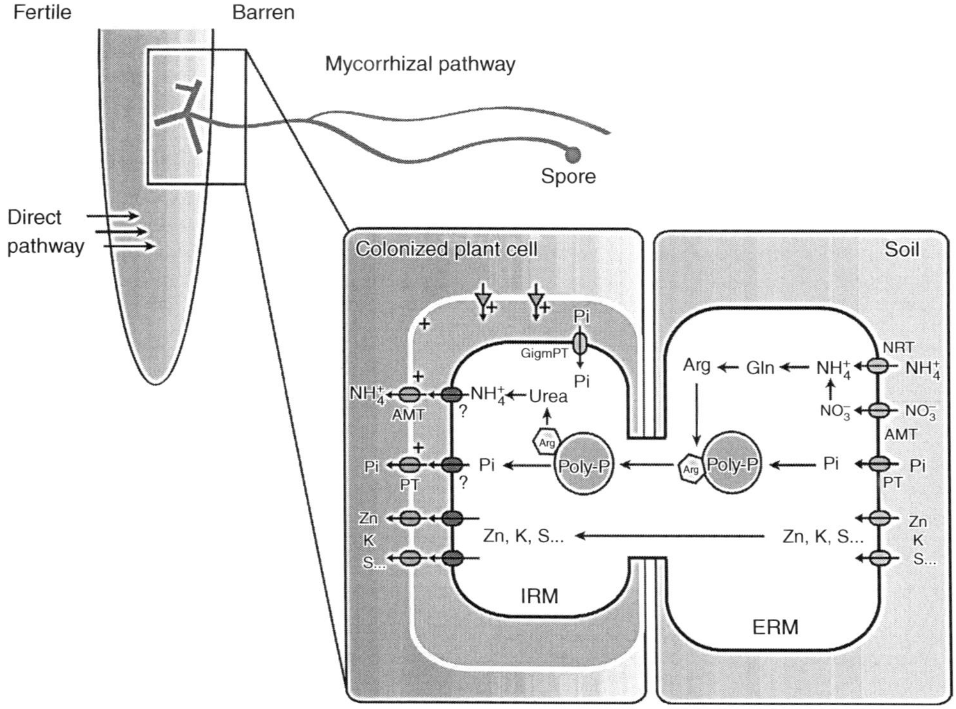

Figure 5.2 The indirect and direct mycorrhizal associated pathways for the uptake and subsequent transformation of zinc (Zn), potassium (K), nitrogen (N), sulfur (S), phosphorous (P), and other essential micronutrients. *Source:* Wang et al. (2017)/With Permission of Elsevier.

stresses and thus acts as a stress mitigating mechanisms in plants. For example, the mycorrhizal symbiosis in plants facing water stress expresses stress coping genes and produces various proteins to mitigate water stress. Additionally, under water stress conditions, the fungus produce long mycorrhizal hyphae for the exploration of water from long distances.

Nitrogen Uptake Mechanisms of Mycorrhizal Symbionts

Nitrogen transmission from rhizosphere to plant through MP remained insignificant for long time on the grounds that plant capable of accumulating inorganic nitrogen in roots in the form of NH_4^+ and NO_3^- appropriately because of their higher mobility and nonavailability of organic nitrogen to mycorrhizal symbionts. Substantial quantity of NH_4^+ and NO_3^- (20–50 μM) is always present in rhizosphere of unfertilized soil due to its higher mobility and nondepleted nature (McDowell et al. 2004; Bücking and Kafle 2015). The nitrogen in the form of NH_4^+ and NO_3^- from rhizosphere was uptake by certain plant species including mycorrhizal symbiotic plants. The mycorrhizal hyphae uptake higher quantity of NH_4^+ as compared to NO_3^- because the NH_4^+ is less mobile and generally absorbed on organic colloids and soil so only small concentration is available for plant uptake

(Marschner and Rimmington 1988; Bücking and Kafle 2015). Various studies showed that leguminous plants inoculated with arbuscular mycorrhizal fungi augmented absorption of total nitrogen from soil significantly as compared with noninoculated leguminous plants (Smith and Read 2008). However, it was also reported by some recent articles that nonleguminous arbuscular mycorrhizal fungi inoculated plants also enhanced the absorption of nitrogen from soil (Johnson 2010). The uptake of nitrogen from soil as also influenced by availability of moisture in soil and uptake of $NH4^+$ is higher in moist soil as compared to NO_3^- due to its higher mobility (Tanaka and Yano 2005).

The inoculation of cucumber with fungus 10% enhanced the uptake of nitrogen, while in tomato 42% increase in nitrogen accumulation was observed (Johansen et al. 1992; Mader et al. 2000). In monoxenic culture of mycorrhizal fungi, ERM uptakes nitrogen and binds it with arginine amino acid and this complex nitrogen form was translocated from ERM to IRM (Govindarajulu et al. 2005; Sosa-Hernández et al. 2019). The absorption of inorganic nitrogen by fungal hyphae can be combined with amino acids through glutamine oxoglutarate aminotransferase cycle (GS/GOGAT), glutamine synthetase, and lastly transformed into amino acid arginine (Jin et al. 2005; Bücking and Kafle 2015). However, Cruz et al. in 2007 reported that before entering in to the root corticular cells, the nitrogen was released from arginine as NH_4^+ (Cruz et al. 2007). Arginine was destroyed into ornithine and urea IRM. During this destruction process, NH_3 released as the by-product of urea hydrolyzation, which is transferred to plant root cells. The transfer of nitrogen from soil to plant root cells takes place through MP (Jin et al. 2005) (Figure 5.2).

The transport of nitrogen in *Rhizophagus irregularis* with mycorrhizal symbiosis was regulated by three major ammonium transporters genes including Gint-AMT1, Gint-AMT2, and Gint-AMT3 (Calabrese et al. 2016; Wang et al. 2017). These genes were activated during low ammonium supply and in response they enhanced the NH_4^+ accumulation from soil by fungi. Additionally, mycorrhizal fungi are also able to accumulate nitrogen by enhancing the enzymatic destruction of soil organic material like plant leaves (Leigh et al. 2009). Mycorrhizal fungal symbiotic Medicago and soybean also contain two ammonium transporters (GmAMT-4.1 and ATM-2.3) in root corticular cells. These nitrogen transporters are specially located on peri-arbuscular membranes of plant cells, which indicate that active NH_4^+ transformation has taken place in arbuscule branches (Breuillin-Sessoms et al. 2015).

Phosphorus Accumulation Mechanisms of Mycorrhizal Fungus

Fresh research on nutrient accumulation mechanisms in plants showed that fungal symbiosis has resilient potential to enhanced mobilization of nutrients from soil in various ecological environments. The presence of mycorrhizal fungal association enhanced the concentration of phosphorous (P) and nitrogen in rhizosphere of soil from natural substrates (Read and Perez-Moreno 2003). The concentration of inorganic P in soil is typically below 10 μM at low and higher levels of pH because of its adsorption with Al, Ca, and Fe (Schachtman et al. 1998). The mobility of P from soil to root epidermis is lower due to its less availability. Different plant taxons have variable root systems for nutrients absorption, and some plant species are completely dependent upon mycorrhiza for phosphorous uptake from soil (Ortas 2012). The plants dependent on mycorrhizae for P accumulation also required mycorrhizal symbiosis. The plant nutrients requirement and soil fertility are regulators of fungal symbiosis in plant roots. The plants growing under deficiency of soil Zn and P conditions were treated with numerous Zn and P levels together with inoculation of mycorrhizal fungi. It was clearly evident from the obtained results that significant increase in mycorrhizal growth and P accumulation were observed under low applied concentrations of P (Ortas 2012). However, under higher levels of applied P fertilizer, no significant variation was observed in both inoculated and noninoculated plants in term of P and growth increase. This increased accumulation of Zn and P may be due to widespread growth of extra radical fungal hyphae retrieving nutrients wide outside from the rhizospheric area of plants roots under inoculated conditions. Few plants were not be able to directly uptake P with root hairs and root epidermis from soils via direct uptake pathway (Figure 5.2). Various mechanisms were found in plants for P uptake including solubilization of P, assessment of soil outside rhizospheric zone, and absorption by mycorrhizal hyphae (Hinsinger 2001). Due to hyphal penetration in soil, the P present outside rhizospheric zone becomes reachable to the roots of inoculated plants. If required threshold levels for phosphorous absorption was reduced, mycorrhizal attraction for soil phosphorous was increased and transfer of phosphorous into mycorrhizal hyphae increased (Huang et al. 2020). The different forms of N have strong effect on rhizospheric soil pH and in response this change in pH might alter the accumulation of minerals in plant tissues in both fungal inoculated or noninoculated plant species (Ortas et al. 1996). The organic acids produced by different P-solubilizing microbes residing in soil enhanced the free P present in soil, which was taken up and transported to the host plant roots via fungal mycelium (Bolan 1991). The soil present in surrounding of mycorrhizal mycelium showed decrease in rate of P reduction from soil (Silberbush and Barber 1983; Koide 1991). The inorganic P up taken by MP imitates through transporting to the ERM in contradiction of electrochemical potential gradient, the required energy was provided by H^+-ATPase (Bucher 2007). Polyphosphate was prepared by buffering cytoplasmic inorganic phosphorous, and it was stored in fungal hyphae for further transportation to plant root cells (Hijikata et al. 2010). Polyphosphate formation depends upon the concentration of available P, and its behavior also vary from soluble to insoluble range with flexible chain length (Viereck et al. 2004). The inorganic P and polyphosphate both were containing negative charge, and plants balanced their charge by accumulation of Mg^{+2} and K^+ from soil; however, in plants in case of arbuscular mycorrhizal association plants generates Arg^+ with polyphosphate to balance positive charge (Ryan et al. 2007). The inorganic phosphorous membrane transporters of plants and fungi were similar and both these are involved in efflux of inorganic P (Preuss et al. 2010; Stuart and Plett 2020). Early studies presented that mycorrhizal symbiosis does not have any influence on P accumulation through direct pathway, and the presence of MP acts as additional contributor in fungal symbiotic plants. So, the uptake of P in mycorrhizal containing was found more than noninoculated plants. The MP was established as highly responsive for P accumulation in fungal symbiotic plants including onion, leek, and clover. In a recent research, fungal-inoculated tomato plant direct pathway was found totally inactive, while MP transported 100 % plant required P (Smith et al. 2004). This research demonstrates the functional variation of mycorrhizal symbiotic association in phosphate accumulation and transportation through both metabolic paths (Facelli et al. 2010). The

Rhizophagus irregularis showed augmented accretion of phosphate in inoculated tomato plant as associated with *Gigaspora margarita* through MP (Figure 5.2). Similarly, the P uptake in inoculated tomato plant was altered in different mycorrhizal symbiotic species (Ortas et al. 2013). Here significant results were distinguished that total absorption of phosphate was equal through MP and direct pathway (Facelli et al. 2010).

In mycorrhizal fungi, the transport of inorganic P across plasma membrane of ERM is facilitated by various reported P transporters including GvPT, GmosPT, and GintPT (Fiorilli et al. 2013). The current research revealed that *Gigaspora margarita* contains a specific transporter for phosphate (GigmPT), which is responsible for cross membrane transport of P. They also elaborated that GigmPT works as a high affinity transporter ($K_m = 1.8$ mM) for transport of P (Xie et al. 2016). The optimum pH required for the work of GigmPT is near to MtPT4 transporter situated near to the PAM (Conrad et al. 2014). GigmPT can facilitate phosphate accumulation at the interface of fungus–soil under short supply of inorganic P and its transcription in extra-radical fungal hyphae and might be useful in mycorrhizal association inside the roots concerning about transcription GigmPT through IRM (Fiorilli et al. 2013).

Potassium (K) and Sodium (Na) Uptake Mechanisms of Mycorrhizal Fungi

For potassium (K) uptake in fungi, two different groups of membrane transporters present including Trk transporters and high affinity potassium transporters (HAK). Both these groups are meant for Na^+ and K^+ accumulation from soil (Corratgé-Faillie et al. 2010; Benito et al. 2011). The most abundant group of K transporter has Trk in genetic database of ectomycorrhizae, while it was not found in endomycorrhizae fungi. In *R. irregularis* instead of Trk, the HAK transporter was present for subsequent transportation of K (Garcia and Zimmermann 2014). These K transporters in both types of fungal symbiosis were localized, specifically expressed and regulated by the fungus and plants. Two specific types of ion channels were responsible for the control and regulation of K transport including SKC (shaker-like potassium ion channel) and TOK (tandem-pore outward potassium ion channel) (Garcia and Zimmermann 2014). In *R. irregularis,* the two SKC-regulating genes were present. While in TOK, ion channels were not found, but they are related to K release in ectomycorrhizae. On the other hand, in plants, the molecular data were still not found about K transport; however, a specific K transporter KUP (putative K uptake) was found that was 44-fold upregulated in mycorrhizal-inoculated plant *Lotus japonicus* (Guether et al. 2009a). The sweet corn growing in soil inoculated with mycorrhiza has higher Na and K accumulation as compared to noninoculated plant concentration (Ortas 2003).

Metabolism of Sulfur in Mycorrhizal Symbiosis

Sulfur (S) acts as a vital macronutrient compulsory for development, growth, stress mitigation, and formation of various S containing complex chemical compounds in plants. The sulfur is only present in the sulfate form, which was directly accumulated by plants from soil through H^+-dependent co-transport mechanism, which was regulated by various membrane transporters of sulfate (Miransari 2013; Stuart and Plett 2020). The leaching capacity of sulfate is manyfolds higher than other known organic forms of sulfur due to which it was not easily available for plant uptake (Casieri et al. 2012). The mycorrhizal symbiosis without changing the activity of S transporters (Giovannetti et al. 2014) facilitates the plants in S accumulation and provide a sufficient amount of S for normal plant growth (Casieri et al. 2012; Sieh et al. 2013; Berruti et al. 2016). These results were revealed by a study in which *M. truncatula* plant were grown inoculated with *R. irregularis* fungi, and various sulfate levels were applied to understand the S accumulating capacity of mycorrhizal symbiosis (Berruti et al. 2016). Casieri et al. (2012) also examined the gene expression of encoding reputed Medicago sulfate transporters (MtSULTRs) and exposed that fungal association considerably augmented the S uptake from soil. Furthermore, it was also analyzed eight different MtSULTRs were located in leaves of host plant and roots. Recently, in a study on *L. japonicas*, a sulfate transporter (LjSultr1) were indentified that was explicitly tangled with the accumulation of S from fungal arbuscules (Giovannetti et al. 2014). However, in this study, it was also seen that a single gene plant S transporter (LjSultr1) was used in both indirect and direct pathways for S accumulation from soil and *L. japonicus*. On the other hand, the S accumulation by plant roots directly depends upon P concentration present in soil, and the accumulation of higher S in roots takes place when only small concentration of phosphate present in soil (Sieh et al. 2013; Behie and Bidochka 2014; Ortaş and Rafique 2017) (Figure 5.2). Additionally, the mycorrhizal symbiosis-based uptake of S also depends upon the release of root exudates, over production of enzymes, enhanced fungal hyphal growth, and increased action of other soil residing microbes like *Thiobacillus*. All these mechanisms enhanced the acidification of rhizospheric soil, which ultimately enhanced the S availability for host plant (Miransari 2013).

Role of Mycorrhizal Lipid Metabolism in Nutrients Accumulation

Glomeromycetes or oleogenic fungi are an important group of fungi because of almost 25% of their body dry mass contains lipids or lipid by-products (Malbreil et al. 2014; Stuart and Plett 2020). Various studies have exposed that lipid mechanism acts as a regulator of definite regulation mechanism in fungal symbiosis. In which carbon is attained from host plant as glucose or hexose then mostly stored in fungi as triacyl glycerol (TAG_ (a C storage form of lipid that allows translocation at long distances) mostly in fungal spores and sometimes in hyphae and is provided at the time of its requirement (Malbreil et al. 2014). According to different studies, the formation of palmitic acid through fatty acid synthesis initially takes place in IRM, and it is then used in the germination of spores, IRM, and ERM (Trepanier et al. 2005; Bhandari and Garg 2017). Additionally, Tisserant et al. (2012) exposed that *R. irregularis* contains all the genetic materials incorporated on their DNA for the formation of fatty acids, and fungal symbiosis is not required for any extra supply of lipids from host plant. Still, posttranscriptional regulation was required for the synthesis of fatty acids. Numerous genes connected to fatty acid metabolism like lipase and desaturase were upregulated in plants (Tisserant et al. 2013; Malbreil et al. 2014).

Mechanism of Micronutrients and Heavy Metal Uptake in Mycorrhizae

Some heavy metals present in soil (Ni^{2+}, Fe^{2+}, Cu^{2+}, Mn^{2+}, Co^{2+}, and Zn^{2+}) in small concentration act as vital micronutrients for normal growth, development, and metabolism of plants, but in higher concentrations, these can be converted into highly toxic ion. For example, plants closely monitor Fe homeostasis as shortage and excess of Fe are instantly revealed in plants owing to its fast reactivity through Fenton reaction (Morrissey and Guerinot 2009; Liu et al. 2020). The root exudates also have varied chemical nature from all other organic (carbohydrates, amino acids, phenolic, organic acids, carboxylate, anions, enzymes and hormones) and inorganic compounds (phosphates, protons) (Liu et al. 2020). These exudates not only change the chemical conformation of rhizospheric soil but also affected the micronutrients availability and concentration (Javed et al. 2018). The use of mycorrhizal fungi in association with plants to overcome various kinds of environmental stresses and bioremediation of soils highly polluted with heavy metals is a newly reputable tactic. The use of fungal symbionts in contaminated soils supports the plants in improving growth and reducing toxic heavy metal transport to roots and shoots (Kamal et al. 2010). Recently, numerous researchers showed that several mycorrhizal fungi like *G. mosseae* and *R. irregularis* improved the transport of essential heavy metals in shoot with higher efficacy (Zaefarian et al. 2013; Ali et al. 2015). In addition, mycorrhizal fungi also involved in the regulation of heavy metals accomodation by bearing in mind the health of plant and facilitate the crop plants in sidestepping the metalic noxiousness (Mn, Cd and Zn) inside the plant tissues (Kamal et al. 2010; Diagne et al. 2020), though some upsurge in metal concentration was also pragmatic (Liao et al. 2003). Micronutrients like zinc and copper have limited diffusion in soil–plant system during inoculated with mycorrhizal fungi. In a previous research, it was experienced that coffee plant growing in Zn and Cu contamination has higher growth under inoculated conditions with fungal hyphae, and it was also decided that mycorrhizal fungi shields plant sprout from metal contamination (Andrade et al. 2010). Plant experienced heavy metal contamination through alteration in three diverse molecular mechanisms on the base of their availability and physical and chemical characteristics like (i) over production of ROS by a series of chemical reactions of Fenton cycle or selfoxidation of transition metals (Fe and Cu), (ii) the obstructing vital functional groups in complex biomolecules through activity of non-redox heavy metals (Cd^{2+} and Hg^{2+}), and (iii) relocating vital metalic ions in complex biomolecules. Plants facing contamination of Fe^{2+} and Cu^{2+} enhanced oxidative stress demage through enhancing generation of H_2O_2, oxidative burst, and lipid peroxidation (Tanwir et al. 2015; Javed et al. 2017; Abbas et al. 2020).

Toxic heavy metals, mainly Cd^{2+}, causes inhibition of oxidative enzymes, especially the action of glutathione reductase. The contamination of Cd^{2+} might enhanced redox regulator of plant tissues, which in response start a series of action terminating the growth interruption of plants. Furthermore, exciting various secondary metabolic reactions, lignification, and eventually result in cell death of plants (Tanwir et al. 2021). Some other studies revealed that mycorrhizal fungi-inoculated plants showed nonsensitive behavior toward Cd^{2+} as compared to those noninoculated with fungal hyphae plants. Probably, interface of mycorrhizal fungal partner and crop plants also activates *Paxillus-pinus*-accumulated phenolic defence system. Mostly, the Cd tempted variations in retort of phenolics in roots of the fungal symbiosis are buffered. Mycorrhizal fungi also resists structural variations in Cd contamination through infected roots (Figure 5.2). Various mechanisms of plants were used by mycorrhizal fungi to save plant root injury from Cd contamination by stimulating metal chelation, limiting Cd availability in soil, limiting excess of Cd to extra- and entracellular sites, and also by activating defence systems (Schützendübel and Polle 2002).

Carbons-Based Triggering of Nutrients Accumulation in Mycorrhizal Symbiosis

Carbon acts as a chief component in mutual association between host plant and mycorrhizal fungal partner, though the transmission of carbon to mycorrhizal interaction was first reported in 1960; however, its complete and comprehensive molecular machinery persisted in an imperative question for a long period of time (Smith and Read 2010). Around 25% of the total C produced through photosynthesis were transported to mycorrhizal fungi (Douds et al. 2000; Stuart and Plett 2020). The mycorrhizal interaction and its rhizobium inoculation in roots of plants caused doubling of carbon distribution in nodules, and C distribution in roots of plants occupied by mycorrhizal symbiosis was higher (12% of fixed C) than that for mycorrhizal colonized plants (4%) (Kucey and Paul 1982).

The mycorrhizal host plants transmits carbon in sucrose (hexose sugar) form to fungi after processing by invertase or acid sucrose synthase and accumulated by a highly attracting monosaccharide transporter (Schaarschmidt et al. 2006; Helber et al. 2011; Liu et al. 2020). Mycorrhizal fungi encourage plant acid invertase expression because sucrose sugar cannot be able to use C as a sugar source (Schaarschmidt et al. 2006; Begum et al. 2019). Transfer of carbon and phosphorous between fungi and plant host is equal in both and also help each other. In addition, carbon also facilitates the accumulation and transporatation of fungal nitrogen by acting as triggering agent for the expression of specific genes present in fungi (Fellbaum et al. 2012a). The transportation of phosphorous and carbon among mycorrhizal fungi and plant host cells was also controlled by gene expression through mycorrhizal induceable plant phosphorous transporter (Pt-4) and mycological monosaccharide transporter (MST-2) (Helber et al. 2011). The successful formation of symbiotic association was preconditioned on the basis of P transport through the expression of Pt4 (Javot et al. 2011; Frey 2019). The degreation and reformation of arbuscules also take place under nitrogen deficiency due to the expression of Pt4 mutants. It was evident that C, N, and P can jointly assist each other when controlling intracellular colonization (Fellbaum et al. 2012b). The plant-based carbon was utilized in the formation of new arbuscules, increase in biomass production of mycorrhiza-associated fungi, and extension of extraradical hyphae and in energy-utilizing reactions. The IRM directly accumulate glucose from the linking sites of fungus and plant symbiosis; however, the supply of C and P accumulation from mycorrhizal fungi was directly proportional (Woolhouse 1975; Kiers et al. 2011). In a confirmatory research, it was revealed that the foliar application of different levels of glucose enhances the efflux of inorganic P from IRM of *Gigaspora margarita* (Solaiman and Saito 2001). Furthermore, the expression of fungal genes also regulates the accumulation and transport of N from mycorrhizal fungi through the continuous supply C supply (Dietz et al. 2011). It effects the N accumulation through biosynthesis of arginine in ecto-radical mycelium and transferred to IRM where it enhanced arginase and urease activity by enhancing NH^+ concentration in IRM to enable N removal from mycorrhizae (Bücking and Kafle 2015). The increase in the transport of C also enhanced the arginine and N uptake by stimulating genetic modifications (Fellbaum et al. 2012a).

Conclusion

The terrestrial plants form mycorrhizal symbiosis through roots with various species of fungi, which facilitate host plants in intake of nutrients. It was evident that different forms of membrane transporters are involved, which regulate the smooth flow of nutrients traffic between soil, host plant, and mycorrhizal fungi. The carbon from host moved to mycorrhizal fungi in triacyl glycerides form. Various other transporters are still not identified, particularly the symbiotic transporters that regulate the flow of sugars and lipids in fungi and host plant. The mycorrhizal symbiosis inproved the plant uptake capacity of S, P, N, K, and micronutrients in exchange of C from its host plant. The release of organic acid exudates, hormones, enzymes, and hyphal growth are also other factors that not only help in establishing mycorrhizal association but also promote the availability and uptake of vital nutrients. In future, further efforts should be required to emphasis on investigative approach toward the nutrient exchange regulation mechanism between fungal partner and host. The mycorrhizal symbiosis frolicked a dynamic role in environmental and agricultural development.

Further understanding of the pheonemonon behind integrated environmental and developmental signals also required that produce varied, robust, and dynamic responses in host plants in mycorrhizal symbiosis.

References

Abbas, S., Javed, M.T., Shahid, M. et al. (2020). Acinetobacter sp. SG-5 inoculation alleviates cadmium toxicity in differentially Cd tolerant maize cultivars as deciphered by improved physio-biochemical attributes, antioxidants and nutrient physiolcgy. *Plant Physiol. Biochem.* 155: 815–827.

Akiyama, K. and Hayashi, H. (2006). Strigolactones: chemical signals for fungal symbionts and parasitic weeds in plant roots. *Ann Bot (Lond)* 97: 925–931.

Ali, N., Masood, S., Mukhtar, T. et al. (2015). Differential effects of cadmium and chromium on growth, photosynthetic activity, and metal uptake of Linum usitatissimum in association with Glomus intraradices. *Environ. Monit. Assess.* 187 (6): 311.

Andrade, S.A.L., Silveira, A.P.D., and Mazzafera, P. (2010). Arbuscular mycorrhiza alters metal uptake and the physiological response of *Coffea arabica* seedlings to increasing Zn and Cu concentrations in soil. *Sci. Total Environ.* 408: 5381–5391.

Aroca, R., Ruiz-Lozano, J.M., Zamarreno, A.M. et al. (2013). Arbuscular mycorrhizal symbiosis influences strigolactone production under salinity and alleviates salt stress in lettuce plants. *J. Plant Physiol.* 170: 47–55.

Barea, J.M., Pozo, M.J., Azcon, R., and Azcon-Aguilar, C. (2013). Microbial interactions in the rhizosphere. In: *Molecular Microbial Ecology of the Rhizosphere* (ed. F. de Bruijn), 29–44. Hoboken, NJ: Wiley-Blackwell.

Barea, J.M., Pozo, M.J., Lopez-Raez, J.A. et al. (2014). Arbuscular mycorrhizas and their significance in promoting soil-plant system sustainability against environmental stresses. In: *Beneficial Plant-Microbial Interactions Ecology and Applications* (ed. M.B. Rodelas and J. Gonzalez-Lopez), 353–387. Boca Raton, FL: CRC, Taylor & Francis.

Begum, N., Qin, C., Ahanger, M.A. et al. (2019). Role of arbuscular mycorrhizal fungi in plant growth regulation: implications in abiotic stress tolerance. *Front. Plant Sci.* 10: 1068.

Behie, S.W. and Bidochka, M.J. (2014). Ubiquity of insect-derived nitrogen transfer to plants by endophytic insect-pathogenic fungi: an additional branch of the soil nitrogen cycle. *Appl. Environ. Microbiol.* 80 (5): 1553–1560.

Benito, B., Garciadeblas, B., Fraile-Escanciano, A., and Rodrıguez-Navarro, A. (2011). Potassium and sodium uptake systems in fungi. The transporter diversity of *Magnaporthe oryzae*. *Fungal Genet. Biol.* 48: 812–822.

Berbee, M.L., James, T.Y., and Strullu-Derrien, C. (2017). Early diverging fungi: diversity and impact at the dawn of terrestrial life. *Annu. Rev. Microbiol.* 71: 41–60.

Berruti, A., Lumini, E., Balestrini, R., and Bianciotto, V. (2016). Arbuscular mycorrhizal fungi as natural biofertilizers: let's benefit from past successes. *Front. Microbiol.* 6: 1559.

Berta, G., Trotta, A., Fusconi, A. et al. (1995). Arbuscular mycorrhizal induced changes to plant-growth and root-system morphology in *Prunus cerasifera*. *Tree Physiol.* 15: 557–557.

Bhandari, P., Garg,N. (2017). Dynamics of Arbuscular Mycorrhizal Symbiosis and Its Role in Nutrient Acquisition: An Overview. In: Varma, A., Prasad, R., Tuteja, N. (eds) Mycorrhiza - Nutrient Uptake, Biocontrol, Ecorestoration. Springer, Cham. https://doi.org/10.1007/978-3-319-68867-1-2.

Bolan, N. (1991). A critical review on the role of mycorrhizal fungi in the uptake of phosphorus by plants. *Plant Soil* 134: 189–207.

Bonfante, P. and Desiro, A. (2015). Arbuscular mycorrhizas: the lives of beneficial fungi and their plant host. In: *Principles of Plant-Microbe Interactions* (ed. B. Lugtenberg), 235–245. Cham: Springer.

Bonfante, P. and Genre, A. (2010). Mechanisms underlying beneficial plant–fungus interactions in mycorrhizal symbiosis. *Nat. Commun.* 1 (1): 1–11.

Breuillin-Sessoms, F., Floss, D.S., Gomez, S.K. et al. (2015). Suppression of arbuscule degeneration in *Medicago truncatula* phosphate transporter4 mutants is dependent on the ammonium transporter 2 family protein AMT2;3. *Plant Cell* 27: 1352–1366.

Brundrett, M.C. and Tedersoo, L. (2018). Evolutionary history of mycorrhizal symbioses and global host plant diversity. *New Phytol.* 220 (4): 1108–1115.

Bucher, M. (2007). Functional biology of plant phosphate uptake at root and mycorrhiza interfaces. *New Phytol.* 173: 11–26.

Bücking, H. and Kafle, A. (2015). Role of arbuscular mycorrhizal fungi in the nitrogen uptake of plants: current knowledge and research gaps. *Agronomy* 5: 587–612.

Bücking, H., Liepold, E., and Ambilwade, P. (2012). The role of the mycorrhizal symbiosis in nutrient uptake of plants and the regulatory mechanisms underlying these transport processes. *Plant Sci.* 4: 108–132.

Calabrese, S., Perez-Tienda, J., Ellerbeck, M. et al. (2016). GintAMT3-a low-affinity ammonium transporter of the arbuscular mycorrhizal *Rhizophagus irregularis*. *Front. Plant Sci.* 7: 679.

Carbonnel, S. and Gutjahr, C. (2014). Control of arbuscular mycorrhiza development by nutrient signals. *Front. Plant Sci.* 5 (462).

Casieri, L., Gallardo, K., and Wipf, D. (2012). Transcriptional response of Medicago truncatula sulphate transporters to arbuscular mycorrhizal symbiosis with and without sulphur stress. *Planta* 235: 1431–1447.

Conrad, M., Schothorst, J., Kankipati, H.N. et al. (2014). Nutrient sensing and A Phosphate Transceptor in AM Fungus Molecular Plantsignaling in the yeast *Saccharomyces cerevisiae*. *FEMS Microbiol. Rev.* 38: 254–299.

Corratgé-Faillie, C., Jabnoune, M., Zimmermann, S. et al. (2010). Potassium and sodium transport in non-animal cells: the Trk/Ktr/HKT transporter family. *Cell. Mol. Life Sci.* 67: 2511–2532.

Cruz, C., Egsgaard, H., Trujillo, C. et al. (2007). Enzymatic evidence for the key role of arginine in nitrogen translocation by arbuscular mycorrhizal fungi. *Plant Physiol.* 144: 782–792.

Desirò, A., Naumann, M., Epis, S. et al. (2013). Mollicutes-related endobacteria thrive inside liverwort-associated arbuscular mycorrhizal fungi. *Environ. Microbiol.* 15 (3): 822–836.

Diagne, N., Ngom, M., Djighaly, P.I. et al. (2020). Roles of arbuscular mycorrhizal fungi on plant growth and performance: importance in biotic and abiotic stressed regulation. *Diversity* 12 (10): 370.

Dietz, S., von Bülow, J., Beitz, E., and Nehls, U. (2011). The aquaporin gene family of the ectomycorrhizal fungus Laccaria bicolor: lessons for symbiotic functions. *New Phytol.* 190: 927–940.

Douds, D.D.J., Pfeffer, P.E., and Shachar, H.Y. (2000). Carbon partitioning, cost, and metabolism of arbuscular mycorrhizas. In: *Arbuscular Mycorrhizas: Physiology and Function* (ed. Y. Kapulnik and D.D.J. Douds). Dordrecht: Kluwer Academic.

Facelli, E., Smith, S.E., Facelli, J.M. et al. (2010). Underground friends or enemies: model plants help to unravel direct and indirect effects of arbuscular mycorrhizal fungi on plant competition. *New Phytol.* 185: 1050–1061.

Fellbaum, C.R., Gachomo, E.W., Beesetty, Y. et al. (2012a). Carbon availability triggers fungal nitrogen uptake and transport in arbuscular mycorrhizal symbiosis. *Proc. Natl. Acad. Sci. U. S. A.* 109: 2666–2671.

Fellbaum, C.R., Mensah, J.A., Pfeffer, P.E. et al. (2012b). The role of carbon in fungal nutrient uptake and transport: implications for resource exchange in the arbuscular mycorrhizal symbiosis. *Plant Signal. Behav.* 7: 1509–1512.

Field, K.J., Rimington, W.R., Bidartondo, M.I. et al. (2016). Functional analysis of liverworts in dual symbiosis with Glomeromycota and Mucoromycotina fungi under a simulated Palaeozoic CO_2 decline. *ISME J* 10 (6): 1514–1526.

Fiorilli, V., Lanfranco, L., and Bonfante, P. (2013). The expression of GintPT, the phosphate transporter of *Rhizophagus irregularis*, depends on the symbiotic status and phosphate availability. *Planta* 237 (5): 1267–1277.

Frey, S.D. (2019). Mycorrhizal fungi as mediators of soil organic matter dynamics. *Annu. Rev. Ecol. Evol. Syst.* 50: 237–259.

Garcia, K. and Zimmermann, S.D. (2014). The role of mycorrhizal associations in plant potassium nutrition. *Front. Plant Sci.* 5: 337.

Garg, N. and Chandel, S. (2010). Arbuscular mycorrhizal networks: process and functions. A review. *Agron. Sustain. Dev.* 30: 581–599.

Genre, A. (2012). Signalling and the re-structuring of plant cell architecture in am symbiosis. In: *Signaling and Communication in Plant Symbiosis, Signaling and Communication in Plants*, vol. 11 (ed. S. Perotto and F. Baluška), 51–71. Berlin: Springer.

Genre, A. and Bonfante, P. (2010). The making of symbiotic cells in arbuscular mycorrhizal roots. In: *Arbuscular Mycorrhizas: Physiology and Function* (ed. H. Koltai and Y. Kapulnik), 57–71. Dordrecht: Springer.

Giovannetti, M., Tolosano, M., Volpe, V. et al. (2014). Identification and functional characterization of a sulfate transporter induced by both sulfur starvation and mycorrhiza formation in *Lotus japonicus*. *New Phytol.* 204: 609–619.

Govindarajulu, M., Pfeffer, P.E., Jin, H. et al. (2005). Nitrogen transfer in the arbuscular mycorrhizal symbiosis. *Nature* 435: 819–823.

Guether, M., Balestrini, R., Hannah, M. et al. (2009a). Genome wide reprogramming of regulatory networks, transport, cell wall and membrane biogenesis during arbuscular mycorrhizal symbiosis in *Lotus japonicus*. *New Phytol.* 182: 200–212.

Guether, M., Neuhauser, B., Balestrini, R. et al. (2009b). Amycorrhizal-specific ammonium transporter from *Lotus japonicus* acquires nitrogen released by arbuscular mycorrhizal fungi. *Plant Physiol.* 150: 73–83.

Gutjahr, C. and Parniske, M. (2013). Cell and developmental biology of the arbuscular mycorrhiza symbiosis. *Annu. Rev. Cell Dev. Biol.* 29: 593–617.

Harrison, M.J. (2005). Signaling in the arbuscular mycorrhizal symbiosis. *Annu. Rev. Microbiol.* 59: 19–42.

He, J.D., Dong, T., Wu, H.H. et al. (2019). Mycorrhizas induce diverse responses of root TIP aquaporin gene expression to drought stress in trifoliate orange. *Sci. Hortic.* 243: 64–69.

Helber, N., Wippel, K., Sauer, N. et al. (2011). A versatile monosaccharide transporter that operates in the arbuscular mycorrhizal fungus Glomus sp is crucial for the symbiotic relationship with plants. *Plant Cell* 23: 3812–3823.

Hijikata, N., Murase, M., Tani, C. et al. (2010). Polyphosphate has a central role in the rapid and massive accumulation of phosphorus in extraradical mycelium of and arbuscular mycorrhizal fungus. *New Phytol.* 186: 285–289.

Hinsinger, P. (2001). Bioavailability of soil inorganic P in the rhizosphere as affected by rootinduced chemical changes: a review. *Plant Soil* 237: 173–195.

Huang, G.M., Zou, Y.N., Wu, Q.S. et al. (2020). Mycorrhizal roles in plant growth, gas exchange, root morphology, and nutrient uptake of walnuts. *Plant Soil Environ.* 66 (6): 295–302.

Javed, M.T., Akram, M.S., Habib, N. et al. (2018). Deciphering the growth, organic acid exudations, and ionic homeostasis of *Amaranthus viridis* L. and *Portulaca oleracea* L. under lead chloride stress. *Environ. Sci. Pollut. Res.* 25: 2958–2971.

Javed, M.T., Akram, M.S., Tanwir, K. et al. (2017). Cadmium spiked soil modulates root organic acids exudation and ionic contents of two differentially Cd tolerant maize (*Zea mays* L.) cultivars. *Ecotoxicol. Environ. Saf.* 141: 216–225.

Javot, H., Penmetsa, R.V., Breuillin, F. et al. (2011). Medicago truncatula mtpt4 mutants reveal a role for nitrogen in the regulation of arbuscule degeneration in arbuscular mycorrhizal symbiosis. *Plant J.* 68: 954–965.

Jeffries, P. and Barea, J.M. (2012). Arbuscular mycorrhiza – a key component of sustainable plant-soil ecosystems. In: *The Mycota* (ed. B. Hock), 51–75. Berlin: Springer.

Jin, H., Pfeffer, P., Douds, D. et al. (2005). The uptake, metabolism, transport and transfer of nitrogen in an arbuscular mycorrhizal symbiosis. *New Phytol.* 168: 687–696.

Johansen, A., Jakobsen, I., and Jensen, E. (1992). Hyphal transport of ^{15}N-labelled nitrogen by a vesiculararbuscular mycorrhizal fungus and its effect on depletion of inorganic soil N. *New Phytol.* 122: 281–288.

Johnson, N.C. (2010). Resource stoichiometry elucidates the structure and function of arbuscular mycorrhizas across scales. *New Phytol.* 185: 631–647.

Kamal, S., Prasad, R., and Varma, A. (2010). Soil microbial diversity in relation to heavy metals. In: *Soil Heavy Metals* (ed. I. Sherameti and A. Varma), 31–64. Berlin: Springer.

Kiers, E.T., Duhamel, M., Beesetty, Y. et al. (2011). Reciprocal rewards stabilize cooperation in the mycorrhizal symbiosis. *Science* 333: 880–882.

Koide, R.T. (1991). Nutrient supply, nutrient demand and plant-response to mycorrhizal infection. *New Phytol.* 117: 365–386.

Koltai, H. and Kapulnik, Y. (2009). Effect of arbuscular mycorrhizal symbiosis on enhancement of tolerance to abiotic stresses. In: *Defensive Mutualism in Microbial Symbiosis* (ed. J.F. White and M.S. Torres), 217–234. Boca Raton, FL: CRC, Taylor & Francis.

Kucey, R.M.N. and Paul, E.A. (1982). Carbon flow, photosynthesis, and N_2 fixation in mycorrhizal and nodulated faba beans (*Vicia-faba* L.). *Soil Biol. Biochem.* 14: 407–412.

Kumar, A., Dames, J.F., Gupta, A. et al. (2015). Current developments in arbuscular mycorrhizal fungi research and its role in salinity stress alleviation: a biotechnological perspective. *Crit. Rev. Biotechnol.* 35: 461–474.

Leigh, J., Hodge, A., and Fitter, A.H. (2009). Arbuscular mycorrhizal fungi can transfer substantial amounts of nitrogen to their host plant from organic material. *New Phytol.* 181: 199–207.

Liao, J.P., Lin, X.G., Cao, Z.H. et al. (2003). Interactions between arbuscular mycorrhizae and heavy metals under sand culture experiment. *Chemosphere* 50: 847–853.

Liu, Y., Li, X., and Kou, Y. (2020). Ectomycorrhizal fungi: participation in nutrient turnover and community assembly pattern in forest ecosystems. *Forests* 11 (4): 453.

Lopez-Bucio, J., Cruz-Ramırez, A., and Herrera-Estrella, L. (2003). The role of nutrient availability in regulating root architecture. *Curr. Opin. Plant Biol.* 6: 280–287.

Lopez-Raez, J.A. (2016). How drought and salinity affect arbuscular mycorrhizal symbiosis and strigolactone biosynthesis. *Planta* 243 (6): 1375–1385.

Lopez-Raez, J.A., Bouwmeester, H., and Pozo, M.J. (2012). Communication in the rhizosphere, a target for pest management. In: *Agroecology and Strategies for Climate Change, Sustainable Agriculture Reviews*, vol. 8 (ed. E. Lichtfouse), 109–133. Dordrecht: Springer.

Lopez-Raez, J.A. and Pozo, M.J. (2013). Chemical signalling in the arbuscular mycorrhizal symbiosis: biotechnological applications. In: *Symbiotic Endophytes, Soil Biology*, vol. 37 (ed. R. Aroca), 215–232. Berlin: Springer.

Mader, P., Vierheilig, H., Streitwolf-Engel, R. et al. (2000). Transport of ^{15}N from a soil compartment separated by a polytetrafluoroethylene membrane to plant roots via the hyphae of arbuscular mycorrhizal fungi. *New Phytol.* 146: 155–161.

Maillet, F., Poinsot, V., Andre, O. et al. (2011). Fungal lipochitooligosaccharide symbiotic signals in arbuscular mycorrhiza. *Nature* 469: 58–63.

Malbreil, M., Tisserant, E., Martin, F., and Roux, C. (2014). Genomics of arbuscular mycorrhizal fungi. Out of the shadows. *Adv. Bot. Res.* 70: 259–290.

Marschner, H. and Rimmington, G. (1988). Mineral nutrition of higher plants. *Plant Cell Environ.* 11: 147–148.

Martin, F., Kohler, A., Murat, C. et al. (2016). Unearthing the roots of ectomycorrhizal symbioses. *Nat. Rev. Microbiol.* 14 (12): 760–773.

Martin, F.M., Uroz, S., and Barker, D.G. (2017). Ancestral alliances: plant mutualistic symbioses with fungi and bacteria. *Science* 356 (6340): eaad4501.

McDowell, W.H., Magill, A.H., Aitkenhead-Peterson, J.A. et al. (2004). Effects of chronic nitrogen amendment on dissolved organic matter and inorganic nitrogen in soil solution. *Forest Ecol. Manag.* 196: 29–41.

Miransari, M. (2013). Arbuscular mycorrhizal fungi and uptake of nutrients. In: *Symbiotic Endophytes, Soil Biology*, vol. 37 (ed. R. Aroca), 253–270. Berlin: Springer.

Mohanta, T.K. and Bae, H. (2015). Functional genomics and signaling events in mycorrhizal symbiosis. *J. Plant Interact.* 10 (1): 21–40.

Montesinos-Navarro, A., Valiente-Banuet, A., and Verdú, M. (2019). Mycorrhizal symbiosis increases the benefits of plant facilitative interactions. *Ecography* 42 (3): 447–455.

Morrissey, J. and Guerinot, M.L. (2009). Iron uptake and transport in plants: the good, the bad, and the ionome. *Chem. Rev.* 109: 4553–4567.

Nasim, G. (2013). Host allelopathy and arbuscular mycorrhizal fungi. In: *Allelopathy* (ed. Z.A. Cheema, M. Farooq and A. Wahid), 429–450. Berlin: Springer.

Nath, M., Bhatt, D., Prasad, R. et al. (2016). Reactive oxygen species generation-scavenging and signaling during plant-arbuscular mycorrhizal and *Piriformospora indica* interaction under stress condition. *Front. Plant Sci.* 7: 1574.

Oldroyd, G.E.D. (2013). Speak, friend, and enter: signalling systems that promote beneficial symbiotic associations in plants. *Nat. Rev. Microbiol.* 11: 252–263.

Orfanoudakis, M., Wheeler, C.T., and Hooker, J.E. (2010). Both the arbuscular mycorrhizal fungus Gigaspora rosea and Frankia increase root system branching and reduce root hair frequency in *Alnus glutinosa*. *Mycorrhiza* 20: 117–126.

Ortas, I. (2003). Effect of selected mycorrhizal inoculation on phosphorus sustainability in sterile and non-sterile soils in the Harran Plain in South Anatolia. *J. Plant Nutr.* 26: 1–17.

Ortas, I. (2012). Do maize and pepper plants depend on mycorrhizae in terms of phosphorus and zinc uptake. *J. Plant Nutr.* 35: 1639–1656.

Ortas, I., Harris, P.J., and Rowell, D.L. (1996). Enhanced uptake of phosphorus by mycorrhizal sorghum plants as influenced by forms of nitrogen. *Plant Soil* 184: 255–264.

Ortaş, I. and Rafique, M. (2017). The mechanisms of nutrient uptake by arbuscular mycorrhizae. In: *Mycorrhiza-Nutrient Uptake, Biocontrol, Ecorestoration*, 1–19. Cham: Springer.

Ortas, I., Sari, N., Akpinar, C., and Yetisir, H. (2013). Selection of arbuscular mycorrhizal fungi species for tomato seedling growth, mycorrhizal dependency and nutrient uptake. *Eur. J. Hortic. Sci.* 78: 209–218.

Pozo, M.J., López-Ráez, J.A., Azcón, C., and García-Garrido, J.M. (2015). Phytohormones as integrators of environmental signals in the regulation of mycorrhizal symbioses. *New Phytol.* 205: 1431–1436.

Prasad, R., Bhola, D., Akdi, K. et al. (2017). Introduction to mycorrhiza: historical development. In: *Mycorrhiza* (ed. A. Varma, R. Prasad and N. Tuteja), 1–7. Cham: Springer.

Pressel, S., Bidartondo, M.I., Field, K.J. et al. (2016). Pteridophyte fungal associations: current knowledge and future perspectives. *J. Syst. Evol.* 54 (6): 666–678.

Preuss, C.P., Huang, C.Y., Gilliham, M., and Tyerman, S.D. (2010). Channel-like characteristics of the low-affinity barley phosphate transporter PHT1; 6 when expressed in *Xenopus oocytes. Plant Physiol.* 152: 1431–1441.

Pumplin, N., Mondo, S.J., Topp, S. et al. (2010). Medicago truncatula Vapyrin is a novel protein required for arbuscular mycorrhizal symbiosis. *Plant J.* 61: 482–494.

Read, D.J. and Perez-Moreno, J. (2003). Mycorrhizas and nutrient cycling in ecosystems – a journey towards relevance? *New Phytol.* 157: 475–492.

Rimington, W.R., Pressel, S., Duckett, J.G., and Bidartondo, M.I. (2015). Fungal associations of basal vascular plants: reopening a closed book? *New Phytol.* 205 (4): 1394–1398.

Rosewarne, G.M., Smith, F.A., Schachtman, D.P., and Smith, S.E. (2007). Localization of proton-ATPase genes expressed in arbuscular mycorrhizal tomato plants. *Mycorrhiza* 17: 249–258.

Rouphael, Y., Franken, P., Schneider, C. et al. (2015). Arbuscular mycorrhizal fungi act as biostimulants in horticultural crops. *Sci. Hortic.* 196: 91–108.

Ryan, M.H., McCully, M.E., and Huang, C.X. (2007). Relative amounts of soluble and insoluble forms of phosphorus and other elements in intraradical hyphae and arbuscules of arbuscular mycorrhizas. *Funct. Plant Biol.* 34: 457–464.

Schaarschmidt, S., Roitsch, T., and Hause, B. (2006). Arbuscular mycorrhiza induces gene expression of the apoplastic invertase LIN6 in tomato (*Lycopersicon esculentum*) roots. *J. Exp. Bot.* 57: 4015–4023.

Schachtman, D.P., Reid, R.J., and Ayling, S.M. (1998). Phosphorus uptake by plants: from soil to cell. *Plant Physiol.* 116: 447–453.

Schützendübel, A. and Polle, A. (2002). Plant responses to abiotic stresses: heavy metal-induced oxidative stress and protection by mycorrhization. *J. Exp. Bot.* 53: 1351–1365.

Shah, M.A. (2014). *Mycorrhizas: Novel Dimensions in the Changing World*, 5. New Delhi: Springer.

Shao, Y.D., Zhang, D.J., Hu, X.C. et al. (2018). Mycorrhiza-induced changes in root growth and nutrient absorption of tea plants. *Plant Soil Environ.* 64: 283–282.

Sieh, D., Watanabe, M., Devers, E.A. et al. (2013). The arbuscular mycorrhizal symbiosis influences sulfur starvation responses of *Medicago truncatula. New Phytol.* 197: 606–616.

Silberbush, M. and Barber, S. (1983). Sensitivity of simulated phosphorus uptake to parameters used by a mechanistic-mathematical model. *Plant Soil* 74: 93–100.

Singh, S., Katzer, K., Lambert, J. et al. (2014). CYCLOPS, a DNA-binding transcriptional activator, orchestrates symbiotic root nodule development. *Cell Host Microbe* 15: 139–152.

Smith, S. and Read, D. (2008). *Mycorrhizal Symbiosis*, 3e. Cambridge: Academic, Elsevier.

Smith, S.E. and Read, D.J. (2010). *Mycorrhizal Symbiosis*. San Diego, CA: Academic.

Smith, S.E. and Smith, F.A. (2011). Roles of arbuscular mycorrhizas in plant nutrition and growth: new paradigms from cellular to ecosystem scales. *Annu. Rev. Plant Biol.* 62: 227–250.

Smith, S.E., Smith, F.A., and Jakobsen, I. (2004). Functional diversity in arbuscular mycorrhizal (AM) symbioses: the contribution of the mycorrhizal P uptake pathway is not correlated with mycorrhizal responses in growth or total P uptake. *New Phytol.* 162: 511–524.

Solaiman, M. and Saito, M. (2001). Phosphate efflux from intraradical hyphae of *Gigaspora margarita* in vitro and its implication for phosphorus translocation. *New Phytol.* 151: 525–533.

Song, Y.Y., Zeng, R.S., Xu, J.F. et al. (2010). Inter plant communication of tomato plants through underground common mycorrhizal networks. *PLoS One* 5 (10): e13324.

Sosa-Hernández, M.A., Leifheit, E.F., Ingraffia, R., and Rillig, M.C. (2019). Subsoil arbuscular mycorrhizal fungi for sustainability and climate-smart agriculture: a solution right under our feet? *Front. Microbiol.* 10: 744.

Strullu-Derrien, C., Kenrick, P., Pressel, S. et al. (2014). Fungal associations in H orneophyton ligneri from the R hynie C hert (c. 407 million year old) closely resemble those in extant lower land plants: novel insights into ancestral plant–fungus symbioses. *New Phytol.* 203 (3): 964–979.

Stuart, E.K. and Plett, K.L. (2020). Digging deeper: in search of the mechanisms of carbon and nitrogen exchange in ectomycorrhizal symbioses. *Front. Plant Sci.* 10: 1658.

Tanaka, Y. and Yano, K. (2005). Nitrogen delivery to maize via mycorrhizal hyphae depends on the form of N supplied. *Plant Cell Environ.* 28: 1247–1254.

Tang, N., San Clemente, H., Roy, S. et al. (2016). A survey of the gene repertoire of *Gigaspora rosea* unravels conserved features among Glomeromycota for obligate biotrophy. *Front. Microbiol.* 7: 233.

Tanwir, K., Akram, S.A., Masood, S. et al. (2015). Cadmium-induced rhizospheric pH dynamics modulated nutrient acquisition and physiological attributes of maize (*Zea mays* L.). *Environ. Sci. Pollut. Res.* 22 (12): 9193–9203.

Tanwir, K., Javed, M.T., Abbas, S. et al. (2021). Serratia sp. CP-13 alleviates Cd toxicity by morpho-physio-biochemical improvements, antioxidative potential and diminished Cd uptake in *Zea mays* L. cultivars differing in Cd tolerance. *Ecotoxicol. Environ. Saf.* 208: 111584.

Tedersoo, L. and Brundrett, M.C. (2017). Evolution of ectomycorrhizal symbiosis in plants. In: *Biogeography of Mycorrhizal Symbiosis* (ed. L. Tedersoo), 407–467. Cham: Springer.

Tisserant, E., Kohler, A., Dozolme-Seddas, P. et al. (2012). The transcriptome of the arbuscular mycorrhizal fungus Glomus intraradices (DAOM 197198) reveals functional tradeoffs in an obligate symbiont. *New Phytologist* 193 (3): 755–769.

Tisserant, E., Malbreil, M., Kuoc, A. et al. (2013). Genome of an arbuscular mycorrhizal fungus provides insight into the oldest plant symbiosis. *Proc. Natl. Acad. Sci. U. S. A.* 110: 20117–20122.

Trepanier, M., Becard, G., Moutoglis, P. et al. (2005). Dependence of arbuscular-mycorrhizal fungi on their plant host for palmitic acid synthesis. *Appl. Environ. Microbiol.* 71: 5341–5347.

Van der Heijden, M.G., Martin, F.M., Selosse, M.A., and Sanders, I.R. (2015). Mycorrhizal ecology and evolution: the past, the present, and the future. *New Phytol.* 205 (4): 1406–1423.

Viereck, N., Hansen, P.E., and Jakobsen, I. (2004). Phosphate pool dynamics in the arbuscular mycorrhizal fungus *Glomus intraradices* studied by in vivo ^{31}P NMR spectroscopy. *New Phytol.* 162: 783–794.

Vogel, J.T., Walter, M.H., Giavalisco, P. et al. (2010). SlCCD7 controls strigolactone biosynthesis, shoot branching and mycorrhiza-induced apocarotenoid formation in tomato. *Plant J.* 61: 300–311.

Wang, W., Shi, J., Xie, Q. et al. (2017). Nutrient exchange and regulation in arbuscular mycorrhizal symbiosis. *Mol. Plant* 10: 1147–1158.

Woolhouse, H. (1975). Membrane structure and transport problems considered in relation to phosphorus and carbohydrate movements and the regulation of endotrophic mycorrhizal associations. *Endomycorrhizas; Proceedings of a symposium*, Woolhouse, University of Leeds, West Yorkshire, England.

Xie, X., Lin, H., Peng, X. et al. (2016). Arbuscular mycorrhizal symbiosis requires a phosphate transceptor in the *Gigaspora margarita* fungal symbiont. *Mol. Plant* 9: 1583–1608.

Yamamoto, A., DeWald, D.B., Boronenkov, I.V. et al. (1995). Novel PI (4) P 5-kinase homologue, Fab1p, essential for normal vacuole function and morphology in yeast. *Mol. Biol. Cell* 6 (5): 525–539.

Yoneyama, K., Xie, X., Kim, H. et al. (2012). How do nitrogen and phosphorus deficiencies affect strigolactone production and exudation? *Planta* 235: 1197–1207.

Zaefarian, F., Rezvani, M., Ardakani, M.R. et al. (2013). Impact of mycorrhizae formation on the phosphorus and heavy-metal uptake of Alfalfa. *Commun. Soil Sci. Plant Anal.* 44: 1340–1352.

Zhang, F., Zou, Y.N., Wu, Q.S., and Kuča, K. (2020). Arbuscular mycorrhizas modulate root polyamine metabolism to enhance drought tolerance of trifoliate orange. *Environ. Exp. Bot.* 171: 103926.

6

Nutrient Availability Regulates Root System Behavior

Salar Farhangi-Abriz and Kazem Ghassemi-Golezani

Department of Plant Eco-physiology, Faculty of Agriculture, University of Tabriz, Tabriz, Iran

Introduction

Roots are important plant organs that absorb nutrients and water from the rhizosphere and transport them to plant shoots (Glinski 2018). Roots also provide physical support to crops and synthesize some hormones such as cytokinin that affects many biochemical and physiological processes associated with development and growth. Strong root networks are needed for the growth of vigorous plants with a great yield potential. Increased knowledge of root architecture and root development dynamics could help to improve crop productivity in agroecosystems. A better understanding of root architecture and growth dynamics of annual crops may lead to a more efficient use of applied nutrients and water (Fageria 2012).

Plants need various macro and micronutrients for better growth and development in their life cycles (Meena et al. 2017). The bioavailability of nutrients in the soil solution may determine root growth, proliferation, and specific functional responses that depend on the prevailing nutrient status of the plant. Availability or unavailability of different nutrients for plants alters morphological and physiological characteristics of their roots (Shahzad and Amtmann 2017) (Figure 6.1). In early studies on barley and maize plants, researchers found that low nitrogen, phosphorus, and potassium in growth media significantly reduced the lengths of main roots, while the lateral root length was unaffected (Drew and Saker 1975; Mollier and Pellerin 1999). Nutrients deficit not only changes the main root growth of plants but also modifies root anatomical specifications. For example, root epidermal cells in *Arabidopsis thaliana* formed root hairs under low levels of phosphorus in the rhizosphere (Ma et al. 2001). Low level of phosphorous in growth media also increased the density of root hair (up to five-fold), compared to roots grown under adequate level of phosphorous in soil. In a study by Schmidt and Schikora (2001), iron deficit increased ethylene production in *Arabidopsis* roots and formed a great number of root hairs on the root surface. The positive effects of sufficient levels of nutrients on root branching and growth are confirmed in many reports (Forde and Lorenzo 2001; Fageria

ORCID 0000-0001-9560-1869 (Kazem Ghassemi-Golezani)

Plant Ionomics: Sensing, Signaling, and Regulation, First Edition.
Edited by Vijay Pratap Singh and Manzer H. Siddiqui.

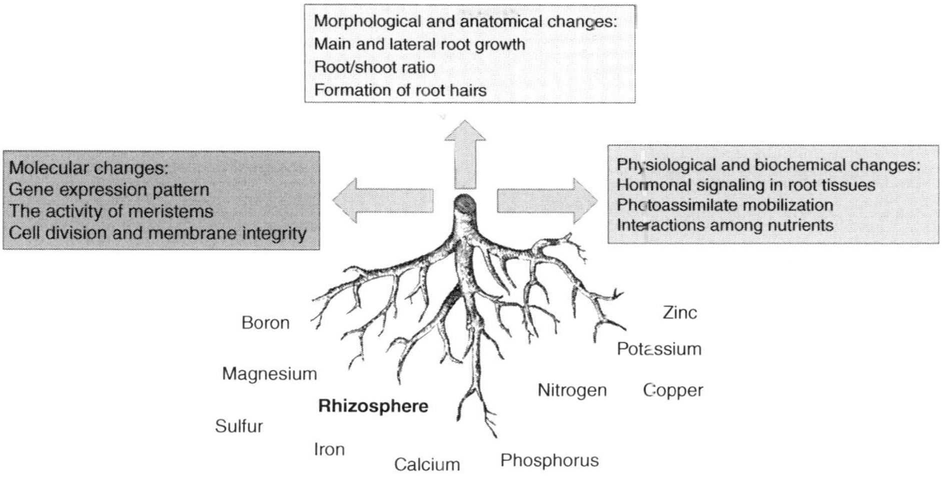

Figure 6.1 Responses of plant roots to various nutrients.

and Moreira 2011; de Kroon et al. 2012). Roots growth and branching increase in response to adequate amounts of nutrients (especially nitrogen and phosphorous). This morphological reaction helps plant to enhance the allocation of resources to root growth within the region of the soil that will yield the most benefit in terms of nutrient capture (López-Bucio et al. 2003). Adequate amounts of different nutrients such as magnesium and calcium are critical for root growth. For example, Li et al. (2020) found that magnesium promotes root growth of *Arabidopsis* under normal and heavy metal stress, or in another study Shabala et al. (2003) stated that optimum levels of calcium are required for better root development in barley seedlings.

Similar to morphological and anatomical changes in plant roots in response to various nutritional conditions, physiological characteristics of plant roots are also affected by nutrient availability. For example, nitrate availability in soil noticeably improves auxin biosynthesis in *Arabidopsis* roots and improves cell elongation (Zhang and Forde 1998). In a recent study, Ghassemi-Golezani and Abdoli (2021) reported that enhancement of iron content in root cells of ajowan plants considerably improves root growth, H^+-ATPase and H^+-PPase activities, and ATP content under normal and saline conditions. In another study, Farhangi-Abriz and Ghassemi-Golezani (2021) stated that increasing magnesium and manganese in soil can enhance water content and osmotic adjustment under saline and nonsaline conditions. Potassium availability controls cell elongation and expansion in different plant species. Sustr et al. (2019) showed a great correlation between potassium content of soil and physiological changes in plant roots. Based on previous reports, plant root growth and development are very sensitive to nutrients availability in the rhizosphere. Hence, in this book chapter, the possible physiological and morphological changes of plant roots in response to nutrients bioavailability in rhizosphere will be discussed in detail.

Nutrients Importance in Root Growth and Development

Nutrients bioavailability in rhizosphere is a critical factor for plant growth and productivity. Changes in nutrient content in the rhizosphere determine root growth, development, proliferation, and specific physiological and biochemical responses, depending on the principal nutrient status of the plant (Forde and Lorenzo 2001). Plant physiologists proved that about 17 nutrients (C, H, O, N, P, K, Ca, Mg, S, Zn, Cu, Mn, Fe, B, Mo, Cl, and Ni) are critical for plant growth and development (Fageria 2016). Absence of these nutrients creates some abnormalities in plant growth and productivity. Plants can absorb carbon, hydrogen, and oxygen from atmosphere and soil water, while other nutrients must be absorbed from the soil. Carbon, hydrogen, and oxygen are the main structural nutrients that build 95% of dry mass in plants and the lasting 5% is the other 14 nutrients. Some of the nutrients such as carbon, hydrogen, oxygen, nitrogen, phosphorous, potassium, calcium, magnesium, and sulfur are mandatory by plants in large amounts and we know these nutrients as macronutrients. Some other nutrients including zinc, copper, manganese, iron, boron, molybdenum, chlorine, and nickel are required in small amounts by plants and scientists calling them micronutrients. Typically, plants absorb chlorine in large amounts, but it is required in only small amounts. Hence, scientists classify chlorine as a micronutrient (Fageria 2016).

All of the micro and macronutrients are equally critical for successful growth of plants. The absence of these macro or micronutrients limits the plant shoot and root growth. However, some nutrients have a heavier burden in adjusting plant root growth. Nitrogen, phosphorus, potassium, calcium, iron, and sulfur are among the nutrients that have important roles in root growth and developmental processes (Fageria and Moreira 2011). Root growth of various plant species was reduced dramatically in response to absence of nitrogen, phosphorous, and potassium in the rhizosphere (Walch-Liu et al. 2006; Hansel et al. 2017; Song et al. 2018). In an earlier study, Lund (1970) reported that calcium ions have important roles in improving root growth of soybean. This researcher indicated that calcium ion not only enhances root growth of soybean but also have some interaction with other nutrients (especially potassium and magnesium) in adjusting root growth and development. Similarly, Emanuelsson (1984) reported similar results about increasing root length and dry weight by rising calcium availability in rhizosphere. In recent years, by improving molecular techniques in plant science, scientists found that calcium ion by engaging plasma membrane calcineurin B-like calcium-ion sensor proteins regulates root growth in plants (Chu et al. 2021). Moreover, it seems that calcium acts as an important signaling molecule in adjusting root growth in plants. For example, Wu et al. (2017) stated that a secreted chitinase-like protein (OsCLP) controls root growth through calcium signaling in rice plants.

Sulfate is another important nutrient that controls root growth and architect. Sulfate has some complex roles in adjusting root growth. Kutz et al. (2002) in an in vitro study observed that sulfate availability (SO_4^{2-}) increased main root elongation of *Arabidopsis* plants, and lateral roots developed at some distance from the root tip, whereas limitation of sulfate reduced main root length and formed a branched root system in *Arabidopsis* plants. Moreover, in sulfate-depleted *Arabidopsis*, lateral roots are shaped closer to the root tip and at augmented density. The increment of lateral root growth of *Arabidopsis* plants under sulfate-deficit condition is related to the transcriptional activation of the *NITRILASE3*

(NIT3) gene. The *NITRILASE3* is part of nitrilase gene family, which can convert indole-3-acetonitrile to indole-3-acetic acid. Plants grown in sulfate-deficit conditions also have increased auxin levels in root tissues, suggesting a main role for *NITRILASE3* in auxin biosynthesis and lateral root growth. Based on these results, plants have different physiological and morphological root responses under various nutritional conditions. A noticeable variation among crop species in nutrient acquisition and metabolism shows modifications in root physiology and morphology that either support or avoid ion movement into the root (O'toole and Bland 1987). Plants root responses to availability or unavailability of nutrients are not equal, hence in next topics, the impacts of individual nutrients on adjusting plant root growth will be discussed in detail.

Morpho-Physiological Responses of Plant Roots to Nutrients Availability

Macronutrients

Nitrogen

Nitrogen is one of the most important nutrients for crop production in agriculture. Crop producers use nitrogen in large amount, and this nutrient is so critical for successful growth and productivity of crops (Blumenthal et al. 2008; Farhangi-Abriz and Ghassemi-Golezani 2016). Nitrogen is frequently added to soil as a fertilizer and is essential for all kinds of plant species. Among the different plant species, only legume plants have the ability to fix their own nitrogen, and other plant species need nitrogen fertilizers as a nitrogen source. Among the essential nutrients, nitrogen availability has the superior impact on rising growth and development of crops, compared to other nutrients (Fageria 2016). Nitrogen plays a critical role in numerous physiological and biochemical pathways in plants. This nutrient is a component of some vital biomolecules such as proteins and nucleic acids. Nitrogen is a constituent of chlorophyll, and an adequate amount of this nutrient is essential for photosynthetic activities in different plant species. Nitrogen deficiency diminishes cell growth, leaf expansion, and shoot biomass production in various types of plants (Leghari et al. 2016). Nitrogen availability not only affects leaf and shoot growth but also controls root growth and expansion (Stober et al. 2000).

In earlier studies, Bengough and Mullins (1990) and Eghball and Maranville (1993) showed that root morphology was affected by soil physical conditions and nitrogen availability. Eghball and Maranville (1993) by testing maize plants under nitrogen-deficit condition reported that unavailability of nitrogen reduces root growth in maize plants. In next studies, other researchers such as Costa et al. (2002) reported that the nitrogen application rate is so vital for optimum growth of plant roots. Costa et al. (2002) showed that the highest root length and root surface area in maize plants were obtained under 128 kg N ha^{-1} (moderate rate), compared with either the control (nonapplied nitrogen fertilizer) or the higher rate of nitrogen application (255 kg N ha^{-1}). In another study, Sainju et al. (2001), by testing tomato plants under various concentrations of nitrogen in soil, stated that the nitrogen bioavailability is critical for the optimum growth of tomato roots. Baligar et al. (1998)

illustrated that the absence of nitrogen in growth media of rice, bean, maize, and soybean plants reduces root dry mass by about 38, 56, 35, and 11%, respectively. In a recent study, Chen et al. (2018) showed that sufficient levels of nitrogen in soil noticeably improve cotton root growth.

Nitrogen sources and the timing of application are the other important parameters that control root growth and development in plants. In a comprehensive study, Fageria et al. (2014) by testing root growth of 20 upland rice genotypes under different nitrogen sources reported a noticeable interaction between nitrogen application rates and its sources. Fageria et al. (2014) illustrated that the root dry weight of rice genotypes was increased by nitrogen application up to 400 mg kg^{-1} in soil. However, these enhancements depended on nitrogen sources. For example, the highest amount of root dry weight in response to urea application was obtained under 281 mg N kg^{-1}, while the highest root growth of rice plants in response to ammonium sulfate was achieved under high rates of nitrogen application rate (400 mg kg^{-1} of soil). These scientists also reported that the application of ammonium sulfate produced more vigorous roots, compared to other types of nitrogen sources and this might be related to noticeable content of sulfur in this fertilizer. Moreover, Fageria et al. (2014) showed that the timing of nitrogen application can also affect the root growth of plants. These scientists reported that the highest root length of rice plants was achieved when the total nitrogen applied at sowing time, while the split application of nitrogen (1/2 nitrogen applied at initiation of tillering + 1/2 nitrogen applied panicle initiation) had a better effect on rising root dry weight of rice plants, compared to other kinds of application methods. In a recent study, Ötvös et *al.* (2021) by testing the ammonium and nitrate fertilizers on changing *Arabidopsis* root growth stated that the nitrate has a better effect than ammonium on the rising root growth of *Arabidopsis* plants. Based on this report, this superiority of nitrate application in rising root growth is related to the enhancement of root cell elongation and the better adjustment of auxin metabolism.

Nitrogen availability also controls root density and branching in different crop species. Hoad et al. (2001) reported that the nitrogen availability in cereal plants (wheat, barley, and oat) noticeably improves root density. The nitrate content of soil changes the root morphology and branching in various crops. In barley and *Arabidopsis* roots, nitrate availability (moderate and low levels) enhanced the number of lateral roots and their lengths (Drew and Saker 1975; Linkohr et al. 2002). Zhang et al. (1999) stated that nitrate stimulates lateral root growth by enhancing the number of cells in the root tips directly exposed to the nitrate signaling. Root branching in response to nitrate availability completely depends on the nitrate content in root cells. For example, lateral root growth of *Arabidopsis* plants was reduced under high content of nitrate in growth media, but low nitrate concentration improves lateral root growth (Zhang and Forde 1998). Based on available reports, low content of nitrate can stimulate the activity of lateral root meristems and consequently improves root branching in different plant species. Nitrate signaling by engaging *ANR1* gene, which is a member of the MADS box family of transcription factors, can enhance lateral root growth of plants (Zhang and Forde 1998). Since *ANR1* gene is a positive controller of root branching, downregulation of this gene under adequate amounts of nitrogen shows a possible role for feedback adjustment of root branching and lateral root growth of plants by the nitrogen availability in the rhizosphere.

Most of the available reports in about the nitrogen effects on adjusting plant root system show many positive effects of this nutrient on rising root development, but some scientists believe that the application of nitrogen at a high rate might have some negative effects on developing plant root system. For example, Anderson (1987) and Costa et al. (2002) showed that the application of nitrogen at high rates enhances the root length and root surface area in maize plants, but reduces root mass per unit area. Comfort et al. (1988) showed that higher rates of nitrogen application in soil decrease the depth of rooting in wheat plants. The decrement of root growth under high levels of nitrogen content of soil could be related to modifications in hormonal signaling in root cells. High amount of nitrogen content in growth media can enhance auxin biosynthesis in plant roots, and typically high concentration of auxin in plant cells has a negative effect on developing root systems in plants (Zhang et al. 1999). Also, high concentration of soil nitrate stimulates abscisic acid production in *Arabidopsis* root tissues, which noticeably inhibits lateral root growth and root branching in this plant (Signora et al. 2001). In a recent study, Blaser et al. (2020) by testing nitrate and ammonium impacts on changing root growth of *Vicia faba* and *Hordeum vulgare* showed that there is no tangible difference in root growth of *Vicia faba* under different nitrogen sources and application rates. However, high nitrate concentrations decreased lateral root growth of *Hordeum vulgare*, while high ammonium concentrations enhanced the formation of lateral roots. These results showed that plant response to nitrogen availability of soil might be different among various crop species.

Phosphorus

Phosphorus is an important nutrient for plant growth and productivity. Phosphorus shortage in plants is a common problem in different kinds of the soils, particularly in acidic soils. The deficiency of phosphate in plant tissues is related to low natural content of phosphate in some soils, loss by soil erosion, high fixation capacity in acidic soils, and immobilization of phosphate in these soils and biotic and abiotic stresses during plant life cycle. Dissimilar to nitrogen, which can be found in rhizosphere by fixation from the atmosphere, phosphorus must be supplied from external sources. Because of this, phosphate fertilizers are used by farmers in large amounts (over 30 million metric tons annually) (Fageria 2016).

Phosphorus has numerous roles in plant life cycle, and a sufficient level of this nutrient is critical for plant growth and productivity. Phosphorus is an important nutrient in physiological pathways of plants by adjusting energy storage and transfer in plant cells. For example, adenosine diphosphate (ADP) and adenosine triphosphate (ATP) are the main forms of energy in plant cells, which provide adequate energy for photosynthetic activities (Raghothama 2005). Phosphorus also has a critical role in controlling oxidation and reduction reactions in respiration and photosynthesis pathways. This important duty of phosphorous in oxidation–reduction activities is related to its participation in triphosphopyridine nucleotide structure. Triphosphopyridine nucleotide act as a transporter of hydrogen or electrons in plant cells (Raghothama 2005; Wang et al. 2021).

Phosphorus is a vital nutrient for root growth and development in various types of soils (Anghinoni and Barber 1980; Hill et al. 2006; Richardson et al. 2009). Availability of this nutrient changes the root architecture in different plants. Available reports stated that the plants have various phosphorus requirements to achieve maximum root growth. For example, Baligar et al. (1998) by testing root growth of wheat, dry bean, and cowpea in response to different levels (0–200 mg phosphorus kg^{-1} soil) of phosphorus in rhizosphere reported that the highest root growth in wheat plants was achieved in 152 mg phosphorus kg^{-1} soil, whereas maximum root dry weight for cowpea and dry bean was attained under 159 and 134 mg phosphorus kg^{-1} soil, respectively. These researchers also indicated that the root growth of these legume and cereal plants was reduced either in low or in higher phosphorus level of soil. Most of the published reports elucidate the importance of phosphorus availability in rising root growth of plants under various soil conditions. For example, Fageria et al. (2006) by examination of root growth of plants in Brazilian Oxisol reported that the absence of phosphorus in growth media reduces root growth of rice, maize, soybean, and common bean by about 62, 50, 21, and 74% (respectively), compared to adequate phosphorus level in soil. There are various reports showing the importance of phosphorus on controlling root growth of plants. In a study by Fageria et al. (2015), the impact of phosphorus availability in controlling root growth of upland rice genotypes was studied. They observed no significant interaction between rice genotypes and phosphorous availability in rhizosphere in adjusting root growth, but in general the highest root growth of rice plants was recorded under low levels of phosphate (25 mg kg^{-1}), compared to high level of phosphate (200 mg kg^{-1}) in soil. Chevalier et al. (2003) by testing 73 *Arabidopsis* ecotypes reported that low content of phosphate limited the primary root growth in half of the ecotypes. These variations in observations show that root growth inhibition in response to phosphate limitation is related to gene expression pattern in plant cells, which is confirmed by molecular investigations (Reymond et al. 2006).

Plant roots are able to sense soil nutrient availability and they can respond to these nutritional signals by changing their root shape. Phosphorus availability not only controls root growth but also adjusts its architecture (Niu et al. 2013). To handle the phosphorus limitations, crops have evolved some multifaceted adaptive mechanisms that include physiological and morphological adjustments in root system. The main impact of phosphorus limitation in *Arabidopsis* plants is a noticeable reduction of primary root cell elongation (Sánchez-Calderón et al. 2005). Johnson et al. (1996) reported that low level of phosphate in soil considerably reduces the main root length and causes formation of root as a proteoid shape. About one year later, Dubrovsky (1997) stated that these proteoid roots after a few days become exhausted and form large numbers of hairs. Production of proteoid roots by plants is an advanced mechanism to enhance phosphate absorption by plants. Proteoid roots provide a larger absorptive surface and also have greater phosphate transporters, compared to regular roots. Williamson et al. (2001) found that low and moderate levels of soil phosphate enhance lateral root growth/primary root growth in *Arabidopsis*.

Anatomical studies have confirmed that plant roots grown under low level of soil phosphate do not have a normal apex, also these roots typically increase the expression of phosphate transporter genes in their cells. However, plant roots grown under high concentration of soil phosphate have a great mitotic activities and auxin concentration in root meristem cells (Abel 2011). Moreover, the activity of phosphate transporters is very low in this type of root (López-Bucio et al. 2003). These results confirm that proteoid roots provide a competitive superiority for crops when growing under limited phosphate availability. Such helpful mechanisms in root growth adjustments are the results of changes in gene expression pattern in plant roots. For example, expression of *siz1* (SUMO E3 ligase) and *pdr2* (P5-type ATPase) genes enhances root sensitivity to phosphate availability in soil (Ticconi et al. 2004; Miura et al. 2005). The *pnp* (ribonuclease polynucleotide phosphorylase) is the other important gene in plant response to phosphate availability in soil, which is typically overexpressed under low phosphate in soil. Overexpression of this gene increases lateral root formation and elongation in *Arabidopsis* plants (Marchive et al. 2009). The *bhlh32* and *fbx2* are the other genes that stimulate root hair formation under low content of phosphate in soil (Chen et al. 2007, 2008).

Potassium

Potassium is a vital nutrient for plant growth and development that is used more commonly as a fertilizer, compared to other nutrients. Potassium chloride and muriate of potash are the main forms of potassium fertilizers in the world. The chief sources of potassium for plant growth are chemical fertilizers and organic manures, while the main depletion causes of potassium in the farmlands are leaching, runoff losses, and removal by crops. The absence of potassium in growth media is harmful for plant growth, since it has many critical roles in crop plants (Fageria 2016). For example, potassium increases shoot and root growth of plants and involved in the opening and closing of the stomata. It is also required to activate several enzymes (at least 60 different enzymes) in physiological pathways (Pandey and Mahiwal 2020).

Potassium has some important roles in controlling growth and architecture of plant roots. Adequate amount of potassium is required for protein synthesis and enzymes activities in root cells. Potassium improves cell expansion by adjusting turgor pressure in the elongation zones of roots (Prajapati and Modi 2012). Earlier studies showed that potassium deficiency noticeably reduces root growth and crop yield in various soil types. For example, Baligar et al. (1998) by testing root growth of different cereals and legumes showed that inadequate potassium in an Inceptisol reduced root growth by 12% in maize, by 23% in rice, by 30% in common bean and by 11% in soybean. These researchers also reported that low level of potassium in soil reduces root growth of different maize genotypes by about 35% in an Oxisol. Moreover, higher potassium concentration in farmlands slightly decreases root growth of plants. However, the number of root hairs increases in response to high levels of potassium in soil (Baligar et al. 1998).

Plants have different strategies to cope potassium deficit. For example, in potassium-efficient genotypes of rice, potassium uptake and translocation under low level of soil potassium were higher than inefficient genotypes (Jia et al. 2008). Tomato plants have two well-known mechanisms to enhance potassium uptake under insufficient potassium of soil. First, these plants enhance potassium uptake capacity per root area, and second, they promote root growth. Typically, these responses are observed in some genotypes that somewhat resist potassium deficiency of soil (Chen and Gabelman 1995, 2000). Plants can activate high-affinity potassium transporters in their roots, which help better potassium absorption under potassium deficiency. In tobacco and *Arabidopsis* plants, primary root length was reduced, while the number of laterals roots was enhanced in response to low levels of potassium in growth media (Kellermeier et al. 2014; Song et al. 2018). Based on available reports, potassium shortage modifies auxin transport and root apical meristem maintenance pathways by ethylene and nitric oxide signaling, and thus reduces the activity of the root meristems (Song et al. 2018). Low potassium in plant shoots reduces photosynthetic activities and photo-assimilate production, thereby decreasing phloem loading and assimilate mobilization to the roots. Low transport of photo-assimilates to the roots can potentially decrease plant growth (Thompson and Zwieniecki 2005).

The literature review reveals that not only potassium availability in soil but also potassium fertilizer application methods can modify root growth of plants. Subsoil addition of potassium noticeably improves cotton root growth and root density in a fine sandy loam soil (Mullins et al. 1994). Oosterhuis (2002) observed that cotton roots had a higher growth rate when potassium was band applied to a potassium-deficient soil. However, some other researchers such as Hallmark and Barber (1984) and Yibirin et al. (1993) showed that localized application of potassium fertilizer did not have a beneficial impact on root growth of soybean and maize plants.

Calcium

Typically, neutral and alkaline soils have adequate amounts of calcium in their structure. However, calcium deficiency is common in acidic soils, chiefly in Oxisols and Ultisols. The soil with minimum calcium has low cation exchange capacity with high leaching. On the other hand, calcium-deficient soils typically have high aluminum content, which can be toxic for plants in most of the time (Kirkby and Pilbeam 1984; Thor 2019). Additions of lime and gypsum are practical methods for increasing calcium bioavailability in acidic soils (Anderson et al. 2020). Calcium has critical roles in cell elongation, cell division, and keeping cell membrane integrity in an optimum level for plant growth (Fageria 2016). Moreover, bioavailability of calcium controls nutrient balance and reduces heavy metals uptake by the plants (Huang et al. 2017).

The bioavailability of calcium in soil modifies root growth of plants. For example, Fageria and Baligar (2008) reported that addition of gypsum and lime to the soil increased root growth of plants by reducing soil acidity and aluminum toxicity. These scientists stated that the addition of gypsum and lime to the soil increased the leaching of calcium and sulfate ions in the rhizosphere and consequently root growth of plants was improved. Bruce et al. (1988) by testing root growth of soybean plants under acidic soil in southeast Queensland in Australia reported that calcium deficiency is the main harsh impact of low pH in this soil, which noticeably reduces root growth of plants. In another study, Fageria and Zimmermann (1998) evaluated the possible impacts of increasing soil pH by addition of lime on root growth of common bean and wheat plants and found that liming process significantly improves root growth of these plants by elevating soil pH from 4.1 to 7.0. These researchers concluded that liming is a practical way to enhance calcium bioavailability in rhizosphere. The calcium availability in rhizosphere reduces harmful impacts of biotic and abiotic stresses by protecting the plasma membrane integrity (White 2004).

Gonzalez-Erico et al. (1979) examined the root growth and agronomical responses of maize plants to limestone application in an Oxisol and reported that rising calcium level in soil enhances the root growth, water utilization, and grain yield of plants. Similar results were reported by Doss et al. (1979) for maize and cotton plants. In a recent study, Chu et al. (2021) showed that adequate amounts of calcium by engaging plasma membrane calcineurin B-like calcium-ion sensor proteins improve root growth of plants.

Magnesium

Magnesium is a vital macronutrient for plant growth and physiological activities. Adequate amount of magnesium in the rhizosphere is important for normal growth and development of plants. Similar to calcium deficiency, low level of magnesium in farmlands is more common in weathered acid soils with coarse textured (Senbayram et al. 2016). Magnesium has several roles in plant growth and development. For example, magnesium is an enzyme activator in different physiological pathways and has also a structural role in chlorophyll molecules, hence it is actively involved in photosynthetic activities in plant thylakoids. This nutrient has some key roles in energy and phosphate metabolisms, formation of carbohydrates, ribosome structure preservation, and integrity in plant cells. The absence of magnesium in growth media causes drastic damages to plants growth and productivity (Walker and Duffus 1983; Yan and Hou 2018).

Root growth of plants inhibited under low content of magnesium in rhizosphere. Fageria and De Souza (1991) evaluated the root growth of rice, common bean, and cowpea under different levels of magnesium and observed that root growth of rice plants was greater under low levels of magnesium (1 $cmol_c\ kg^{-1}$ soil), compared to moderate and high levels of this nutrient in growth media. However, the highest root growth in common bean and cowpea plants was achieved under moderate levels of magnesium in soil (2.5–3 $cmol_c\ kg^{-1}$ soil). These researchers also indicated that high amount of magnesium in soil (4–7 $cmol_c\ kg^{-1}$ soil) noticeably reduced root growth in all studied plants. In a novel study, Ghassemi-Golezani et al. (2021) and Farhangi-Abriz and Ghassemi-Golezani (2021) observed that magnesium-enriched biochar increased root growth of safflower plants under normal and saline conditions. These researchers found a strong correlation between magnesium availability in soil and safflower root growth. They also reported a strong correlation between magnesium content in root cells by enhancing potassium uptake in both saline and nonsaline conditions.

Niu et al. (2014) by testing *Arabidopsis* root growth under different levels of magnesium reported that lateral root formation and primary root growth of this plant were not affected by low magnesium content. However, high level of magnesium in soil significantly reduced main and lateral root growth. Magnesium availability not only controls root growth but also changes the root/shoot ratio in plants. For example, Damm et al. (2011) showed that low magnesium in growth media increased the root/shoot ratio in rice plants. This impact of magnesium deficiency might be related to allocation of carbon to the leaf tissues instead of the root. In a recent study, Li et al. (2020) showed that adequate amounts of magnesium in soil promoted the root growth of *Arabidopsis* plants by engaging nitric oxide signaling under normal condition and aluminum toxicity.

Sulfur

Sulfur is a macronutrient with different roles in plant growth and development. Since sulfur is a structural element in cysteine and methionine amino acids, it has a main role in protein formation in plant cells. Sulfur activates some important enzymes involving in plant physiological defense mechanisms, and it has also an impact on nodule formation in legume plants (Hawkesford and De Kok 2007). Compared to other macronutrients, few studies have evaluated the effects of sulfur on root growth and development. Availability of sulfur in soil significantly increases roots number and dry weight in soybean in comparison with control treatment (Zhao et al. 2008). Oates and Kamprath (1985) reported that sulfur availability in soil enhances root development in different soils.

Kutz et al. (2002) by investigating the sulfur impacts on root growth of *Arabidopsis* showed that the low sulfur in growth media created a branched root system in plants. Adequate amount of sulfate in growth media increased the main root elongation and formed the lateral roots at some distance from the root tip. However, the absence of sulfur in growth media formed the lateral roots closer to the root tip. Improvements in lateral root growth under low sulfur are related to transcriptional activation of the *NITRILASE3* (*NIT3*) gene. Nitrilases are enzymes that increase auxin content in plant tissues by converting indole-3-acetonitrile to indole-3-acetic acid. The increment of auxin content in root tissues creates a branched root system. Carciochi et al. (2017) observed that sulfur improves root growth at different stages of plant life cycle and improves nitrogen use efficiency in wheat crop. Hitsuda et al. (2005) in a comprehensive study evaluated the sulfur requirement of eight crops at early stages of growth. They examined the root growth of soybean, maize, rice, wheat, sorghum, cotton, sunflower, and field bean under various concentrations of sulfur in growth media and observed that the addition of sulfur to the growth media noticeably improves root growth in all tested crops. Zhao et al. (2008) showed that sulfur availability in growth media increases root growth of soybean (dry weight) up to 34%, in comparison with the control plants. The responses of plant roots to different levels of macronutrients in rhizosphere are shown in Table 6.1.

Micronutrients

The required concentrations of micronutrients in plant tissues are low, compared to the macronutrients. While micronutrients are mandatory in small amounts for crops, their impact could be as great as that of macronutrients in agricultural activities. There are seven main micronutrients (boron, zinc, manganese, iron, copper, molybdenum, and chlorine) that constitute less than 1% of the dry weight of most plants. The absence of these nutrients can create some abnormalities in plant growth (Merchant 2010).

Zinc

Zinc is one of the important micronutrients that needed in small amounts by crop plants. This nutrient has some important roles in plant life cycle. For example, zinc is required for producing carbohydrates and chlorophyll and a better metabolism of nitrogen in plant cells (Hafeez et al. 2013). The absence of this nutrient creates some serious problems in plant growth and development. Fageria (2016) showed that zinc addition to soil up 120 mg kg^{-1} significantly improves root growth of different crops such as maize, dry bean, and soybean in Brazilian Oxisols. Broadley et al. (2010) by testing rice plants under zinc deficit reported that the addition of zinc to the soil improves the number of roots and their weight. Phuphong et al. (2020) reported that foliar application of zinc (0.5% $ZnSO_4$) enhances root growth of rice seedlings under various concentrations of zinc in the growth media.

Root response to zinc in rhizosphere is dependent on its concentration. Based on available reports, low and moderate concentrations of zinc in growth media have a positive impact on root growth, but high concentration of this nutrient reduces root growth in different plant species. In recent research, Feigl et al. (2019) found that low zinc

Table 6.1 Responses of plant roots to low and high amounts of macronutrients in rhizosphere.

Nutrient	Absorption form	Root response	References
Nitrogen	NO_3^- and NH_4^+	The low content of nitrogen reduces main root growth and increases lateral root numbers in plants. Excessive amount of nitrogen in growth media enhances auxins and abscisic acid concentration in root cells, leading to a reduction in lateral root growth and root mass per unit area in plants.	(Drew and Saker 1975; Anderson 1987; Zhang and Forde 1998; Zhang et al. 1999; Signora et al. 2001; Linkohr et al. 2002; Costa et al. 2002)
Phosphorus	$H_2PO_4^-$ and HPO_4^{2-}	Both of the low and high levels of phosphorous in growth media decrease the growth of main root in plants. Moreover, low level of phosphorous in growth media stimulates the lateral root growth in various plant species and creates a proteoid shape in plant root. Roots under low level of phosphorous do not have appropriate cell elongation and mitotic activities.	(Johnson et al. 1996; Sánchez-Calderón et al. 2005; Fageria et al. 2006; Abel 2011)
Potassium	K^+	The low content of potassium in growth media noticeably reduces root growth and modifies auxin transport and root apical meristem maintenance pathways by ethylene and nitric oxide signaling, and thus decreases the activity of the root meristems. Moreover, low content of potassium in plant cells reduces assimilate mobilization to roots and consequently diminishes root growth. High content of potassium slightly decreases root growth of plants and enhances the number of root hairs.	(Baligar et al. 1998; Kellermeier et al. 2014; Song et al. 2018)
Calcium	Ca^{2+}	The unavailability of calcium ions in acidic soil reduces root growth of plants. The adequate amount of calcium by engaging plasma membrane calcineurin B-like calcium-ion sensor proteins improves root growth of plants.	(Fageria and Zimmermann 1998; Chu et al. 2021)
Magnesium	Mg^{2+}	The low and moderate levels of magnesium in growth media improve root growth of plants, but high level of magnesium in soil significantly reduces main and lateral root growth.	(Fageria and De Souza 1991; Niu et al. 2014)
Sulfur	SO_4^{2-}	The low content of sulfur in growth media enhances the accumulation of auxins in root cells and decreases main root growth and forms a branched root system in plants.	Kutz et al. 2002; Hitsuda et al. 2005

supplementation modifies protein nitration pattern and stimulates root growth in *Brassica napus*. While, high zinc concentration increases nitrosative stress (overproduction of nitric oxide) and consequently inhibits root growth. These scientists proved that nitrosative processes have a critical role in Zn-induced root growth responses. Similarly, Duan et al. (2015) studied the interaction of nitric oxide and reactive oxygen species in regulating root growth in wheat seedlings under zinc stress and reported that zinc toxicity reduces root cell viability and elongation.

Boron

Boron is a critical micronutrient for crops. This nutrient is vital for protein, nucleic acids, seed, and cell wall formation. Boron deficiency depresses plant growth and causes shortening of the internodes (Camacho-Cristóbal et al. 2008). Deficiency of boron typically occurs in arid and semiarid regions with sandy, light texture, and acidic soils (Gupta 1993). Boron requirement is different among crop species. For example, maximum root growth of rice plants is achieved under 0.4 mg B kg^{-1} soil, whereas maximum root growth of bean plant is attained at 1.9 mg B kg^{-1} soil. The highest growth of root in wheat plants occurred under 4 mg B kg^{-1} soil. In general, application of boron at lower rates on most of the plant species increases root growth, while boron application in high amounts noticeably reduces root growth (Fageria 2000).

Lukaszewski and Blevins (1996) stated that root growth inhibition under boron deficit might be related to a low level of ascorbate metabolism in root cells. Boron has a critical role in controlling hormonal signaling in plant cells. Li et al. (2016) in a molecular study showed that boron deficiency noticeably reduces indole-3-acetic acid production in root cells by diminishing expression of *TAA1, TAR2, YUC3,* and *YUC8* genes. Indole-3-acetic acid is one of the main phytohormones in adjusting root growth of plants. Donghua et al. (2000) indicated that the boric acid has a stimulatory effect on root growth at concentrations of 10^{-6}~10^{-3} M and an inhibitory effect at higher concentrations. They showed that low and moderate levels of boron are required for successful cell division in root tips of broad bean. Further investigations by Poza-Viejo et al. (2018) showed that these improvements in cell division are related to enhancement of cytokinin concentration in root meristems.

Copper

Copper is absorbed by plants in cation form Cu^{2+} and has critical roles in adjusting plant growth. Copper is involved in the process of photosynthesis and has some roles in carbohydrate metabolism and cell wall formation in plant cells (Yruela 2005). The deficiency of copper in plants is common in soils inherently low in copper or has a high amount of organic matter content. Available reports show that copper in low content stimulates, but in high amount inhibits root growth.

Fageria (2002) reported that low level of copper (2 mg kg^{-1}) in growth media improves root growth of wheat and dry bean plants. However, high amounts of copper inhibit root growth in both plants. Arduini et al. (1995) showed that high amount of copper (5 μM $CuSO_4$) in growth media reduces, but low levels of copper increase root growth. Similar results have been reported for maize plants (Jiang et al. 2001). Jiang et al. (2001) showed that high content of copper in soil modifies chromosomal morphology and anaphase bridges in cell division process of plant roots and consequently reduces plant root growth.

Groppa et al. (2008) reported similar results about the root growth inhibition in sunflower roots under copper toxicity.

Iron

Iron is an important micronutrient in plant life cycle that has some vital roles in physiological pathways. This nutrient is a component of many enzymes and has an important role in electron transfer between the photosystems. Moreover, this nutrient as an electron transporter is involved in oxidation–reduction reactions (Miller et al. 1995). Iron deficiency is common in most crop species, and its absence in growth media creates numerous abnormalities in plant growth and productivity. Sun et al. (2017) showed that iron deficiency reduces root growth by manipulating auxin balance and nitric oxide signaling in rice plants. Zhang et al. (2019) reported that iron deficiency seriously inhibited root growth and caused abnormal growth of roots in different plant species. In comparison, Zhang et al. (2018) reported that excessive amounts of iron in growth media noticeably reduces root growth of *Arabidopsis* plants. These researchers illustrated that excessive amounts of iron decrease root growth by inhibition of potassium metabolism in root cells. Similar reports about the root growth inhibitions in response to iron toxicity have been reported by Fageria et al. (2008) in rice plants. The summary of the responses of plant root to different levels of some micronutrients in rhizosphere is shown in Table 6.2.

Table 6.2 Responses of plant roots to low and high levels of some micronutrients in rhizosphere.

Nutrient	Absorption form	Root response	References
Zinc	Zn^{2+}	The low and moderate concentrations of zinc in soil have a positive impact on root growth, but a high concentration of this nutrient reduces root growth. High zinc concentration in growth media increases nitrosative stress (overproduction of nitric oxide) and consequently inhibits root growth. Moreover, high level of zinc in growth media reduces root cell viability and elongation.	(Duan et al. 2015; Feigl et al. 2019)
Boron	Boric acid	Boron at lower rates increases root growth in most of the plant species, while boron application in high amounts noticeably reduces root growth. This nutrient at low level stimulates the root cell elongation and division. Boron controls hormonal signaling in plant roots and in lower consecration enhances cytokinin and auxin biosynthesis in plant roots.	(Fageria 2000; Li et al. 2016; Poza-Viejo et al. 2018)
Copper	Cu^{2+}	The low level of copper stimulates root growth, but the high level of this nutrient in soil modifies chromosomal morphology and anaphase bridges in cell division process and consequently reduces root growth.	(Arduini et al. 1995; Jiang et al. 2001)
Iron	Ferric ions	Iron deficiency reduces root growth by manipulating auxin balance and nitric oxide signaling. Excessive amounts of iron ions in growth media decrease root growth by inhibiting the uptake of other nutrients such as potassium from the growth media.	(Sun et al. 2017; Zhang et al. 2018)

Nano Nutrients and Root System Modifications

Nano-fertilizers contain nutrients in nano-dimensions ranging from 30 to 100 nm and capable of preserving great number of ions due to their high surface area. Nano nutrients interact with different physiological and morphological pathways in plants. The general impacts of nano nutrients on plant growth are similar to their regular form. However, due to better absorption capacity of nano nutrients, they have a higher use efficiency in plants, compared to regular forms (Subramanian et al. 2015; Qureshi et al. 2018). Typically, low and moderate contents of nano nutrients have a positive impact on root growth, but nano nutrients in high amounts have some drastic effects on plant root growth, due to ion toxicity (El-Ramady et al. 2018). For example, Zuverza-Mena et al. (2016) reported that application of nanosized silver in growth media of radish plants significantly reduced root growth of this plant. Although nano nutrients can have a toxic impact in higher doses on root growth and development, most effects of these nutrients on root growth are positive (Siddiqui et al. 2015).

Wang et al. (2012b) and Cañas et al. (2008) reported that multiwalled and single-walled carbon nanotubes noticeably improved root elongation in wheat and onion plants, respectively. Prasad et al. (2012) illustrated that the application of zinc in nano form significantly increased root growth of peanuts. In another study, Salama (2012) showed that application of silver in nano form enhanced root growth in maize and bean plants. Lee et al. (2010) by testing titanium oxide nanoparticle on changing root growth of *Arabidopsis* plants reported a superior impact of this particle in improving root growth. A similar report is available about the impacts of aluminum oxide nanoparticles on rising root growth of *Arabidopsis* plants (Jin et al. 2017). Kim et al. (2014) indicated that application nanoscale zero-valent iron particles increased the root growth of *Arabidopsis* plants. Wu et al. (2012) by applying cobalt oxide nanoparticle reported that this nanosized material enhanced root growth of radish plants. Wang et al. (2012a) applied the hydroxyapatite suspension on growing lettuce plants and observed that root growth was promoted in response to this nano form material.

Management Strategies for Maximizing Root Systems

Maximizing root growth of plants is an important goal in rising nutrient use efficiency and crop productivity in various environments. Researchers suggested some practical ways to improve root growth of plants (Gregory 1994), which can be summarized as (i) soil management and (ii) plant management.

Soil Management

Management of soil conditions is a practical way to improve root growth and nutrient uptake by plants. There are various soil management strategies that must be applied by the farmers to enhance plant root growth. For example, preserving and increasing soil organic matter and adjusting soil pH are the most important ways to enhance root growth and nutrient absorption rate in different crop species. Soil organic matter improves water-holding capacity and aggregation of soil, increases nutrients bioavailability, and decreases nutrients leaching (Tiessen et al. 1994). The increment in nutrients bioavailability may produce strong root systems and higher crop yields. Nutrients application methods and effective sources of nutrients are the other management strategies to improve root growth of plants. As we explained in the previous sections, application methods of nutrients and their effective sources modify the root growth of plants, so farmers must pay more attention to these factors. Appropriate tillage system in each environment is another important factor that controls root growth and nutrient uptake by the plants (Anderson 1988). Typically, deep plowing of soil by breaking compacted layers that roots cannot readily penetrate and increases the root growth of plants. However, deep plowing reduces organic matter content and increases nutrients leaching in soil, which are negative factors for root growth in long term. It seems that choosing proper method of soil plowing depends on general soil and climate conditions such as, soil texture, soil organic matter, and precipitation (Foy 1992). Adjusting soil pH can also have a superior impact on rising root growth and nutrient uptake in plants. Soil pH controls the bioavailability of nutrients and changes the root growth and architecture. Application of lime, gypsum, and compost could be helpful methods in adjustment of soil pH under various environments.

Plant Management

Plant management strategies are the practical ways to maximize root growth of plants. Different plant management strategies such as selecting the cultivars with strong root system could be helpful in maximizing root growth of plants and nutrient uptake in farmlands. There are extensive variations in root growth of plant genotypes in response to different environmental situations. Nutrient deficiency induces significant differences in root growth and morphology of plants, depending on plant species and genotypes. Hence, selecting the right genotype of a crop in each environment is critical for better nutrient use efficiency and crop productivity. Providing suitable condition for plant growth is another important plant management strategy in controlling root growth. For example, sowing crops in an appropriate density helps the plants to expand their roots in soil more properly (Khan et al. 2017; Hecht et al. 2019). Appropriate sowing density prevents the competition of roots for absorption of mineral nutrients and promotes their growth.

Conclusions and Future Perspectives

The study of plant roots' behavior to bioavailability of nutrients is one of the most interesting, but least explored, aspects of plant science. Nutrients bioavailability noticeably interacts with root growth and its behavior. Excessive amount or the absence of a nutrient causes physiological and morphological modifications in root growth. Moreover, appropriate amounts of nutrients for maximum root growth of plants depend on crop species and application methods. The important findings about the interaction of plant roots with different nutrients should be presented by the researchers to the farmers. These findings could be helpful in rising nutrients use efficiency and crop productivity under various growth conditions of plants. We noticed that the number of studies that were conducted to evaluate the roots' interactions with nutrient bioavailability (especially micronutrients) are few. This may be related to physical location and growth habits of roots. It would be meaningful, if future investigations are directed to evaluate molecular responses of plant roots to nutrients bioavailability in various soil and climate conditions.

References

Abel, S. (2011). Phosphate sensing in root development. *Curr. Opin. Plant Biol.* 14: 303–309.

Anderson, E.L. (1987). Corn root growth and distribution as influenced by tillage and nitrogen fertilization. *Agron. J.* 79: 544–549.

Anderson, E.L. (1988). Tillage and N fertilization effects on maize root growth and root: shoot ratio. *Plant Soil* 108: 245–251.

Anderson, G.C., Pathan, S., Easton, J. et al. (2020). Short-and long-term effects of lime and gypsum applications on acid soils in a water-limited environment: 2. Soil chemical properties. *Agronomy* 10: 1987.

Anghinoni, I.A. and Barber, S.A. (1980). Phosphorus influx and growth characteristics of corn roots as influenced by phosphorus supply. *Agron. J.* 72: 685–688.

Arduini, I., Godbold, D.L., and Onnis, A. (1995). Influence of copper on root growth and morphology of *Pinus pinea* L. and *Pinus pinaster* Ait. seedlings. *Tree Physiol.* 15: 411–415.

Baligar, V.C., Fageria, N.K., and Elrashidi, M.A. (1998). Toxicity and nutrient constraints on root growth. *HortScience* 33: 960–965.

Bengough, A.G. and Mullins, C.E. (1990). Mechanical impedance to root growth: a review of experimental techniques and root growth responses. *J. Soil Sci.* 41: 341–358.

Blaser, S.R., Koebernick, N., Spott, O. et al. (2020). Dynamics of localised nitrogen supply and relevance for root growth of *Vicia faba* ('fuego') and *Hordeum vulgare* ('Marthe') in soil. *Sci. Rep.* 10: 1–16.

Blumenthal, J.M., Baltensperger, D.D., Cassman, K.G. et al. (2008). Importance and effect of nitrogen on crop quality and health. In: *Nitrogen in the Environment* (ed. L. Hatfield and R.F. Follett), 51–70. Academic Press.

Broadley, M.R., Rose, T., Frei, M. et al. (2010). Response to zinc deficiency of two rice lines with contrasting tolerance is determined by root growth maintenance and organic acid exudation rates, and not by zinc-transporter activity. *New Phytol.* 186: 400–414.

Bruce, R.C., Warrell, L.A., Edwards, D.G., and Bell, L.C. (1988). Effects of aluminium and calcium in the soil solution of acid soils on root elongation of *Glycine max* cv. Forrest. *Aust. J. Agric. Res.* 39: 319–338.

Camacho-Cristóbal, J.J., Rexach, J., and González-Fontes, A. (2008). Boron in plants: deficiency and toxicity. *J. Integr. Plant Biol.* 50: 1247–1255.

Cañas, J.E., Long, M., Nations, S. et al. (2008). Effects of functionalized and nonfunctionalized single-walled carbon nanotubes on root elongation of select crop species. *Environ. Toxicol. Chem.* 27: 1922–1931.

Carciochi, W.D., Divito, G.A., Fernández, L.A., and Echeverría, H.E. (2017). Sulfur affects root growth and improves nitrogen recovery and internal efficiency in wheat. *J. Plant Nutr.* 40: 1231–1242.

Chen, J. and Gabelman, W.H. (1995). Isolation of tomato strains varying in potassium acquisition using a sand-zeolite culture system. *Plant Soil* 176: 65–70.

Chen, J. and Gabelman, W.H. (2000). Morphological and physiological characteristics of tomato roots associated with potassium-acquisition efficiency. *Sci. Hortic.* 83: 213–225.

Chen, Z.H., Nimmo, G.A., Jenkins, G.I., and Nimmo, H.G. (2007). BHLH32 modulates several biochemical and morphological processes that respond to Pi starvation in *Arabidopsis*. *Biochemist* 405: 191–198.

Chen, Z.H., Jenkins, G.I., and Nimmo, H.G. (2008). Identification of an F-box protein that negatively regulates Pi starvation responses. *Plant Cell Physiol.* 49: 1902–1906.

Chen, J., Liu, L., Wang, Z. et al. (2018). Determining the effects of nitrogen rate on cotton root growth and distribution with soil cores and minirhizotrons. *PLoS One* 13: e0197284.

Chevalier, F., Pata, M., Nacry, P. et al. (2003). Effects of phosphate availability on the root system architecture: large-scale analysis of the natural variation between *Arabidopsis* accessions. *Plant Cell Environ.* 26: 1839–1850.

Chu, L.C., Offenborn, J.N., Steinhorst, L. et al. (2021). Plasma membrane calcineurin B-like calcium-ion sensor proteins function in regulating primary root growth and nitrate uptake by affecting global phosphorylation patterns and microdomain protein distribution. *New Phytol.* 229: 2223–2237.

Comfort, S.D., Malzer, G.L., and Busch, R.H. (1988). Nitrogen fertilization of spring wheat genotypes: influence on root growth and soil water depletion. *Agron. J.* 80: 114–120.

Costa, C., Dwyer, L.M., Zhou, X. et al. (2002). Root morphology of contrasting maize genotypes. *Agron. J.* 94: 96–101.

Damm, C., Mallembakam, M.R., Voronov, A., and Peukert, W. (2011). Production of filled hydrogels by mechanochemically induced polymerization. *J. Appl. Polym. Sci.* 120: 799–807.

Donghua, L., Jiang, W., Zhang, L., and Lufang, L.I. (2000). Effects of boron ions on root growth and cell division of broadbean (*Vicia faba* L.). *Isr J Plant Sci* 48: 47–51.

Doss, B.D., Dumas, W.T., and Lund, Z.F. (1979). Depth of lime incorporation for correction of subsoil acidity. *Agron. J.* 71: 541–544.

Drew, M.C. and Saker, L.R. (1975). Nutrient supply and the growth of the seminal root system in barley: II. Localized, compensatory increases in lateral root growth and rates op nitrate uptake when nitrate supply is restricted to only part of the root system. *J. Exp. Bot.* 26: 79–90.

Duan, X., Li, X., Ding, F. et al. (2015). Interaction of nitric oxide and reactive oxygen species and associated regulation of root growth in wheat seedlings under zinc stress. *Ecotoxicol. Environ. Saf.* 113: 95–102.

Dubrovsky, J.G. (1997). Determinate primary root growth in *Stenocereus gummosus* (Cactaceae). In: *Biology of Root Formation and Development* (ed. A. Altman and Y. Waisel), 13–20. Boston, MA: Springer.

Eghball, B. and Maranville, J.W. (1993). Root development and nitrogen influx of corn genotypes grown under combined drought and nitrogen stresses. *Agron. J.* 85: 147–152.

El-Ramady, H., Abdalla, N., Alshaal, T. et al. (2018). Plant nano-nutrition: perspectives and challenges. In: *Nanotechnology, Food Security and Water Treatment* (ed. K.M. Gothandam, S. Ranjan, N. Dasgupta, et al.), 129–161. Cham: Springer.

Emanuelsson, J. (1984). Root growth and calcium uptake in relation to calcium concentration. *Plant Soil* 78: 325–334.

Fageria, N.K. (2000). Adequate and toxic levels of boron for rice, common bean, corn, soybean and wheat production in cerrado soil. *Rev Bras Eng Agric Ambient* 4: 57–62.

Fageria, N.K. (2002). Micronutrients' influence on root growth of upland rice, common bean, corn, wheat, and soybean. *J. Plant Nutr.* 25: 613–622.

Fageria, N.K. (2012). *The Role of Plant Roots in Crop Production*. CRC Press.

Fageria, N.K. (2016). *The Use of Nutrients in Crop Plants*. CRC press.

Fageria, N.K. and Baligar, V.C. (2008). Ameliorating soil acidity of tropical Oxisols by liming for sustainable crop production. *Adv. Agron.* 99: 345–399.

Fageria, N.K. and De Souza, C.M.R. (1991). Upland rice, common bean, and cowpea response to magnesium application on an Oxisol. *Commun. Soil Sci. Plant Anal.* 22: 1805–1816.

Fageria, N.K. and Moreira, A. (2011). The role of mineral nutrition on root growth of crop plants. *Adv. Agron.* 110: 251–331.

Fageria, N.K. and Zimmermann, F.J.P. (1998). Influence of pH on growth and nutrient uptake by crop species in an Oxisol. *Commun. Soil Sci. Plant Anal.* 29: 2675–2682.

Fageria, N.K., Baligar, V.C., and Clark, R. (2006). *Physiology of Crop Production*. CRC Press.

Fageria, N.K., Santos, A.B., Barbosa Filho, M.P., and Guimarães, C.M. (2008). Iron toxicity in lowland rice. *J. Plant Nutr.* 31: 1676–1697.

Fageria, N.K., Carvalho, M.C.S., and Dos Santos, F.C. (2014). Root growth of upland rice genotypes as influenced by nitrogen fertilization. *J. Plant Nutr.* 37: 95–106.

Fageria, N.K., Santos, A.B., and Carvalho, M.C.S. (2015). Agronomic evaluation of phosphorus sources applied to upland and lowland rice. *Commun. Soil Sci. Plant Anal.* 46: 1097–1111.

Farhangi-Abriz, S. and Ghassemi-Golezani, K. (2016). Improving amino acid composition of soybean under salt stress by salicylic acid and jasmonic acid. *J. Appl. Bot. Food Qual.* 89: 243–248.

Farhangi-Abriz, S. and Ghassemi-Golezani, K. (2021). Changes in soil properties and salt tolerance of safflower in response to biochar-based metal oxide nanocomposites of magnesium and manganese. *Ecotoxicol. Environ. Saf.* 211: 111904.

Feigl, G., Molnár, Á., Szőllősi, R. et al. (2019). Zinc-induced root architectural changes of rhizotron-grown *B. napus* correlate with a differential nitro-oxidative response. *Nitric Oxide* 90: 55–65.

Forde, B. and Lorenzo, H. (2001). The nutritional control of root development. *Plant Soil* 232: 51–68.

Foy, C.D. (1992). Soil chemical factors limiting plant root growth. In: *Limitations to Plant Root Growth* (ed. J.L. Hatfield and B.A. Stewart), 97–149. New York, NY: Springer.

Ghassemi-Golezani, K. and Abdoli, S. (2021). Improving ATPase and PPase activities, nutrient uptake and growth of salt stressed ajowan plants by salicylic acid and iron-oxide nanoparticles. *Plant Cell Rep.* 40: 559–573.

Ghassemi-Golezani, K., Farhangi-Abriz, S., and Abdoli, S. (2021). How can biochar-based metal oxide nanocomposites counter salt toxicity in plants? *Environ. Geochem. Health* 43: 2007–2023.

Glinski, J. (2018). *Soil Physical Conditions and Plant Roots*. CRC Press.

Gonzalez-Erico, E., Kamprath, E.J., Naderman, G.C., and Soares, W.V. (1979). Effect of depth of lime incorporation on the growth of corn on an Oxisol of Central Brazil. *Soil Sci. Soc. Am. J.* 43: 1155–1158.

Gregory, P.J. (1994). Root growth and activity. In: *Physiology and Determination of Crop Yield* (ed. K.J. Boote, J.M. Bennett, T.R. Sinclair and G.M. Paulsen), 65–93. USA: American Society of Agronomy.

Groppa, M.D., Zawoznik, M.S., Tomaro, M.L., and Benavides, M.P. (2008). Inhibition of root growth and polyamine metabolism in sunflower (*Helianthus annuus*) seedlings under cadmium and copper stress. *Biol. Trace Elem. Res.* 126: 246–256.

Gupta, U.C. (1993). *Boron and its Role in Crop Production*. CRC Press.

Hafeez, B., Khanif, Y.M., and Saleem, M. (2013). Role of zinc in plant nutrition-a review. *J. Exp. Agric. Int.* 374–391.

Hallmark, W.B. and Barber, S.A. (1984). Root growth and morphology, nutrient uptake, and nutrient status of early growth of soybeans as affected by soil P and K1. *Agron. J.* 76: 209–212.

Hansel, F.D., Amado, T.J., Ruiz Diaz, D.A. et al. (2017). Phosphorus fertilizer placement and tillage affect soybean root growth and drought tolerance. *Agron. J.* 109: 2936–2944.

Hawkesford, M.J. and De Kok, L.J. (2007). *Sulfur in Plants*, vol. 6. London: Springer.

Hecht, V.L., Temperton, V.M., Nagel, K.A. et al. (2019). Plant density modifies root system architecture in spring barley (*Hordeum vulgare* L.) through a change in nodal root number. *Plant Soil* 439: 179–200.

Hill, J.O., Simpson, R.J., Moore, A.D., and Chapman, D.F. (2006). Morphology and response of roots of pasture species to phosphorus and nitrogen nutrition. *Plant Soil* 286: 7–19.

Hitsuda, K., Yamada, M., and Klepker, D. (2005). Sulfur requirement of eight crops at early stages of growth. *Agron. J.* 97: 155–159.

Hoad, S.P., Russell, G., Lucas, M.E., and Bingham, I.J. (2001). The management of wheat, barley, and oat root systems. *Adv. Agron.* 74: 193–246.

Huang, D., Gong, X., Liu, Y. et al. (2017). Effects of calcium at toxic concentrations of cadmium in plants. *Planta* 245: 863–873.

Jia, Y.B., Yang, X.E., Feng, Y., and Jilani, G. (2008). Differential response of root morphology to potassium deficient stress among rice genotypes varying in potassium efficiency. *J. Zhejiang Univ. Sci. B* 9: 427–434.

Jiang, W., Liu, D., and Liu, X. (2001). Effects of copper on root growth, cell division, and nucleolus of *Zea mays*. *Biol. Plant.* 44: 105–109.

Jin, Y., Fan, X., Li, X. et al. (2017). Distinct physiological and molecular responses in *Arabidopsis thaliana* exposed to aluminum oxide nanoparticles and ionic aluminum. *Environ. Pollut.* 228: 517–527.

Johnson, J.F., Vance, C.P., and Allan, D.L. (1996). Phosphorus deficiency in *Lupinus albus* (altered lateral root development and enhanced expression of phosphoenolpyruvate carboxylase). *Plant Physiol.* 112: 31–41.

Kellermeier, F., Armengaud, P., Seditas, T.J. et al. (2014). Analysis of the root system architecture of *Arabidopsis* provides a quantitative readout of crosstalk between nutritional signals. *Plant Cell* 26: 1480–1496.

Khan, A., Najeeb, U., Wang, L. et al. (2017). Planting density and sowing date strongly influence growth and lint yield of cotton crops. *Field Crop Res.* 209: 129–135.

Kim, J.H., Lee, Y., Kim, E.J. et al. (2014). Exposure of iron nanoparticles to *Arabidopsis thaliana* enhances root elongation by triggering cell wall loosening. *Environ. Sci. Technol.* 48: 3477–3485.

Kirkby, E.A. and Pilbeam, D.J. (1984). Calcium as a plant nutrient. *Plant Cell Environ.* 7: 397–405.

de Kroon, H., Hendriks, M., van Ruijven, J. et al. (2012). Root responses to nutrients and soil biota: drivers of species coexistence and ecosystem productivity. *J. Ecol.* 100: 6–15.

Kutz, A., Müller, A., Hennig, P. et al. (2002). A role for nitrilase 3 in the regulation of root morphology in sulphur-starving *Arabidopsis thaliana*. *Plant J.* 30: 95–106.

Lee, C.W., Mahendra, S., Zodrow, K. et al. (2010). Developmental phytotoxicity of metal oxide nanoparticles to *Arabidopsis thaliana*. *Environ. Toxicol. Chem.* 29: 669–675.

Leghari, S.J., Wahocho, N.A., Laghari, G.M. et al. (2016). Role of nitrogen for plant growth and development: a review. *Adv. Environ. Biol.* 10: 209–219.

Li, Q., Liu, Y., Pan, Z. et al. (2016). Boron deficiency alters root growth and development and interacts with auxin metabolism by influencing the expression of auxin synthesis and transport genes. *Biotechnol. Biotechnol. Equip.* 30: 661–668.

Li, D., Ma, W., Wei, J. et al. (2020). Magnesium promotes root growth and increases aluminum tolerance via modulation of nitric oxide production in *Arabidopsis*. *Plant and Soil* 457 (1): 83–95.

Linkohr, B.I., Williamson, L.C., Fitter, A.H., and Leyser, H.O. (2002). Nitrate and phosphate availability and distribution have different effects on root system architecture of *Arabidopsis*. *Plant J.* 29: 751–760.

López-Bucio, J., Cruz-Ramırez, A., and Herrera-Estrella, L. (2003). The role of nutrient availability in regulating root architecture. *Curr. Opin. Plant Biol.* 6: 280–287.

Lukaszewski, K.M. and Blevins, D.G. (1996). Root growth inhibition in boron-deficient or aluminum-stressed squash may be a result of impaired ascorbate metabolism. *Plant Physiol.* 112: 1135–1140.

Lund, Z.F. (1970). The effect of calcium and its relation to several cations in soybean root growth. *Soil Sci. Soc. Am. J.* 34: 456–459.

Ma, Z., Bielenberg, D.G., Brown, K.M., and Lynch, J.P. (2001). Regulation of root hair density by phosphorus availability in *Arabidopsis thaliana*. *Plant Cell Environ.* 24: 459–467.

Marchive, C., Yehudai-Resheff, S., Germain, A. et al. (2009). Abnormal physiological and molecular mutant phenotypes link chloroplast polynucleotide phosphorylase to the phosphorus deprivation response in *Arabidopsis*. *Plant Physiol* 151: 905–924.

Meena, V.S., Meena, S.K., Verma, J.P. et al. (2017). Plant beneficial rhizospheric microorganism (PBRM) strategies to improve nutrients use efficiency: a review. *Ecol. Eng.* 107: 8–32.

Merchant, S.S. (2010). The elements of plant micronutrients. *Plant Physiol* 154: 512–515.

Miller, G.W., Huang, I.J., Welkie, G.W., and Pushnik, J.C. (1995). Function of iron in plants with special emphasis on chloroplasts and photosynthetic activity. In: *Iron Nutrition in Soils and Plants* (ed. J. Abadía), 19–28. Dordrecht: Springer.

Miura, K., Rus, A., Sharkhuu, A. et al. (2005). The *Arabidopsis* SUMO E3 ligase SIZ1 controls phosphate deficiency responses. *Proc. Natl. Acad. Sci. U. S. A.* 102: 7760–7765.

Mollier, A. and Pellerin, S. (1999). Maize root system growth and development as influenced by phosphorus deficiency. *J. Exp. Bot.* 50: 487–497.

Mullins, G.L., Reeves, D.W., Burmester, C.H., and Bryant, H.H. (1994). In-row subsoiling and potassium placement effects on root growth and potassium content of cotton. *Agron. J.* 86: 136–139.

Niu, Y.F., Chai, R.S., Jin, G.L. et al. (2013). Responses of root architecture development to low phosphorus availability: a review. *Ann. Bot.* 112: 391–408.

Niu, Y., Jin, G., and Zhang, Y.S. (2014). Root development under control of magnesium availability. *Plant Signal. Behav.* 9: e29720.

Oates, K.M. and Kamprath, E.J. (1985). Sulfur fertilization of winter wheat grown on deep sandy soils. *Soil Sci. Soc. Am. J.* 49: 925–927.

Oosterhuis, D.M. (2002). Potassium management of cotton. In: *Potassium for Sustainable Crop Production*, 331–346.

O'toole, J.C. and Bland, W.L. (1987). Genotypic variation in crop plant root systems. *Adv. Agron.* 41: 91–145.

Ötvös, K., Marconi, M., Vega, A. et al. (2021). Modulation of plant root growth by nitrogen source-defined regulation of polar auxin transport. *EMBO J.* 40: e106862.

Pandey, G.K. and Mahiwal, S. (2020). *Role of Potassium in Plants*. Springer.

Phuphong, P., Cakmak, I., Yazici, A. et al. (2020). Shoot and root growth of rice seedlings as affected by soil and foliar zinc applications. *J. Plant Nutr.* 43: 1259–1267.

Poza-Viejo, L., Abreu, I., González-García, M.P. et al. (2018). Boron deficiency inhibits root growth by controlling meristem activity under cytokinin regulation. *Plant Sci.* 270: 176–189.

Prajapati, K. and Modi, H.A. (2012). The importance of potassium in plant growth–a review. *Indian J. Plant Sci.* 1: 177–186.

Prasad, T., Sudhakar, P., Sreenivasulu, Y. et al. (2012). Effect of nanoscale zinc oxide particles on the germination, growth and yield of peanut. *J. Plant Nutr.* 35: 905–927.

Qureshi, A., Singh, D.K., and Dwivedi, S. (2018). Nano-fertilizers: a novel way for enhancing nutrient use efficiency and crop productivity. *Int. J. Curr. Microbiol. App. Sci.* 7: 3325–3335.

Raghothama, K.G. (2005). Phosphorus and plant nutrition: an overview. In: *Phosphorus: Agriculture and the Environment*, vol. 46 (ed. J.T. Sims and A.N. Sharpley), 353–378. USA: American Society of Agronomy.

Reymond, M., Svistoonoff, S., Loudet, O. et al. (2006). Identification of QTL controlling root growth response to phosphate starvation in *Arabidopsis thaliana*. *Plant Cell Environ.* 29: 115–125.

Richardson, A.E., Hocking, P.J., Simpson, R.J., and George, T.S. (2009). Plant mechanisms to optimise access to soil phosphorus. *Crop Pasture Sci.* 60: 124–143.

Sainju, U.M., Singh, B.P., and Whitehead, W.F. (2001). Comparison of the effects of cover crops and nitrogen fertilization on tomato yield, root growth, and soil properties. *Sci. Hortic.* 91: 201–214.

Salama, H.M. (2012). Effects of silver nanoparticles in some crop plants, common bean (*Phaseolus vulgaris* L.) and corn (*Zea mays* L.). *Int. Res. J. Biotechnol.* 3: 190–197.

Sánchez-Calderón, L., López-Bucio, J., Chacón-López, A. et al. (2005). Phosphate starvation induces a determinate developmental program in the roots of *Arabidopsis thaliana*. *Plant Cell Physiol.* 46: 174–184.

Schmidt, W. and Schikora, A. (2001). Different pathways are involved in phosphate and iron stress-induced alterations of root epidermal cell development. *Plant Physiol.* 125: 2078–2084.

Senbayram, M., Gransee, A., Wahle, V., and Thiel, H. (2016). Role of magnesium fertilisers in agriculture: plant–soil continuum. *Crop Pasture Sci.* 66: 1219–1229.

Shabala, S., Shabala, L., and Van Volkenburgh, E. (2003). Effect of calcium on root development and root ion fluxes in salinised barley seedlings. *Funct. Plant Biol.* 30: 507–514.

Shahzad, Z. and Amtmann, A. (2017). Food for thought: how nutrients regulate root system architecture. *Curr. Opin. Plant Biol.* 39: 80–87.

Siddiqui, M.H., Al-Whaibi, M.H., Firoz, M., and Al-Khaishany, M.Y. (2015). Role of nanoparticles in plants. In: *Nanotechnology and Plant Sciences* (ed. M.H. Siddiqui, M.H. Al-Whaibi and F. Mohammad), 19–35. Cham: Springer.

Signora, L., De Smet, I., Foyer, C.H., and Zhang, H. (2001). ABA plays a central role in mediating the regulatory effects of nitrate on root branching in *Arabidopsis*. *Plant J.* 28: 655–662.

Song, W., Xue, R., Song, Y. et al. (2018). Differential response of first-order lateral root elongation to low potassium involves nitric oxide in two tobacco cultivars. *J. Plant Growth Regul.* 37: 114–127.

Stober, C., George, E., and Persson, H. (2000). Root growth and response to nitrogen. In: *Carbon and Nitrogen Cycling in European Forest Ecosystems*, 99–121. Berlin, Heidelberg: Springer.

Subramanian, K.S., Manikandan, A., Thirunavukkarasu, M., and Rahale, C.S. (2015). Nano-fertilizers for balanced crop nutrition. In: *Nanotechnologies in Food and Agriculture* (ed. M. Rai, C. Ribeiro, L. Mattoso and N. Duran), 69–80. Cham: Springer.

Sun, H., Feng, F., Liu, J., and Zhao, Q. (2017). The interaction between auxin and nitric oxide regulates root growth in response to iron deficiency in rice. *Front. Plant Sci.* 8: 2169.

Sustr, M., Soukup, A., and Tylova, E. (2019). Potassium in root growth and development. *Plan. Theory* 8: 435.

Thompson, M.V. and Zwieniecki, M.A. (2005). The role of potassium in long distance transport in plants. In: *Vascular Transport in Plants* (ed. N.M. Holbrook and M.A. Zwieniecki), 221–240. Academic Press.

Thor, K. (2019). Calcium – nutrient and messenger. *Front. Plant Sci.* 10: 440.

Ticconi, C.A., Delatorre, C.A., Lahner, B. et al. (2004). *Arabidopsis* pdr2 reveals a phosphate-sensitive checkpoint in root development. *Plant J.* 37: 801–814.

Tiessen, H., Cuevas, E., and Chacon, P. (1994). The role of soil organic matter in sustaining soil fertility. *Nature* 371: 783–785.

Walch-Liu, P.I.A., Ivanov, I.I., Filleur, S. et al. (2006). Nitrogen regulation of root branching. *Ann. Bot.* 97: 875–881.

Walker, G.M. and Duffus, J.H. (1983). Magnesium as the fundamental regulator of the cell cycle. *Dent. Mag.* 2: 1–16.

Wang, M., Chen, L., Chen, S., and Ma, Y. (2012a). Alleviation of cadmium-induced root growth inhibition in crop seedlings by nanoparticles. *Ecotoxicol. Environ. Saf.* 79: 48–54.

Wang, X., Han, H., Liu, X. et al. (2012b). Multi-walled carbon nanotubes can enhance root elongation of wheat (*Triticum aestivum*) plants. *J. Nanopart. Res.* 14: 1–10.

Wang, Y., Chen, Y.F., and Wu, W.H. (2021). Potassium and phosphorus transport and signaling in plants. *J. Integr. Plant Biol.* 63: 34–52.

White, P.J. (2004). Calcium signals in root cells: the roles of plasma membrane calcium channels. *Biologia* 59: 77–83.

Williamson, L.C., Ribrioux, S.P., Fitter, A.H., and Leyser, H.O. (2001). Phosphate availability regulates root system architecture in *Arabidopsis*. *Plant Physiol* 126: 875–882.

Wu, S.G., Huang, L., Head, J. et al. (2012). Phytotoxicity of metal oxide nanoparticles is related to both dissolved metals ions and adsorption of particles on seed surfaces. *J. Pet. Environ. Biotechnol.* 3: 126.

Wu, J., Wang, Y., Kim, S.G. et al. (2017). A secreted chitinase-like protein (OsCLP) supports root growth through calcium signaling in *Oryza sativa*. *Physiol. Plant.* 161: 273–284.

Yan, B. and Hou, Y. (2018). Effect of soil magnesium on plants: a review. *IOP Conference Series: Earth and Environmental Science* (Vol. 170, No. 2, p. 022168). IOP Publishing.

Yibirin, H., Johnson, J.W., and Eckert, D.J. (1993). No-till corn production as affected by mulch, potassium placement, and soil exchangeable potassium. *Agron. J.* 85: 639–644.

Yruela, I. (2005). Copper in plants. *Braz. J. Plant Physiol.* 17: 145–156.

Zhang, H. and Forde, B.G. (1998). An *Arabidopsis* MADS box gene that controls nutrient-induced changes in root architecture. *Science* 279: 407–409.

Zhang, H., Jennings, A., Barlow, P.W., and Forde, B.G. (1999). Dual pathways for regulation of root branching by nitrate. *Proc. Natl. Acad. Sci. U. S. A.* 96: 6529–6534.

Zhang, L., Li, G., Wang, M. et al. (2018). Excess iron stress reduces root tip zone growth through nitric oxide-mediated repression of potassium homeostasis in *Arabidopsis*. *New Phytol.* 219: 259–274.

Zhang, X., Zhang, D., Sun, W., and Wang, T. (2019). The adaptive mechanism of plants to iron deficiency via iron uptake, transport, and homeostasis. *Int. J. Mol. Sci.* 20: 2424.

Zhao, Y., Xiao, X., Bi, D., and Hu, F. (2008). Effects of sulfur fertilization on soybean root and leaf traits, and soil microbial activity. *J. Plant Nutr.* 31: 473–483.

Zuverza-Mena, N., Armendariz, R., Peralta-Videa, J.R., and Gardea-Torresdey, J.L. (2016). Effects of silver nanoparticles on radish sprouts: root growth reduction and modifications in the nutritional value. *Front. Plant Sci.* 7: 90.

7

Potassium Transport Systems at the Plasma Membrane of Plant Cells. Tools for Improving Potassium Use Efficiency of Crops

Jesús Amo, Almudena Martínez-Martínez*, Vicente Martínez, Manuel Nieves-Cordones, and Francisco Rubio*

Departamento de Nutrición Vegetal, Centro de Edafología y Biología Aplicada del Segura-CSIC, Murcia, Spain

Potassium (K^+) as a Macronutrient for Plants

Functions of K^+ and Its Concentration in Plant Cells

Plant growth and development requires efficient acquisition of essential elements. Potassium (K^+) is the most abundant intracellular cation in all living organisms, and its homoeostasis is completely essential. In plant tissues, K^+ comprises up to 10% of a plant's dry weight (White and Karley 2010). K^+ has a high mobility in the plant, and it is translocated between the root and the aerial part by the xylem (root to shoot) and phloem (shoot to root), as well as within individual cells and within tissues (White 2012a, 2012b).

The concentration of K^+ in soil solution may vary widely between 0.1 and 1 mM (White and Karley). This contrasts with its concentrations in the cytosol, where is maintained around 100 mM (Walker et al. 1996). This concentration is essential for K^+ to fulfill all its physiological functions. K^+ deficiency has very negative effects for the plant because it cannot be fully replaced by other inorganic cations. In subcellular compartments, the concentration of K^+ varies depending on its availability in the external solution and the physiological state of the plant. In the vacuole, the concentration usually varies between 10 and 200 mM, and it can reach 500 mM (Luan et al. 2017). Upon K^+ deficiency, vacuolar K^+ ions are released to the cytosol, maintaining the cytoplasmic K^+ concentration in the range of 100 mM. Even so, with prolonged K^+ starvation, cytosolic concentration of K^+ declines, causing first visible symptoms of deficiency such as brown spots at leaf and chlorosis and eventually wilting and necrosis (Hawkesford et al. 2012a).

K^+ takes part in a series of biophysical and biochemical processes that are very significant for the plant, carrying out vital functions in cell metabolism, adaptation to stress, and plant growth. Since K^+ is positively charged and so abundant, it participates in the

*contributed equally to this work

Plant Ionomics: Sensing, Signaling, and Regulation, First Edition.
Edited by Vijay Pratap Singh and Manzer H. Siddiqui.

stabilization of negatively charged molecules, for instance nucleic acids and proteins. Moreover, K^+ is required for the activation of a considerable number of enzymes such as pyruvate kinase, phosphofructokinase, ADP-glucose starch synthase, membrane-bound proton-pumping ATPases, and vacuolar pyrophosphatase isoforms (Nieves-Cordones et al. 2016c). This cation binds to enzymes and induces conformational changes that increase the rate of catalytic reaction (V_{max}) and in some cases the affinity for the substrate (K_m). These effects are highly specific, and K^+ cannot be replaced by other similar ions such as Na^+ or Li^+ (Hawkesford et al. 2012a).

K^+ also plays an essential role in the synthesis of several proteins among which ribulose bisphosphate carboxylase (RuBisCo) and nitrate reductase stand out. The effect on the latter is reflected in the accumulation of soluble N components in K^+-deficient plants (Hawkesford et al. 2012a). In addition, plant metabolism may be affected through transcriptional and posttranscriptional regulations of metabolic enzymes by K^+ deficiency.

K^+ affects photosynthesis in different ways. It is required for the flow of protons (H^+) through the membrane of thylakoids (Tester and Blatt 1989) and also for the establishment of the pH gradient that drives ATP synthesis through photophosphorylation. Furthermore, K^+ is required for an optimal RuBisCo activity, maintaining the high pH in the stroma.

One of the most important functions of K^+ is the maintenance of turgor pressure that drives plant and organ growth and processes that depend on it such as opening of the stomata, pollen tube growth, leaf movement, and osmocontractility (Hawkesford et al. 2012a).

The K^+ requirement for optimal plant growth is 20–50 g kg^{-1} in vegetative parts, fleshy fruits, and tubers (Hawkesford et al. 2012a). When K^+ is limiting, growth is slow and leaves and roots are short-lived. Stems are weak, and seeds and fruits are small and dehydrated. The physiological symptoms of K^+ deficiency include damage in phloem transport, increased leaf carbohydrate concentrations, a reduction in chlorophyll concentrations and photosynthetic capacity, decreased water content, decreased turgor, impaired stomatal regulation, and reduced transpiration (White and Karley 2010). Organs become chlorotic and necrotic. K^+-deficient plants are more susceptible to abiotic and biotic stresses because more reactive oxygen species (ROS) are produced. In K^+-deficient plants, photosynthesis and RuBisco activity are significantly reduced as well as photorespiration. When K^+ is deficient, there is also an increase of stomatal resistance to CO_2 and dark respiration (Hawkesford et al. 2012a). In K^+-deficient plants, there is an accumulation of reducing sugars and positively charged amino acids, and a depletion of nitrate, organic acids, negatively charged amino acids, and pyruvate (Amtmann et al. 2008)

Concentrations of K^+ in Soil, K^+-Deficient Soils, and Presence of Environmental Conditions that Affect K^+ Nutrition

K^+ represents approximately 2.9% of the earth's crust, being among the most abundant elements present in the upper continental crust. However, only a proportion of the total K^+ amount in soil can be taken up and utilized by plants.

Three types of K^+ pools can be identified in soil depending on their availability to be absorbed by the plant. About 90–98% of total K^+ is found in the structural form (Rengel and Damon 2008). This pool is principally comprised of K^+ bearing primary minerals and is considered nonavailable to plants. The second pool (1–10%) is "non-exchangeable" where

K^+ is associated with clay lattices. The third and last pool is "exchangeable" K^+ (1–2%). It comprises the K^+ dissolved in the soil solution and weakly bound to soil minerals, being able to be absorbed easily by the plant.

As mentioned earlier, the concentration of K^+ in soil solution is widely variable and in most soils lies in the 0.1–1 mM range. It depends on factors such as soil water content, soil depth, pH, cation-exchange capacity, redox potential, quantity of soil organic matter and microbial activity, season, and fertilizer application (Marschner and Rengel 2012). When the K^+ concentration decreases in the soil solution, the exchangeable K^+ pool is released quickly, being immediately available to the plant. Particularly in soils with low in exchangeable K^+, a considerable proportion of the K^+ taken up by the plant comes from the non-exchangeable fraction.

In spite of K^+ abundance in earth's crust, there are large areas of K^+-deficient soils which may limit agricultural production. One-quarter and two-thirds of arable soil are K^+-deficient in China and Australia, respectively (Zörb et al. 2014). Intense K^+ fertilization is required in these deficient areas. Moreover, to increase crop yield, fertilization is a widely used agricultural practice worldwide, which results in important economic and environmental costs. This is of special concern especially for a future agriculture facing an important increase in world's population under a scenario of climate change that requires sustainable agriculture.

K^+ Transport Systems

Acquisition of K^+ from the soil solution is achieved by plant roots, in particular by epidermal and cortical cells. Then, K^+ is loaded into the xylem vessels for its distribution to the rest of the plant (White 2012a). Within the root, the suberized endodermis that defines the central cylinder containing the xylem and phloem (stele) impedes direct transfer of K^+ from the soil solution (apoplast) to the xylem. Thus, to enter the stele and reach the xylem, K^+ has to enter into the symplast, which is composed by the continuum of cytoplasm of cells connected via plasmodesmata. Once in the symplast, K^+ crosses the endodermis and it is released into the xylem. The important idea to keep in mind regarding this process is that crossing a plasma membrane of an epidermal or cortical cell is the first point of control in K^+ acquisition by the plant. Once into the cell, K^+ can be accumulated into the vacuole or other organelles, which also involves crossing a membrane through transport systems. From the xylem, K^+ is distributed to the plant, and in this process it is first released to the shoot apoplast and then transported into the shoot symplast by membrane transport systems (White 2012b). From shoot, K^+ is transferred to other organs through phloem, which also involves membrane transport systems, since at some points plasmodesmata are not present or K^+ diffusion through them is too slow (White 2012b). In conclusion, K^+ transport systems located at plasma membrane of different cells are key pieces for the control of K^+ homeostasis within the plant and proper K^+ nutrition. In the following, the families containing the most relevant plasma membrane K^+ transport systems are described from a phylogenetic point of view, and information regarding their physiological roles is also provided. Transport systems located in other cell membranes are not described here and are reviewed elsewhere (Sze and Chanroj 2018; Ragel et al. 2019).

HAK/KT/KUP Transporters

The members of the HAK/KT/KUP family show homology with bacterial KUP and fungal HAK transporters. This family (High-Affinity K^+/K^+ Transporter/K^+ UPtake) has been widely studied and is present in all plant genomes, indicating that its members are very important for K^+ homeostasis (Santa-María et al. 2018). Although K^+ transport is the main feature of this family, recent studies show that some family members are involved in Na^+ transport too (Zhang et al. 2019; Wang et al. 2020; Benito et al. 2012). The ion movement through the systems can be bidirectional (Garciadeblas et al. 2002), and the role of the transporter in the plant depends on whether it is involved in cation influx into cells cation efflux from them.

One characteristic of the transporters of this family is that they do not highly discriminate among K^+, Rb^+, or Cs^+ and that they are inhibited by NH_4^+. These pharmacological properties have been very instrumental to study the relevance of these transporters for root K^+ uptake. Thus, Rb^+ has been used as a tracer for HAK-mediated K^+ uptake and NH_4^+ as a specific inhibitor of this type of transporters (Rubio et al. 2008). Besides, their capacity to mediate Cs^+ transport makes some members of this family of transporters especially relevant in the accumulation of this toxic cation (Nieves-cordones and Rubio 2021).

In plants, these systems are composed of 10–15 transmembrane domains (TM) (Véry et al. 2014). Based on their phylogeny, the members of this family can be classified in five clusters (clusters I to V), with two subgroups in cluster I (Ia and Ib) and three in cluster II (IIa, IIb, and IIC) (Nieves-Cordones et al. 2016b).

Cluster I: This group is divided into two clusters Ia and Ib. Cluster Ia includes sequences from dicotyledonous and monocotyledonous, while in cluster Ib only sequences from dicotyledonous are found. In general, members of cluster I are K^+–H^+ symporters. Clade Ia includes the well-characterized high-affinity K^+ transporters from barley, rice, tomato, Arabidopsis, or *Eutrema salsuginea* (HvHAK1, OsHAK1, SlHAK5, AtHAK5, and EsHAK5 respectively) that operate at the plasma membrane of root cells to mediate the high-affinity K^+ uptake observed in K^+-starved plants (Nieves-Cordones et al. 2016b). Additional functions for some members of this clade have been recently shown. Thus, OsHAK1, OsHAK5, or AtHAK5 are involved in K^+ translocation from root to shoot (Yang et al. 2014; Chen et al. 2015; Nieves-Cordones et al. 2019c). Other members like OsHAK21 are involved in the distribution of K^+ to the aerial parts of plants growing under salt stress (Shen et al. 2015). Moreover, SlHAK5 is involved in pollen K^+ uptake, which is a determinant for pollen tube elongation and therefore for pollen viability (Nieves-Cordones et al. 2020a).

Regarding cluster Ib, well-characterized systems such as DmHAK5 from *Dionaea muscipula* and VvKUPI/VvHAK1 from *Vitis vinifera*. DmHAK5 is expressed in the trap and mediates K^+ uptake from the digested prey (Scherzer et al. 2015). VvKUP1/VvHAK1 is targeted to the berry skin during the early fruit development stages (pre-veraison), where the transporter regulates berry size and berry K^+ amount through K^+ uptake (Davies et al. 2006).

Cluster II: this cluster is divided into three subgroups, IIa, IIb, and IIc. This cluster is composed by systems that differ in their transport properties and can mediate high- or low-affinity K^+ transport and may be involved in different processes such as hormone transport or development. For example, within cluster Ia, AtKUP4/TRH1 has been reported to transport auxin, being involved in root hair growth and gravitropic responses (Rigas et al. 2001, 2013; Vicente-Agullo et al. 2004). AtKUP6 and AtKUP8, in clade IIc, function as low-affinity K^+ efflux systems, involved in growth regulation, osmotic stress adaptation, and cell expansion through auxin and ABA signaling (Osakabe et al. 2013). Also within this cluster, K^+ transporters involved in tolerance to salinity stand out, as for example PhaHAK2 from *Phragmites australis* (Takahashi et al. 2007) and McHAK3 from *Mesembryanthemum crystallinum* (Su et al. 2001). One of the first characterized HAK/KUP/KT transporters, AtKUP1/KT1, belongs to cluster IIb and has been characterized as a dual-affinity K^+ transport system, mediating both high- and low-affinity K^+ transport (Fu and Luan 1998; Kim et al. 1998). A clear role for this transporter in the plant has not been described.

KUP9 and KUP7 are representative members of cluster III. They fulfill different roles and are targeted to different membranes. AtKUP9 is expressed in the endoplasmic reticulum membrane where is involved in root growth under low K^+ stress (Zhang et al. 2020). KUP7 is located at the plasma membrane of root cells and is involved in K^+ acquisition from the external solution and its translocation from root to shoot (Han et al. 2016).

Members of cluster IV have been described as high-affinity Na^+ transporters. Two of them, ZmHAK4 and SlHAK20, mediate Na^+ transport and are involved in maintaining K^+/Na^+ homeostasis under salt stress in maize and tomato plants, respectively (Zhang et al. 2019; Wang et al. 2020). Such Na^+ transport capacity is also observed in the *Physcomitrella patens* transporter PpHAK13, which mediates high-affinity Na^+ uptake (Benito et al. 2012). Importantly, PpHAK13 does not transport K^+ and is inhibited by the presence of high Na^+ concentrations (Benito et al. 2012). Thus, Na^+ transport capacity by HAK transporters seems to be originated early in land plant evolution.

Within cluster V, AtKUP5 and PpHAK1 have been described. AtKUP5 has been barely characterized, while more information is available for PpHAK1. This latter one is targeted to the plasma membrane and is involved in K^+ uptake in a wide range of concentrations (Garciadeblas et al. 2007). PpHAK1 does not discriminate between K^+, Rb+, and Cs^+ and is a crucial to control K^+ content especially under long-term K^+ starvation.

Voltage-Gated K^+ Channels

Contrarily to the HAK family, voltage-gated K^+ channels are present in all kingdoms, plants, animals, fungi, and bacteria (Jegla et al. 2018). The members of this family are constituted by multimeric proteins involved in K^+ transport through plasma membrane. These channels show a transmembrane core which allows the specific movement of K^+ through a pore region. They also have a long cytosolic tail with different domains: C-linker domain, cyclic nucleotide binding domain, ankyrin domain (in some subunits), and K_{HA} domain (Jegla et al. 2018). Based on phylogenetic studies, the voltage-gated K^+ channels can be classified into five groups (Pilot et al. 2003).

Group I is formed by channels involved in K^+ influx, also known as inward rectifiers. Members of this group have in their structure an ankyrin domain. The AKT1 channel, one of the best characterized K^+ channels in plants, is found in this group. AKT1 is targeted to the plasma membrane of root cells and plays a crucial role in maintaining K^+ nutrition because it constitutes the main route for K^+ entry from the external solution at concentrations above $20\,\mu M$ K^+ (Hirsch et al. 1998). Other representatives of this group are SPIK and

AKT6. They are both expressed in flower tissues (Lacombe et al. 2000; Mouline et al. 2002). SPIK is expressed in pollen grains and regulates their germination and growth through K^+ uptake (Mouline et al. 2002).

Unlike those of Group I, members of Group II do not show an ankyrin domain. However, all of them are also involved in K^+ influx. In this group, KAT1 and KAT2 are two well-characterized members. They are expressed in guard cells and are involved in turgor control and stomatal movements mediating their opening (Pilot et al. 2001; Lebaudy et al. 2010).

Regarding Group III, there is only one member identified, AKT2. Structurally, AKT2 presents the ankyrin domain as the Group I members do. AKT2 is a weakly rectifier channel that can mediate K^+ influx or efflux (Lacombe et al. 2000). AKT2 presents two transport states: inward rectifier and leak-like channel. Interconversion between both states seems to be regulated by phosphorylation and plays an important role for solute transport under energy-limiting conditions (Michard et al. 2005; Gajdanowicz et al. 2011). AKT2 is expressed in the phloem and in the mesophyll where it forms heteromeric channels with KAT2 (belonging to Group I) or homomeric channels (Xicluna et al. 2007). In grapevine, VvK3.1, which is an AKT2 homolog, is expressed in the pulvinus and in berry phloem (Nieves-Cordones et al. 2019a). Pulvinus is involved in leaf movements, and the berry phloem serves as the main pathway for solute transport in late stages of berry development. Thus, AKT2-like channels seem to have additional physiological functions in plant species other than Arabidopsis.

As in the previous one, Group IV is only represented by one channel, KC1. This channel does not present an ankyrin domain and its subunits do not form functional homotetramers as other members of the family do. Therefore, it is often denominated a silent subunit. However, KC1 forms heterotetramers by interacting with members of others groups, in particular with subunits that form inward or weakly rectifying channels (Véry et al. 2014). KC1 is expressed in root cells where it interacts with AKT1 to modulate its activity, regulating K^+ acquisition from the soil (Reintanz et al. 2002; Honsbein et al. 2009; Wang et al. 2010).

Finally, in Group V, the SKOR- and GORK-like channels are found. Group V subunits form outward-rectifying channels involved in K^+ efflux from plant cells. The ankyrin domain is present in both channels. SKOR is targeted to pericycle and xylem parenchyma cells where it contributes to loading K^+ into the xylem (Gaymard et al. 1998). GORK is targeted to root hairs, where it mediates K^+ efflux to the external solution, and in guard cells, where it regulates the osmotic status of the root cells and plays an important role in stomatal closing (Ivashikina et al. 2001; Hosy et al. 2003). Recent work has shown that GORK-like channels take part in nodule formation in leguminous plants (Drain et al. 2020). MtGORK contributes to membrane repolarization by mediating K^+ efflux in the Nod factor membrane signaling.

HKT Transporters

This family has no representatives in animals, and its members come from an ancestral K^+ channel subunit which is related to the fungal Trk/Ktr family (Corratgé-Faillie et al. 2010). HKT representatives are found in fungi, bacteria, and plants. Unlike bacteria, where HKT transport systems are multimeric complexes, in fungi and plants they are single-subunit systems (Corratgé-Faillie et al. 2010).

Studies in plants indicate that HKT transporters have substrate diversity depending on their ion selectivity. Thereby, some family members are involved in Na^+ transport, while others are involved in K^+ mobilization. According to this, members of the HKT family can be classified in three types: K^+-selective transporters, Na^+-selective transporters, and Na^+–K^+ symporters (Véry et al. 2014).

Based on phylogenetic studies, the HKT transporters can be classified into two subfamilies: subfamily 1 and subfamily 2 (Platten et al. 2006). Members of subfamily 1 have been identified in most higher plants (Platten et al. 2006). In Arabidopsis and rice, species where this family has been studied the most, the HKT transporters of this subfamily are involved in Na^+ transport. Some representative examples of this subfamily are AtHKT1, OsHKT1, OsHKT1.3, OsHKT1.5, and OsHKT1.4. AtHKT1 is targeted to root xylem parenchyma and phloem where it contributes to salinity tolerance regulating the amount of Na^+ that is transported to the shoot through the xylem sap and back to the root through the phloem (Berthomieu et al. 2003; Sunarpi et al. 2005; Davenport et al. 2007).

The transporters belonging to subfamily 2 are only present in monocotyledonous plants (Platten et al. 2006). Unlike members of subfamily 1, subfamily 2 transporters can mediate transport of Na^+ and/or K^+. Despite the large body of evidence showing the physiological role of Na^+ transport mediated by HKT proteins in the plant, only one article has reported a physiological function for HKT-mediated K^+ transport in plants. In maize, ZmHKT2 is a K^+-preferring transporter that is involved in K^+ reabsorption from the xylem sap under salt stress (Cao et al. 2019). Interestingly, ZmHKT2 activity under salt stress is counterproductive, since it gives rise to higher Na^+/K^+ ratios in the xylem sap and shoots which negatively impacts on shoot growth (Cao et al. 2019). Thus, maize KO mutants in ZmHKT2 exhibited better performance than wild-type plants under salt stress.

Cyclic Nucleotide Gated Channels

Nonselective cation channels have been proposed to form a pathway for K^+ transport in plants (Demidchik and Maathuis 2007). However, the large variability of the nonselective conductance in plant cells and their unknown genetic identity have hampered a profound understanding of their roles and transport mechanisms in plant cells. Several transport system families have been proposed to fall in the category of nonselective cation channels that transport K^+. Cyclic nucleotide gated channel (CNGC) is one of the best characterized one so far (Kaplan et al. 2007). CNGCs are also present in animal cells and, similarly to voltage-gated K^+ channels, they form tetrameric channels (either homo- or heteromeric). They have a cyclic nucleotide binding domain in the C-terminus, but the role of cyclic nucleotides as their regulators has been called into question (Dietrich et al. 2020). Calmodulin seems to be a regulator of these channels (Dietrich et al. 2020). Four phylogenetic groups can be distinguished, being Group IV divided in groups IVa and IVb (Mäser et al. 2001). In fact, CNGCs and voltage-gated K^+ channels are evolutionarily-related (Jegla et al. 2018). CNGCs have been shown to be permeable to K^+ (Demidchik and Maathuis 2007), and several KO mutants showed K^+-related phenotypes. For example, Arabidopsis *cngc3* mutants exhibited reduced K^+ content at high external K^+ concentrations when compared to wild-type plants (Gobert et al. 2006). Besides, Arabidopsis CNGC10 antisense lines showed reduced K^+ content in comparison with wild-type plants (Li et al. 2005). Both

CNGC3 and 10 belong to Group I, but the K^+ transport capacity does not seem to be exclusive of this group since CNGC2 (Group IVb) forms K^+-permeable channels (Leng et al. 1999). Thus, it remains to be clarified whether the separation into different phylogenetic groups is related to different transport properties or physiological functions.

Key Points for K^+ Homeostasis and Transport Systems Involved

The movement of K^+ from the external solution (apoplast) to the root cells (symplast) takes place against a steep concentration gradient. While the K^+ concentration in the cytosol is maintained constant around 100 mM, in the external solution it may be around 1–5 mM and as low as 1–10 μM. To concentrate K^+, plant cells use the energy stored in the electrochemical gradient of H^+ across the plasma membrane, which is created by the plasma membrane H^+-ATPase (Rodríguez-Navarro 2000). This enzyme uses the energy released from ATP hydrolysis to pump H^+ outside the cell. The asymmetric distribution of H^+ results in a H^+ gradient (ΔpH), acidic outside, and an electrical gradient (ΔVm), negative inside (Nieves-Cordones et al. 2014).

Two types of proteins involved in K^+ transport across membranes, channels and transporters, can be found at the plasma membrane. K^+ channels are uniports through which the cation crosses the membrane alone. In channels involved in uptake from the apoplast, K^+ enters the symplast down the electrical gradient of the plasma membrane. The concentrative capacity of these transport systems is determined by the magnitude of the electrical gradient (Rodríguez-Navarro 2000). In symporters, a type of transport systems, K^+ transport is coupled to the transport of other solutes, mainly H^+. These systems use both the H^+ and the electrical gradients to allow K^+ concentration inside the cell (Rodríguez-Navarro 2000). Symporters may have different stoichiometries, for example, coupling one K^+ to one or two H^+. At a given membrane potential, symporters have a higher concentrative capacity than channels because K^+ moves uphill its concentration gradient coupled to H^+ that move downhill both the electrical and the H^+ gradients. It is important to take into account that the electrical membrane potential is affected by the K^+ status of the plant, because the plasma membrane is highly permeable to K^+ (Rubio et al. 2020). When there is an increase in the external K^+ concentration, a large amount of this cation crosses the plasma membrane, causing its rapid depolarization, whereas a decrease in the external concentration of K^+ produces hyperpolarization (Britto and Kronzucker 2008). This dependency of the cell membrane potential on external and internal K^+ concentrations indirectly affects the transport of other solutes that rely on cell polarization to enter or exit the plant cell (Nieves-Cordones et al. 2016c). At the end, the value of the electric potential is determined by both the operation of the H^+-ATPase (hyperpolarizing) and the movements of K^+ inwardly (depolarizing) and outwardly (hyperpolarizing). Thus, in K^+-sufficient plants, the electrical potential is less steep than in K^+-starved plants.

K^+ enters into the root cells via AKT1-type channels, or via AtHAK5-type transporters, which are K^+-H^+ symporters (Scherzer et al. 2015; Nieves-Cordones et al. 2016a). Traditionally, low-affinity K^+ uptake, which is dominant in K^+-sufficient plants, has been ascribed to channels and high-affinity K^+ uptake, dominant in K^+-deficient plants, to symporters (Rodríguez-Navarro and Rubio 2006). In K^+-sufficient plants, AKT1 channels would be sufficient to mediate K^+ uptake, but, in K^+-starved plants, AtHAK5-like transporters may be required if the electrical potential is not negative enough. Increasing evidences show that K^+ uptake from diluted solutions, within the K^+ concentration range assigned to high-affinity K^+ uptake, can take place through AKT1 channels (Hirsch et al. 1998). In K^+-starved Arabidopsis root cells, electrical gradients more negative than −200 mV and as negative as −236 mV have been recorded (Rubio et al. 2014a). According to Nernst equation, this electrical gradient would allow to concentrate K^+ from concentrations lower than 10 μM. In fact, in Arabidopsis plants starved of K^+, it has been shown that both AKT1 and AtHAK5 contribute to K^+ uptake from diluted solutions, being AtHAK5 able to mediate K^+ uptake at lower concentrations than AKT1 (Nieves-Cordones et al. 2016a). At high external K^+ concentrations, a Ca^{2+}-sensitive channel mediates K^+ uptake in Arabidopsis. Pharmacological analysis of the K^+ transport features of this channel suggested that a CNGC may be the transport system responsible for this K^+ uptake (Caballero et al. 2012).

Once K^+ has entered into the root, it can be returned to the external solution or distributed within the plant via the xylem. K^+ efflux from the root back to the external solutions takes place under conditions that depolarize the membrane potential. In this case, K^+ moves downhill its electrochemical gradient through outward K^+ channels such as GORK (Shabala and Cuin 2008). The release of K^+ into the xylem takes place from the cytoplasm of xylem parenchyma cells to the xylem vessels. This transport to xylem vessels is mediated by multiple transport systems, and their contribution depends on the species. These include SKOR channels, KUP7 transporters, HKT2 transporters, and nonselective cation channels (De Boer 1999; Ragel et al. 2019). Importantly, NRT1.5 a member of the NRT family of transporters, which mediate NO_3^- transport, has been involved in K^+ release into the xylem (Li et al. 2017). In addition, it has recently been shown that the AtHAK5 transporter and the AKT1 channel also contribute to long-distance K^+ transport. While the former contributes to K^+ transport to shoots, the latter plays a negative role in this process by retrieving K^+ from the xylem vessels in plants grown under K^+-sufficient conditions (Nieves-Cordones et al. 2019c). SKOR contributes to massive xylem K^+ load under standard growth conditions (Gaymard et al. 1998) where the membrane potential in xylem parenchyma cells is less negative than the equilibrium potential (De Boer 1999). Thus, SKOR opens and K^+ efflux takes place downhill. When the external K^+ becomes limiting, NRT1.5 enters into scene. It mediates K^+/H^+ antiport and allows K^+ efflux in exchange with H^+ at membrane potentials more negative than the equilibrium potential. Under these conditions, SKOR remains closed. Therefore, xylem K^+ load is maintained in spite of the absence of SKOR-mediated K^+ transport. KUP7 seems also to mediate xylem K^+ load under low external K^+ concentrations (Han et al. 2016); however, the transport mechanism by which K^+ is transferred into xylem vessels is not clear from a thermodynamic point of view. A HKT2 transporter participates in xylem K^+ transport in maize by mediating K^+ resorption from the xylem sap (Cao et al. 2019). It is not clear whether this mechanism is conserved in other plant species as homologs of HKT2 are absent in dicotyledonous species. The role of nonselective channels in xylem K^+ loading remains to be determined, since the genetic identity of these proteins is unknown. They were observed in patch-clamp studies carried out on xylem parenchyma cells (De Boer 1999). It is expected that these channels contribute to xylem K^+ load in cooperation with SKOR under standard growth conditions. It is important to consider that K^+ transport to shoots is still significant in *skor* mutants (Gaymard et al. 1998).

K^+ transport in the phloem has been proposed to exert multiple functions such as providing a pathway for K^+ recirculation from shoot to roots and for K^+ delivery to fruit tissues (Ahmad and Maathuis 2014). Besides, K^+ concentration in the phloem may act as a signal to regulate K^+ uptake by indicating the shoot K^+ status. The best studied K^+ transport system in the phloem is AKT2 channel. This channel operates as a K^+ battery that allows K^+ transport under energy-limiting conditions (Gajdanowicz et al. 2011). The two-state transport modes of AKT2 allow to store energy in the form of a K^+ gradient when the channels are in the inward rectifier mode. When cell energy goes down, the H^+-ATPase activity is reduced and AKT2, by switching to the leak-channel mode, allows K^+ efflux from phloem cells and in turn permits energization of solute transport such as sucrose reloading (Dreyer et al. 2017). In grapevine, the homolog of AKT2, VvK3.1 is expressed in the berry phloem (Nieves-Cordones et al. 2019a). It has been proposed that operation of VvK3.1 in this tissue allows K^+ unload in the berries and supports, by functioning as a K^+ battery, sucrose exchanges between phloem and the berry.

K^+ accumulation into specialized cells such as guard cells or pollen plays critical roles. K^+ is the major cation associated with stomatal movements. K^+ influx leads to increased water uptake. This K^+ influx is triggered by light, for example. Under these conditions, the voltage-gated K^+ channels KAT1 and KAT2 are the main K^+ transport systems involved (Pilot et al. 2001; Lebaudy et al. 2010). At night, or when or water is scarce, stomata close. Stomatal closure is largely dependent on guard cell K^+ efflux. To allow such efflux, voltage-gated K^+ channel GORK and KUP2/6/8 transporters are activated (Hosy et al. 2003; Osakabe et al. 2013). Interestingly, the inward rectifying channel AKT1 seems to negatively affect stomatal closure as the *akt1* mutant close stomata faster than wild-type stomata and allows this mutant to save more water under drought conditions (Nieves-Cordones et al. 2012). In pollen, elongation of its tube is driven by turgor buildup that occurs simultaneously to cell wall softening (Michard et al. 2017). K^+ is the main cation involved in the increase of pollen turgor. K^+ influx is mediated by the voltage-gated K^+ channel SPIK, the tandem-pore K^+ channel TPK4 in Arabidopsis (Becker et al. 2004; Mouline et al. 2002) and, in tomato plants, the SlHAK5 transporter (Nieves-Cordones et al. 2020a). Whereas *tpk4* pollen did not seem to be affected in terms of tube elongation, *spik* and *slhak5* pollen exhibited reduced tube lengths. Thus, HAK transporters and voltage-gated K^+ channels may be regarded as the main actors in K^+ influx in pollen grains. However, species-specific mechanism may exist and additional work is still needed to assess the individual contribution of each transport system type in each species, as it has been done with transport systems involved in K^+ uptake.

Figure 7.1 shows a schematic view of the most relevant K^+ transport systems that have been described earlier and that are involved in K^+ transferred from the external solution to the root stele and K^+ movements in guard cells and pollen grains.

General Mechanisms of Regulation

Transcriptional Regulation

Most of the genes encoding K^+ transport systems do not respond to K^+ supply. Thus, the activity of the systems is mainly regulated at the posttranscriptional level. Exception to this are the genes encoding the high-affinity K^+ transporter HAK5, which are strongly induced

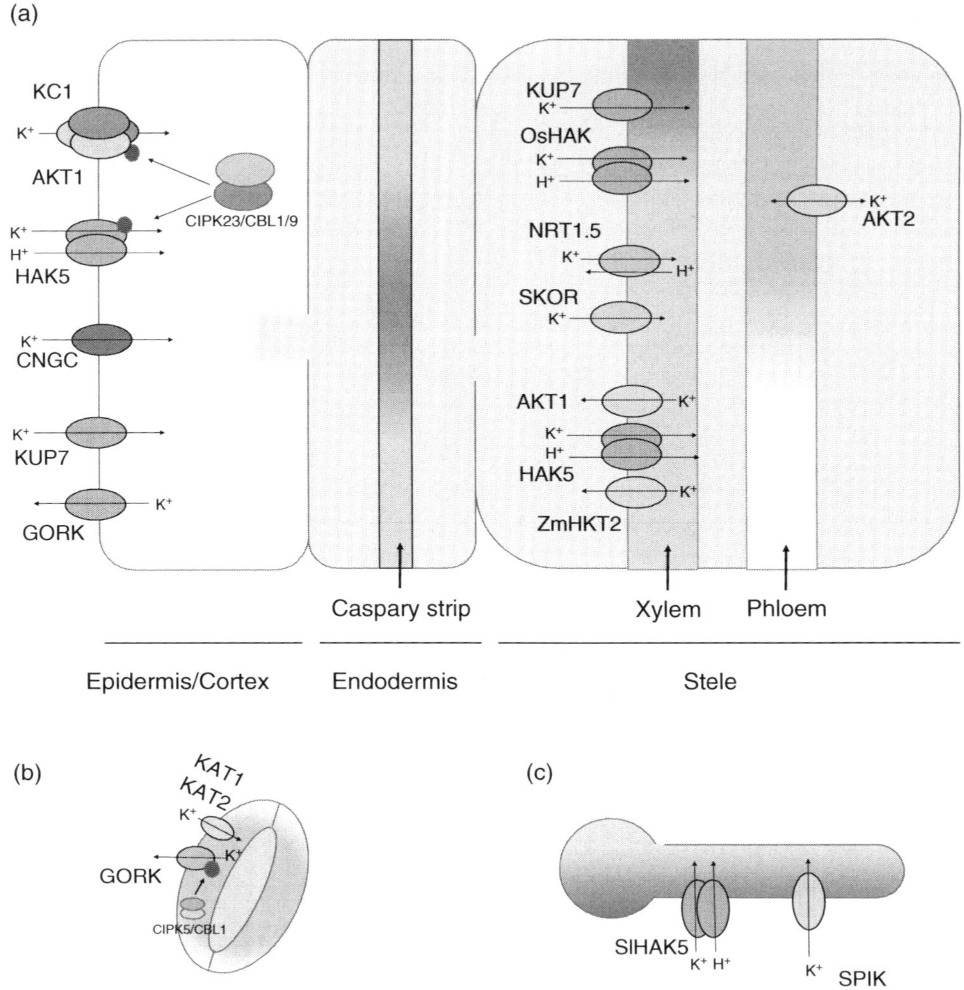

Figure 7.1 K^+ transport systems in roots and specialized cells. (a) K^+ is taken up at the root and enters the symplast of an epidermal or cortical cell by crossing a plasma membrane. Several transport systems mediate this first step of K^+ acquisition and the main actors are depicted in the figure. The inward rectifier channel AKT1 and the HAK5-type high-affinity transporter are the major contributors. When the external K^+ concentration is limiting, both of them are activated by phosphorylation (red circle) mediated by the CIPK23/CBL1-9 complex. Other systems such as KUP7 and unidentified systems, probably of the CNGC family, also participate in root K^+ uptake. K^+ can also be transported back to the external solution via the outward rectifier K^+ channel GORK. Once in the symplast, K^+ moves through plasmodesmata and reaches the stele where it is released in the xylems vessels via the outward rectifier K^+ channel SKOR and the NRT transporter NRT1.5. At this stage, other transport systems have been described to participate in regulating K^+ content into xylem vessels. The AKT1 channel retrieves K^+ from the xylem and AtHAK5 and KUP7 from Arabidopsis, and OsHAK1 and 5 from rice are involved in K^+ release into xylem. The maize ZmHKT2 transporter has been shown to mediate retrieval of K^+ from xylem. From the xylem, K^+ is distributed throughout the plant and it can be remobilized to other organs and back to the root via the phloem where the AKT2 channel mediates phloem K^+ load and unload. Some specialized cells are endowed with K^+ transport systems that are essential for their functionality. Guard cells (b) regulate plant transpiration according to their turgor which determines the aperture of stomatal pores. In order to open, guard cells increase their turgor by taking up K^+ via the inward rectifier channels KAT1 and KAT2, which drives water uptake. For its closing, guard cells lose turgor by moving K^+ out of guard cells via the outward rectifier GORK channel and KUP transporters which is followed by water efflux. The GORK channel is regulated by phosphorylation mediated by the CIPK5/CBL1 complex. Pollen grains (c) elongate their tubes in the pistil because of turgor increase driven by the uptake of K^+ trough the Arabidopsis inward rectifier K^+ channel SPIK and the tomato transporter SlHAK5 which is followed by water uptake.

in roots upon K^+ starvation. The first step of the signal cascade that leads to gene induction is the hyperpolarization of the plasma membrane because of K^+ limitation. Then, an increase in ROS followed by an increase in ethylene result in the upregulation of AtHAK5 (Jung et al. 2009). In addition to ethylene, other hormones such as jasmonic acid (JA), cytokinins, and gibberellins may play a role in the regulation of AtHAK5 transcription. Interestingly, DELLA proteins, known repressors of gibellerin signaling, have been related to the regulation of AtHAK5 (Santa-María et al. 2018). Several transcription factors have been described as involved in gene induction. Apart from AtHAK5, KUP12 expression also responds to K^+ and is downregulated in shoots after K^+ resupply to K^+-starved plants (Armengaud et al. 2004).

PostTranslational Regulation

Multimerization and Regulatory Subunits

Some transport proteins function as multimers. As mentioned, functional channels are homo- or heterotretamers, each monomer contributing with a K^+-selective pore. Heterotetramerization contributes to the regulation of channel activity, because it produces transport entities of different properties. Interestingly, the sensitivity to membrane voltage of AKT1-like K^+ channels is modified by the interaction with an additional subunit of KC1, which basically plays a regulatory role. AKT1 channels also interact with other regulatory proteins as the SNARE SYP121 protein (Nieves-Cordones et al. 2014; Honsbein et al. 2009).

Dimerization of transporters of the HAK family, such as KUP4/TRH1 (Daras et al. 2015), has been reported, and the possible role of dimerization on the regulation of the activity of members of this family is a hypothesis that needs further investigation.

Regulation by Phosphorylation

Ca^{2+} is a well-known second messenger in plant responses to different environments. Upon an external stimulus, the cytosolic Ca^{2+} concentration is increased and the magnitude of this increase as well as its duration and oscillation, the so-called Ca^{2+} signature, is a key piece in determining the specific response of the plant. Importantly, under low K^+ supply, two successive and different Ca^{2+} signals are produced. These two signals differ temporal and spatially. The first one is produced within 1 minute in the postmeristematic stellar tissue of the root elongation zone and the second one takes place several hours later in the elongation and root hair differentiation zones (Behera et al. 2017).

Ca^{2+} signals are perceived by Ca^{2+} sensors and effectors such as Calmodulin (CaM), Calmodulin-like proteins (CML), Ca^{2+}-dependent protein kinases (CDPKs), and calcineruin B-like proteins (CBLs) which operate in coordination with CBL-interacting protein kinase (CIPKs) (Saito and Uozumi 2020). The CBL-CIPK complexes play a central role in the regulation of plant responses to many different stresses, nutritional deficiency included, and especially in K^+ limitation. Thus, it has been shown that the K^+ channel AKT1 is activated upon its phosphorylation mediated by the CIPK23 kinase which interacts with the CBL1 or CBL9 Ca^{2+} sensors. When the plant is subjected to K^+ limitation, a specific Ca^{2+} signature is registered by CBL Ca^{2+} sensors. These Ca^{2+} sensors, located in the PM via myristoylation, have EF-hand motifs that bind Ca^{2+} and recruit CIPK23 to the plasma membrane. Then,

CIPK23 phosphorylates the AKT1 channel by activating it (Xu et al. 2006). Interestingly, the CIPK23 kinase and the KC1 subunit act synergistically to regulate AKT1 (Wang et al. 2016). Interestingly, it has been shown that CBL10 negatively regulates AKT1 by competing with the activating complex CIPK23 to bind AKT1 (Ren et al. 2013).

In addition to AKT1, the same CIPK23/CBL1 complex also activates the high-affinity K^+ transporter AtHAK5 by phosphorylation of a serine residue located in its N-terminus (S35). Phosphorylation of AtHAK5 induces the release of a C-terminal autoinhibitory domain, resulting in a more active transporter (Ródenas et al. 2021). The CIPK23 kinase also interacts with other CBLs such as CBL8, 9, and 10 to activate AtHAK5 (Ragel et al. 2015). Moreover, other kinases such as CIPK1 and CIPK9 have also been shown to activate AtHAK5 (Lara et al. 2020).

It has also been shown that CBL1/CIPK5 regulate the outward rectifier K^+ channel GORK, which is involved in K^+ efflux in stomata and from root cells (Förster et al. 2019)

In conclusion, combinations of CIPK/CBL may form different complexes to regulate root K^+ uptake through AKT1 and AtHAK5 and other systems such as GORK, contributing to maintaining cytosolic K^+ homeostasis.

Agriculture for the Future: K^+ Use Efficiency and Stress Tolerance

During the last decades, crops have been selected according to their yield under optimum conditions. However, current crop demand is moving toward plant species and varieties that require low nutrient inputs and are tolerant to abiotic stresses such as drought, salinity, or waterlogging (Bailey-Serres et al. 2019). Such change in plant breeding is due to side effects of extensive fertilizer use and the higher incidence of weather events associated with climate changes. Interestingly, K^+ has a robust protective effect against abiotic stresses and is a significant fertilizer component (Shabala and Pottosin 2014; Nieves-Cordones et al. 2020c). Thus, a better understanding of how K^+ is accumulated in plants according to the topics described earlier regarding the systems involved in K^+ transport and their regulation will be of strategic importance to breed future crops.

K^+ Use Efficiency

The use of K^+ by plants is conditioned by several factors that could be grouped into two major categories: K^+ transport and K^+ utilization (Moriconi and Santa-María 2013). K^+ transport takes into account the capacity of the plant to take up K^+ from the soil and move it within the plant. K^+ utilization considers the biomass production aspects as a function of K^+ content (Baligar and Fageria 2015).

K^+ transport can be decomposed in several transport processes: root K^+ uptake, long-distance K^+ transport in vascular tissues, and K^+ accumulation and remobilization within plant cells (Baligar and Fageria 2015). K^+ transport systems participating in these transport processes have been described earlier.

Root K^+ uptake is affected by internal (plant) and external (soil). Internal factors include anatomical (root density, root hair length, root depth, etc.) and physiological (such as those involved in the establishment of electrochemical gradients and function transport proteins). Soil factors affect the potassium availability for the plant, including soil moisture, temperature, pH, and composition. Long-distance K^+ transport in vascular tissues is regulated by a greater extent by internal factors (anatomical and physiological) than external ones, since they occur in inner plant tissues.

Coordination between K^+ transport processes is critical for delivering each plant organ. It has been shown that K^+ uptake and translocation are co-regulated (Nieves-Cordones et al. 2019c). Thus, the operation of K^+ transport systems must be balanced to obtain benefits in K^+ accumulation. However, information about this topic is scarce in crops. It should be necessary to identify the rate-limiting K^+ transport processes and the critical transport systems involved. With this regard, transport system features such as K^+ selectivity, affinity, transport rates, and sensitivity to cellular or external factors could be engineered to increase the plant K^+ transport capacity.

The comprehension of potassium uptake mechanisms is far more advanced than that of K^+ utilization ones (Santa-María et al. 2015). K^+ utilization factors include the rate of biochemical reactions at a given K^+ content, K^+ content in structural tissues and molecules, and K^+ substitution by other cations (Baligar and Fageria 2015). The study of ion transport systems can be of interest for K^+ substitution by other cations such as Ca^{2+}, Mg^{2+}, and Na^+. If vacuolar storage of these cations is enhanced, more K^+ will be available for fulfilling other physiological functions such as enzyme activator in the cytosol. Regarding this, several proteins are involved in Ca^{2+}, Mg^{2+}, and Na^+ transport, such as CNGC channels, CAX transporters, MGT/MRS transporters, and HKT transporters (Kaplan et al. 2007; Conn et al. 2011; Kobayashi and Tanoi 2015) and could be studied to understand K^+ substitution by other inorganic cations.

Abiotic Stress Affecting K^+ Homeostasis

Salinity

Salinity involves a high concentration of salts in the soil or the irrigation water. Na^+ is one of the most abundant cations under salinity (Munns and Tester 2008). Due to its chemical similarity to K^+, Na^+ strongly interferes with K^+ transport, and sufficient K^+ accumulation in plant cells prevents excessive Na^+ uptake. This idea can be applied to different plant tissues, particularly those that act as control points in Na^+/K^+ transport.

Root cells in the epidermis and cortex constitute the first control point for K^+ and Na^+ transport. At the protein level, Na^+ inhibits root K^+ uptake through HAK transporters, voltage-gated K^+ channels, and CNGCs (Rubio et al. 2020; Demidchik and Maathuis 2007). However, the inhibitory effect of Na^+ is not of the same magnitude in the three transport systems. HAK and voltage-gated K^+ channels exhibit a high selectivity of K^+ over Na^+, and external Na^+ concentration must be 1000-fold higher than that of K^+ to reduce K^+ uptake through these proteins (Rubio et al. 2020). In the case of CNGCs, permeability to Na^+ and K^+ is comparable in these channels (Demidchik and Maathuis 2007), so the inhibitory effect of Na^+ on K^+ uptake through CNGCs is more significant if present at high external concentrations.

Engineering transport systems with a higher K^+ over Na^+ selectivity will be of great interest to breed salt-tolerant crops. With this regard, the F130S mutation in the K^+ transporter AtHAK5 performed much better in yeast growing at 0.1 mM K^+/400 mM Na^+ than the native AtHAK5 transporter (Aleman et al. 2014).

Na^+ also affects the regulatory mechanisms of K^+ transport systems. Na^+ interferes with the upregulation of HAK5/1 transporters during K^+ deficiency (Nieves-Cordones et al. 2007; Rubio et al. 2008). As described earlier, one of the first effects of K^+ deficiency is the hyperpolarization of the root cells plasma membrane and root cell membrane potential correlates with HAK5/1 expression levels, being its expression higher at more negative membrane potentials (Nieves-Cordones et al. 2008; Rubio et al. 2014b). Therefore, the negative effect of Na^+ seems to be originated by the membrane depolarization induced by Na^+ (Nieves-Cordones et al. 2008; Rubio et al. 2014b). Interestingly, *Eutrema salsuginea,* a salt-tolerant relative of Arabidopsis, is much less affected by the presence of Na^+ than Arabidopsis in terms of HAK5 expression and K^+ uptake (Alemán et al. 2009). Thus, salt-tolerant relatives can be an exciting subject of study for improving the regulation of K^+ transport systems under salt stress.

Besides root K^+ uptake, xylem K^+ loading is another control point with a substantial impact on salinity tolerance. Xylem K^+ concentration positively correlates with plant salt tolerance (Alejandro et al. 2007; Albacete et al. 2009). Thus, the function of K^+ transport systems mediating xylem K^+ loading is of particular interest. As explained previously, SKOR and NRT1.5 are major K^+ transport systems in xylem parenchyma cells in Arabidopsis, and strategies based on their function could be used to breed salt-tolerant crops. Indeed, overexpression of the ortholog SKOR channel from melon improved growth in Arabidopsis under salinity (Long-Tang et al. 2018).

Concerning K^+ transport in the phloem, AKT2-like channels play an essential role in this process. AKT2 function has been proposed to be a relevant candidate to achieve plant salt tolerance by controlling K^+/Na^+ ratios in the phloem sap (Rubio et al. 2020). However, the effects of manipulating AKT2 activity have not been tested under saline conditions yet.

Drought

Due to the critical role that K^+ plays in controlling the cell osmotic potential, K^+ accumulation is of vital relevance for coping with drought conditions (Nieves-Cordones et al. 2019b). It has been shown that root K^+ transporters such as OsHAK1 and OsAKT1 are essential for rice plants under drought conditions (Ahmad et al. 2016; Chen et al. 2017). Overexpression of these transport systems gave rise to drought-tolerant rice plants, whereas loss-of-function mutants are drought-sensitive.

Guard cells are essential players in water saving under drought conditions since ~95% of the water that the plant takes up is lost by transpiration (Kramer 1983). K^+ transport in guard cells assists stomatal opening, and closing and the function of K^+ transport systems have been used to obtain drought-tolerant plants (Kwak et al. 2001; Nieves-Cordones et al. 2012; Papanatsiou et al. 2019). For example, disruption of the *Arabidopsis* inward-rectifying K^+ channel AKT1 allowed akt1 plants to cope better with drought stress (Nieves-Cordones et al. 2012). This phenotype was explained by the response of akt1 guard cells to the stress hormone ABA. *akt1* stomata closed faster than wild-type stomata, thereby reducing water loss during the first stages of drought stress.

Waterlogging

Waterlogging causes hypoxia in roots, and this gives rise to a decrease in cell energy levels (for example, the ATP levels). Among macronutrients, K^+ transport is the most affected by hypoxia (Colmer and Greenway 2011), indicating a high demand for metabolic energy. Plants tend to concentrate K^+ in roots by reducing K^+ transport to shoots to prevent intense K^+ deficiency in roots (Shabala and Pottosin 2014). Therefore, prolonged waterlogging strongly reduced the K^+ content of aerial parts of plants, and this effect seemed to be less pronounced in roots (Shabala and Pottosin 2014).

In Arabidopsis, the protein kinase CIPK25 is upregulated explicitly in the stele by waterlogging and regulates the K^+ channel AKT1 (Tagliani et al. 2020). The loss-of-function mutant *cipk25* was strongly affected by waterlogging (a more significant reduction in plant weight and K^+ content in shoots than WT plants was observed). However, overexpression of CIPK25 did not give rise to higher plant weight and K^+ content in overexpressors than in WT plants (Tagliani et al. 2020). Thus, although CIPK25 constitutes a central player in waterlogging responses by regulating AKT1-mediated transport, additional K^+ transport mechanisms should be taken into account to breed waterlogging-tolerant crops.

Toxic Ions

Several ion pollutants have been shown to be related to K^+ transport. The best-studied case is cesium (Cs^+) contamination (Nieves-cordones and Rubio 2021). Radioactive Cs^+ is released into the environment after accidents in nuclear power plants such as those of Chernobyl or Fukushima (Burger and Lichtscheidl 2018). Cs^+ is chemically similar to K^+, and it permeates through several K^+ transport systems such as HAKs, ABCs, or CNGCs. Thus, radioactive Cs^+ can enter the trophic chain by being taken up by plant roots (Burger and Lichtscheidl 2018). Then, radioactive Cs^+ is translocated to edible organs, and there is a potential health risk if consumed. Plants with a low capacity to accumulate Cs^+ have been developed, and they were obtained by inactivating HAK transporters that allow root Cs^+ uptake (Genies et al. 2017; Nieves-Cordones et al. 2017b, 2020b). However, the mutation in HAK transporters can have side effects on K^+ transport, since they contribute to K^+ nutrition (Genies et al. 2017; Nieves-Cordones et al. 2017b 2020b). Recently, two ABCGs transporters (ABCG33 and ABCG37) from Arabidopsis have been described as K^+-independent Cs^+ transporters (Ashraf et al. 2021). Therefore, they offer alternative transport systems to engineer plants with low or high Cs^+ accumulation capacity without distorting K^+ homeostasis. Additionally, the selectivity of HAKs transporters could be modified to either be more K^+- or Cs^+-selective (for hypo- or hyperaccumulation purposes, respectively) and reduce undesired effects on K^+ nutrition (Nieves-cordones and Rubio 2021).

Excess of heavy metals in soils also imposes a challenge to agriculture. Interestingly, several K^+ transport systems have been related to plant tolerance to heavy metals. Disruption of the HAK/KUP/KT transporter KUP8 from Arabidopsis allowed plants to exhibit improved growth concerning WT plants in a medium with a heavy metal mix of cadmium, chromium, and copper (Sanz-Fernández et al. 2021). Unexpectedly, the mutation in the KUP8 did not affect macro- and micronutrient accumulation in roots and leaves (Sanz-Fernández et al. 2021). Thus, the mechanism by which KUP8 improves plant growth under heavy metal stress remains to be elucidated. KO mutants in the K^+/H^+ antiporter NRT1.5 showed decreased cadmium accumulation in the xylem sap (Chen et al. 2012). Cadmium has not been described as a substrate for NRTs transporters, so the lower cadmium concentration in xylem sap may be originated by an indirect effect caused by the mutation in *NRT1.5*.

Biotic Stress Affecting K^+ Homeostasis

At the physiological level, K^+ nutrition has been linked to the impact of pathogen attacks (Amtmann et al. 2008; Shabala and Pottosin 2014). In some cases, K^+ deficiency can have a positive effect on plant defense mechanisms (Armengaud et al. 2010; Davis et al. 2018), while in others, it has the opposite effect (Davis et al. 2018). Recent studies have shed some light on the specific role of K^+ transport systems during pathogen attacks. In rice, an effector protein called AvrPiz-t from *Magnaporthe oryzae* targets the K^+ channel OsAKT1 (Shi et al. 2018). AvrPiz-t inhibits OsAKT1-mediated K^+ transport, possibly by competing with the regulatory kinase OsCIPK23. Therefore, AvrPiz-t action gives rise to lower K^+ content in plant cells and facilitates pathogen infection.

Stomata are a critical control point for pathogen access to inner plant tissues and for saving water in case of leaf damage. Upon leaf wounding, JA is produced, and it triggered stomatal closure (Förster et al. 2019). The CBL1-CIPK5 Ca^{2+} sensor-kinase complexes are activated by JA signaling and phosphorylated the K^+ channel GORK to give rise to K^+ efflux in guard cells. Guard cells therefore shrink and stomata closed.

More research is needed to better understand the role of K^+ transport systems under abiotic and biotic stress and in nutrient use efficiency in future years.

Biotechnological Approaches and Emerging Techniques for Crop Improvement

Since identifying the first K^+ transport systems in Arabidopsis plants in 1992, AKT1 (Sentenac et al. 1992), and KAT1 (Schachtman et al. 1992), many other K^+ transport systems have been described in a wide range of plant species. Today, we know that plants genomes harbor several gene families composed of many members that encode K^+ transport systems. Besides, many different mechanisms for the regulation of the transport systems have also been described. Scientists now face the challenge of using this great deal of information to generate crops with increased use efficiency of K^+, adapted to produce more with less input and to tolerate abiotic and biotic stress conditions. Recent technological advances allow approaching this challenge with potent tools which will result in improved crop varieties in a short time. In the following, we would like to highlight some of these approaches for this future task.

Models Versus Crops and Translational Research

Many pioneering studies that led to the identification of K^+ transport systems were done in Arabidopsis, as it is an excellent model system. These studies in the model species were fundamental for later identifying homologs in other plant species. It became soon evident that, although most of the characteristics of the Arabidopsis transporters were conserved in other plant species, important differences also existed. Thus, for example, while in Arabidopsis, both the high-affinity K^+ transporter AtHAK5 and the K^+ channel AKT1 contributed to K^+ uptake from diluted solutions, in rice plants, the OsHAK1 transporter dominated over the OsAKT1 channel (Nieves-Cordones et al. 2016a). Another example is provided by the different transcriptional regulation of the Arabidopsis AtHAK5 and its *Eutrema salsuginea* homolog EsHAK5 (Alemán et al. 2009). As mentioned earlier, the presence of Na^+ in the external solution importantly repressed the former but not the latter. The described differences arise from the biodiversity among plant species. Traits of interest can be transferred from one species or variety to a different one for crop improvement. Unfortunately, this variability has been lost through the domestication of wild plants and breeding programs, which have notably reduced the genetic variability of major crops. Thus, traits of interest for tolerance to stress conditions such as K^+ deficiency or salinity may have been lost in major crops, but they may still be present in neglected crops or wild relatives. Moreover, from the knowledge obtained with fundamental research, genetic diversity can be created in the lab by mutagenesis. Ultimately, translational research is required for transferring the alleles of interest, either found in natural variation or generated in the lab, to crops of interest (Jacob et al. 2018) (Figure 7.2).

Natural Variation Exploitation

Classical genetics allowed the identification of the genotype responsible for a phenotype. The identification could be done in a recombinant population through linkage mapping or by association mapping in natural populations. Natural populations can be regarded as experimental designs performed over thousands of years and provide enormous variants for millions of loci. Genotype identification has importantly been enhanced by high-throughput genotyping. The natural variation underlying traits of interest can now be described thanks to massive sequencing and computational methods in Genome-Wide

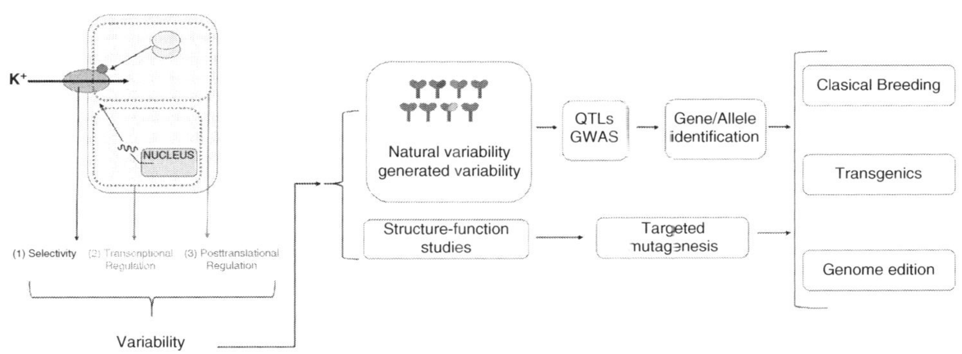

Figure 7.2 Workflow for the generation of crops with increased K^+ use efficiency. Basic research generates knowledge on the systems involved in K^+ transport across cell membranes and their regulation. Important properties such as (1) selectivity, (2) transcriptional regulation of the genes encoding them or mechanisms of regulation of their activity by (3) posttranscriptional regulation are identified. Variability in these three processes may constitute important tools for the improvement of crops. This variability may be present in wild relatives of crops, in neglected crops or in collections generated by crossing parental lines with different properties. Identifying relevant traits and alleles of interest may be then achieved by QTLs study or Genome Wide Association Studies (GWAS). Another approach to generate variability is targeted mutagenesis of functional residues of transport systems and their regulators identified in structure-function studies. Once the allele of interest is identified, it can be introduced into the crop or variety of interest by classical breeding, transgenic breeding, or genome edition.

Association Studies (GWAS) (Figure 7.2). This approach complements classical biparental cross mapping and allows the investigation of multiple traits simultaneously. In addition, high-throughput technologies to determine diversity in gene expression levels (transcriptome), DNA methylation (methylome), or metabolite accumulation (metabolome), which importantly contribute to determining the phenotype, could be used for the identification of traits of interest. All this information will assist scientists to mine natural variation and associate phenotypes with causal genes. This will help develop crops by genome-wide selection (Huang and Han 2014; Liang et al. 2021) (Figure 7.2).

New Alleles Generated in the Lab

As mentioned earlier, genetic diversity can be created in the lab by mutagenesis approaches. In some cases, targeted mutagenesis can be used to create specific alleles (Figure 7.2). For such an approach, previous studies are needed to generate the required information for designing the mutations to be obtained. Either random or site-directed mutagenesis can generate mutants that are later characterized in heterologous systems such as yeast or Xenopus oocytes. An example of this is the structure-function studies performed in the Arabidopsis AtHAK5, which identified protein residues with potential for function improvement. Several domains have been described in AtHAK5, containing residues essential for K^+ transport and selectivity (Ródenas et al. 2021). Mutations in these domains produce transporters with altered selectivity, which may be of interest. An example of this, the F130S mutation in AtHAK5, mentioned earlier, results in a transporter with increased selectivity for K^+ over Na^+ and Cs^+ (Aleman et al. 2014). Other mutations in AtHAK5 partially remove the C-terminal tail of the protein and produce a constitutively active transporter that mediates K^+ transport at maximal rate even in the absence of the regulatory AtCIPK23/AtCBL1 complex (Ródenas et al. 2021). In some cases, mutants that produced nonfunctional transporters may be of interest. This is the case of KO mutants in the gene encoding the K^+ transporter OsHAK1 of rice (Nieves-Cordones et al. 2017a) or SlHAK5 of tomato (Nieves-Cordones et al. 2020b) that result in plants with a lower accumulation of Cs^+. In the case of the *slhak5*, parthenocarpic fruits are also produced.

Genome Editing

Once the allele to be generated has been identified from any of the earlier-described approaches, the next step is to introduce it into a crop or variety of interest. This is a complex and time-consuming task that can be tackled by different approaches such as cross-breeding, mutational breeding, transgenic breeding, and breeding by genome edition. The recent development of the CRISPR-based genome edition technology speeds up this process. CRISPR-based edition allows to specifically modify a DNA sequence to produce the mutant of interest (Figure 7.2). The technology is based on targeted double-strand breaks (DSBs) of genomic DNA generated by the Cas9 endonuclease, which is directed by a single-guide RNA (sgRNA) that base pairs with the DNA target. The cellular machinery then repairs the broken DNA. If a donor DNA containing the desired mutation is provided, DNA repair may occur through homologous recombination (HR), resulting in introducing the specific mutation in the genomic DNA. Unfortunately, HR is rare in plants. In the absence of donor DNA, the broken DNA is repaired by nonhomologous end joining (NHEJ), resulting in insertion and deletions (INDELs) that ultimately modify the reading framing of coding sequences and result in premature stop codons and knockout mutants. Improved editors such as base editors and prime editors have recently been produced. Base editors consist of a modified Cas9 (D10A) and a single-strand deaminase that produce transitions at single-strand breaks generated by the Cas9 at specific points. Prime editors combine a modified Cas9 with a reverse transcriptase that synthesizes a donor DNA from a primer editing guide RNA (pegRNA), leading to introduction of the desired mutation. Gene edition is rapidly evolving, and improved systems are continuously generated. They could be used to generate specific mutants and, for example, circumvent linkage, or for pyramiding traits in a multiplex edition. Attractive approaches are now feasible, such as *de novo* domestication of wild relatives, mutant saturation of a gene of interest, direct protein evolution, or synthetic biology (Hua et al. 2019; Zhu et al. 2020).

References

Ahmad, I. and Maathuis, F.J.M. (2014). Cellular and tissue distribution of potassium: physiological relevance, mechanisms and regulation. *J. Plant Physiol.* 171: 708–714.

Ahmad, I., Mian, A., and Maathuis, F.J.M. (2016). Overexpression of the rice AKT1 potassium channel affects potassium nutrition and rice drought tolerance. *J. Exp. Bot.* 67: 2689–2698.

Albacete, A., MartÍnez-AndÚjar, C., Ghanem, M.E. et al. (2009). Rootstock-mediated changes in xylem ionic and hormonal status are correlated with delayed leaf senescence, and increased leaf area and crop productivity in salinized tomato. *Plant Cell Environ.* 32: 928–938.

Alejandro, S., Rodríguez, P.L., Bellés, J.M. et al. (2007). An Arabidopsis quiescin-sulfhydryl oxidase regulates cation homeostasis at the root symplast-xylem interface. *EMBO J.* 26: 3203–3215.

Aleman, F., Caballero, F., Ródenas, R. et al. (2014). The F130S point mutation in the Arabidopsis high-affinity K^+ transporter AtHAK5 increases K^+ over Na^+ and Cs^+ selectivity and confers Na^+ and Cs^+ tolerance to yeast under heterologous expression. *Front. Plant Sci.* 5.

Alemán, F., Nieves-Cordones, M., Martínez, V., and Rubio, F. (2009). Differential regulation of the HAK5 genes encoding the high-affinity K^+ transporters of thellungiella halophila and Arabidopsis thaliana. *Environ. Exp. Bot.* 65: 263–269.

Amtmann, A., Troufflard, S., and Armengaud, P. (2008). The effect of potassium nutrition on pest and disease resistance in plants. *Physiol. Plant.* 133: 682–691.

Armengaud, P., Breitling, R., and Amtmann, A. (2010). Coronatine-insensitive 1 (COI1) mediates transcriptional responses of Arabidopsis thaliana to external potassium supply. *Mol. Plant* 3: 390–405.

Armengaud, P., Breitling, R., and Amtmann, A. (2004). The potassium-dependent transcriptome of Arabidopsis reveals a prominent role of jasmonic acid in nutrient signaling. *Plant Physiol.* 136: 2556–2576.

Ashraf, M.A., Akihiro, T., Ito, K. et al. (2021). ATP Binding Cassette Proteins ABCG37 and ABCG33 function as potassium- independent cesium uptake carriers in Arabidopsis roots. *Mol. Plant* 14: 664–678.

Bailey-Serres, J., Parker, J.E., Ainsworth, E.A. et al. (2019). Genetic strategies for improving crop yields. *Nature* 575: 109–118.

Baligar, V.C. and Fageria, N.K. (2015). Nutrient use efficiency in plants: an overview. *Nutr. Use Effic. From Basics to Adv.* 32: 1–14.

Becker, D. et al. (2004). AtTPK4, an Arabidopsis tandem-pore K^+ channel, poised to control the pollen membrane voltage in a pH- and Ca^{2+}-dependent manner. *Proc. Natl. Acad. Sci. U. S. A.* 101: 15621–15626.

Behera, S., Long, Y., Schmitz-Thom, I. et al. (2017). Two spatially and temporally distinct Ca^{2+} signals convey Arabidopsis thaliana responses to K^+ deficiency. *New Phytol.* 213: 727–738.

Benito, B., Garciadeblas, B., and Rodriguez-Navarro, A. (2012). HAK transporters from physcomitrella patens and Yarrowia lipolytica mediate sodium uptake. *Plant Cell Physiol.* 53: 1117–1123.

Berthomieu, P. et al. (2003). Functional analysis of AtHKT1 in Arabidopsis shows that Na^+ recirculation by the phloem is crucial for salt tolerance. *EMBO J.* 22: 2004–2014.

De Boer, A.H. (1999). Potassium translocation into the root xylem. *Plant Biol.* 1: 36–45.

Britto, D.T. and Kronzucker, H.J. (2008). Cellular mechanisms of potassium transport in plants. *Physiol. Plant.* 133: 637–650.

Burger, A. and Lichtscheidl, I. (2018). Stable and radioactive cesium: a review about distribution in the environment, uptake and translocation in plants, plant reactions and plants' potential for bioremediation. *Sci. Total Environ.* 618: 1459–1485.

Caballero, F., Botella, M.A., Rubio, L. et al. (2012). A Ca^{2+}-sensitive system mediates low-affinity K^+ uptake in the absence of AKT1 in arabidopsis plants. *Plant Cell Physiol.* 53: 2047–2059.

Cao, Y., Liang, X., Yin, P. et al. (2019). A domestication-associated reduction in K^+ -preferring HKT transporter activity underlies maize shoot K^+ accumulation and salt tolerance. *New Phytol.* 222: 301–317.

Chen, C.Z., Lv, X.F., Li, J.Y. et al. (2012). Arabidopsis NRT1.5 is another essential component in the regulation of nitrate reallocation and stress tolerance. *Plant Physiol.* 159: 1582–1590.

Chen, G., Hu, Q., Luo, L. et al. (2015). Rice potassium transporter OsHAK1 is essential for maintaining potassium-mediated growth and functions in salt tolerance over low and high potassium concentration ranges. *Plant Cell Environ.* 38: 2747–2765.

Chen, G., Liu, C., Gao, Z. et al. (2017). Oshak1, a high-affinity potassium transporter, positively regulates responses to drought stress in rice. *Front. Plant Sci.* 8: 1–17.

Colmer, T.D. and Greenway, H. (2011). Ion transport in seminal and adventitious roots of cereals during O_2 deficiency. *J. Exp. Bot.* 62: 39–57.

Conn, S.J., Gilliham, M., Athman, A. et al. (2011). Cell-specific vacuolar calcium storage mediated by CAX1 regulates apoplastic calcium concentration, gas exchange, and plant productivity in Arabidopsis. *Plant Cell* 23: 240–257.

Corratgé-Faillie, C., Jabnoune, M., Zimmermann, S. et al. (2010). Potassium and sodium transport in non-animal cells: the Trk/Ktr/HKT transporter family. *Cell. Mol. Life Sci.* 67: 2511–2532.

Daras, G., Rigas, S., Tsitsekian, D. et al. (2015). Potassium transporter TRH1 subunits assemble regulating root-hair elongation autonomously from the cell fate determination pathway. *Plant Sci.* 231: 131–137.

Davenport, R.J., Muñoz-Mayor, A., Jha, D. et al. (2007). The Na^+ transporter AtHKT1;1 controls retrieval of Na^+ from the xylem in Arabidopsis. *Plant Cell Environ.* 30: 497–507.

Davies, C., Shin, R., Liu, W. et al. (2006). Transporters expressed during grape berry (Vitis vinifera L.) development are associated with an increase in berry size and berry potassium accumulation. *J. Exp. Bot.* 57: 3209–3216.

Davis, J.L., Armengaud, P., Larson, T.R. et al. (2018). Contrasting nutrient–disease relationships: potassium gradients in barley leaves have opposite effects on two fungal pathogens with different sensitivities to jasmonic acid. *Plant Cell Environ.* 41: 2357–2372.

Demidchik, V. and Maathuis, F.J.M. (2007). Physiological roles of nonselective cation channels in plants: from salt stress to signalling and development. *New Phytol.* 175: 387–404.

Dietrich, P., Moeder, W., and Yoshioka, K. (2020). Plant cyclic nucleotide-gated channels: new insights on their functions and regulation1[OPEN]. *Plant Physiol.* 184: 27–38.

Drain, A., Thouin, J., Wang, L. et al. (2020). Functional characterization and physiological roles of the single Shaker outward K^+ channel in Medicago truncatula. *Plant J.* 102: 1249–1265.

Dreyer, I., Gomez-Porras, J.L., and Riedelsberger, J. (2017). The potassium battery: a mobile energy source for transport processes in plant vascular tissues. *New Phytol.* 216: 1049–1053.

Förster, S. et al. (2019). Wounding-induced stomatal closure requires jasmonate-mediated activation of GORK K^+ channels by a Ca^{2+} sensor-kinase CBL1-CIPK5 complex. *Dev. Cell* 48: 87–99.e6.

Fu, H.H. and Luan, S. (1998). AtKUP1: a dual-affinity K^+ transporter from arabidopsis. *Plant Cell* 10: 63–73.

Gajdanowicz, P., Michard, E., Sandmann, M. et al. (2011). Potassium (K^+) gradients serve as a mobile energy source in plant vascular tissues. *Proc. Natl. Acad. Sci. U. S. A.* 108: 864–869.

Garciadeblas, B., Barrero-Gil, J., Benito, B., and Rodríguez-Navarro, A. (2007). Potassium transport systems in the moss Physcomitrella patens: pphak1 plants reveal the complexity of potassium uptake. *Plant J.* 52: 1080–1093.

Garciadeblas, B., Benito, B., and Rodríguez-Navarro, A. (2002). Molecular cloning and functional expression in bacteria of the potassium transporters CnHAK1 and CnHAK2 of the seagrass Cymodocea nodosa. *Plant Mol. Biol.* 50: 623–633.

Gaymard, F., Pilot, G., Lacombe, B. et al. (1998). Identification and disruption of a plant shaker-like outward channel involved in K^+ release into the xylem sap. *Cell* 94: 647–655.

Genies, L., Orjollet, D., Carasco, L. et al. (2017). Uptake and translocation of cesium by Arabidopsis thaliana in hydroponics conditions: links between kinetics and molecular mechanisms. *Environ. Exp. Bot.* 138: 164–172.

Gobert, A., Park, G., Amtmann, A. et al. (2006). Arabidopsis thaliana cyclic nucleotide gated channel 3 forms a non-selective ion transporter involved in germination and cation transport. *J. Exp. Bot.* 57: 791–800.

Han, M., Wu, W., Wu, W.H., and Wang, Y. (2016). Potassium transporter KUP7 is involved in K^+ acquisition and translocation in Arabidopsis root under K^+-limited conditions. *Mol. Plant* 9: 437–446.

Hawkesford, M., Horst, W., Kichey, T. et al. (2012a). Chapter 6 - Functions of Macronutrients. In: *Marschner's Mineral Nutrition of Higher Plants*, 3e (ed. P. Marschner), 135–189. San Diego: Academic Press.

Hirsch, R.E., Lewis, B.D., Spalding, E.P., and Sussman, M.R. (1998). A role for the AKT1 potassium channel in plant nutrition. *Science* (80) 280: 918–921.

Honsbein, A., Sokolovski, S., Grefen, C. et al. (2009). A tripartite SNARE-K^+ channel complex mediates in channel-dependent K^+ nutrition in arabidopsis. *Plant Cell* 21: 2859–2877.

Hosy, E. et al. (2003). The Arabidopsis outward K^+ channel GORK is involved in regulation of stomatal movements and plant transpiration. *Proc. Natl. Acad. Sci. U. S. A.* 100: 5549–5554.

Hua, K., Zhang, J., Botella, J.R. et al. (2019). Perspectives on the application of genome-editing technologies in crop breeding. *Mol. Plant* 12: 1047–1059.

Huang, X. and Han, B. (2014). Natural variations and genome-wide association studies in crop plants. *Annu. Rev. Plant Biol.* 65: 531–551.

Ivashikina, N., Becker, D., Ache, P. et al. (2001). K^+ channel profile and electrical properties of Arabidopsis root hairs. *FEBS Lett.* 508: 463–469.

Jacob, P., Avni, A., and Bendahmane, A. (2018). Translational research: exploring and creating genetic diversity. *Trends Plant Sci.* 23: 42–52.

Jegla, T., Busey, G., and Assmann, S.M. (2018). Evolution and structural characteristics of plant voltage-gated K^+ channels. *Plant Cell* 30: 2898–2909.

Jung, J.Y., Shin, R., and Schachtman, D.P. (2009). Ethylene mediates response and tolerance to potassium deprivation in arabidopsis. *Plant Cell* 21: 607–621.

Kaplan, B., Sherman, T., and Fromm, H. (2007). Cyclic nucleotide-gated channels in plants. *FEBS Lett.* 581: 2237–2246.

Kim, E.J., Kwak, J.M., Uozumi, N., and Schroeder, J.I. (1998). AtKUP1: an arabidopsis gene encoding high-affinity potassium transport activity. *Plant Cell* 10: 51–62.

Kobayashi, N.I. and Tanoi, K. (2015). Critical issues in the study of magnesium transport systems and magnesium deficiency symptoms in plants. *Int. J. Mol. Sci.* 16: 23076–23093.

Kramer, P.J. (1983). 11 -). Transpiration. In: *Water Relations of Plants* (ed. P.J. Kramer), 291–341. Academic Press.

Kwak, J.M., Murata, Y., Baizabal-Aguirre, V.M. et al. (2001). Dominant negative guard cell K^+ channel mutants reduce inward-rectifying K^+ currents and light-induced stomatal opening in arabidopsis. *Plant Physiol.* 127: 473–485.

Lacombe, B., Pilot, G., Michard, E. et al. (2000). A shaker-like K^+ channel with weak rectification is expressed in both source and sink phloem tissues of Arabidopsis. *Plant Cell* 12: 837–851.

Lara, A., Ródenas, R., Andrés, Z. et al. (2020). Arabidopsis K^+ transporter HAK5-mediated high-affinity root K^+ uptake is regulated by protein kinases CIPK1 and CIPK9. *J. Exp. Bot.* 71: 5053–5060.

Lebaudy, A., Pascaud, F., Véry, A.A. et al. (2010). Preferential KAT1-KAT2 heteromerization determines inward K^+ current properties in Arabidopsis guard cells. *J. Biol. Chem.* 285: 6265–6274.

Leng, Q., Mercier, R.W., Yao, W., and Berkowitz, G.A. (1999). Cloning and first functional characterization of a plant cyclic nucleotide-gated cation channel. *Plant Physiol.* 121: 753–761.

Li, H., Yu, M., Du, X.Q. et al. (2017). NRT1.5/NPF7.3 functions as a proton-coupled H^+/K^+ antiporter for K^+ loading into the xylem in arabidopsis. *Plant Cell* 29: 2016–2026.

Li, X., Borsics, T., Harrington, H.M., and Christopher, D.A. (2005). Arabidopsis AtCNGC10 rescues potassium channel mutants of E. coli, yeast and Arabidopsis and is regulated by calcium/calmodulin and cyclic GMP in E. coli. *Funct. Plant Biol.* 32: 643–653.

Liang, Y., Liu, H.-J., Yan, J., and Tian, F. (2021). Natural variation in crops: realized understanding, continuing promise. *Annu. Rev. Plant Biol.* 72: 1–29.

Long-Tang, H., Li-Na, Z., Li-Wei, G. et al. (2018). Constitutive expression of CmSKOR, an outward K^+ channel gene from melon, in Arabidopsis thaliana involved in saline tolerance. *Plant Sci.* 274: 492–502.

Luan, M., Tang, R.J., Tang, Y. et al. (2017). Transport and homeostasis of potassium and phosphate: limiting factors for sustainable crop production. *J. Exp. Bot.* 68: 3091–3105.

Marschner, P. and Rengel, Z. (2012). Chapter 12 - Nutrient availability in soils. In: *Marschner's Mineral Nutrition of Higher Plants*, 3e (ed. P. Marschner), 315–330. San Diego: Academic Press.

Mäser, P., Thomine, S., Schroeder, J.I. et al. (2001). Phylogenetic relationships within cation transporter families of Arabidopsis. *Plant Physiol.* 126: 1646–1667.

Michard, E., Dreyer, I., Lacombe, B. et al. (2005). Inward rectification of the AKT2 channel abolished by voltage-dependent phosphorylation. *Plant J.* 44: 783–797.

Michard, E., Simon, A.A., Tavares, B. et al. (2017). Signaling with ions: the keystone for apical cell growth and morphogenesis in pollen tubes. *Plant Physiol.* 173: 91–111.

Moriconi, J.I. and Santa-María, G.E. (2013). A theoretical framework to study potassium utilization efficiency in response to withdrawal of potassium. *J. Exp. Bot.* 64: 4289–4299.

Mouline, K., Véry, A.A., Gaymard, F. et al. (2002). Pollen tube development and competitive ability are impaired by disruption of a Shaker K^+ channel in Arabidopsis. *Genes Dev.* 16: 339–350.

Munns, R. and Tester, M. (2008). Mechanisms of salinity tolerance. *Annu. Rev. Plant Biol.* 59: 651–681.

Nieves-Cordones, M., Alemán, F., Martínez, V., and Rubio, F. (2014). K^+ uptake in plant roots. The systems involved, their regulation and parallels in other organisms. *J. Plant Physiol.* 171: 688–695.

Nieves-Cordones, M., Andrianteranagna, M., Cuéllar, T. et al. (2019a). Characterization of the grapevine shaker K^+ channel VvK3.1 supports its function in massive potassium fluxes necessary for berry potassium loading and pulvinus-actuated leaf movements. *New Phytol.* 222: 286–300.

Nieves-Cordones, M., Caballero, F., Martínez, V., and Rubio, F. (2012). Disruption of the arabidopsis thaliana inward-rectifier K^+ channel AKT1 improves plant responses to water stress. *Plant Cell Physiol.* 53: 423–432.

Nieves-Cordones, M., García-Sánchez, F., Pérez-Pérez, J.G. et al. (2019b). Coping with water shortage: an update on the role of K^+, Cl^-, and water membrane transport mechanisms on drought resistance. *Front. Plant Sci.* 10: 1–9.

Nieves-Cordones, M., Lara, A., Ródenas, R. et al. (2019c). Modulation of K^+ translocation by AKT1 and AtHAK5 in Arabidopsis plants. *Plant Cell Environ.* 42: 2357–2371.

Nieves-Cordones, M., Lara, A., Silva, M. et al. (2020a). Root high-affinity K(+) and Cs(+) uptake and plant fertility in tomato plants are dependent on the activity of the high-affinity K(+) transporter SlHAK5. *Plant Cell Environ.* 43: 1707–1721.

Nieves-Cordones, M., Lara, A., Silva, M. et al. (2020b). Root high-affinity K+ and Cs+ uptake and plant fertility in tomato plants are dependent on the activity of the high-affinity K+ transporter SlHAK5. *Plant Cell Environ.* 43: 1707–1721.

Nieves-Cordones, M., Martínez-Cordero, M.A., Martínez, V., and Rubio, F. (2007). An NH4+-sensitive component dominates high-affinity K+ uptake in tomato plants. *Plant Sci.* 172: 273–280.

Nieves-Cordones, M., Martínez, V., Benito, B., and Rubio, F. (2016a). Comparison between Arabidopsis and rice for main pathways of K+ and Na+ uptake by roots. *Front. Plant Sci.* 7.

Nieves-Cordones, M., Miller, A.J., Alemán, F. et al. (2008). A putative role for the plasma membrane potential in the control of the expression of the gene encoding the tomato high-affinity potassium transporter HAK5. *Plant Mol. Biol.* 68: 521–532.

Nieves-Cordones, M., Mohamed, S., Tanoi, K. et al. (2017a). Production of low-Cs(+) rice plants by inactivation of the K(+) transporter OsHAK1 with the CRISPR-cas system. *Plant J.* 92: 43–56.

Nieves-Cordones, M., Mohamed, S., Tanoi, K. et al. (2017b). Production of low-Cs^+ rice plants by inactivation of the K^+ transporter OsHAK1 with the CRISPR-cas system. *Plant J.* 92: 43–56.

Nieves-Cordones, M., Ródenas, R., Chavanieu, A. et al. (2016b). Uneven HAK/KUP/KT protein diversity among angiosperms: species distribution and perspectives. *Front. Plant Sci.* 7: 1–7.

Nieves-cordones, M. and Rubio, F. (2021). The quest for selective Cs^+ transport in plants. *Mol. Plant* 14 (4): 552–554.

Nieves-Cordones, M., Rubio, F., and Santa-María, G.E. (2020c). Editorial: nutrient use-efficiency in plants: an integrative approach. *Front. Plant Sci.* 11: 10–12.

Nieves-Cordones, M., Al Shiblawi, F.R., and Sentenac, H. (2016c). Roles and transport of sodium and potassium in plants BT. In: *The Alkali Metal Ions: Their Role for Life* (ed. A. Sigel, H. Sigel and R.K.O. Sigel), 291–324. Cham: Springer International Publishing.

Osakabe, Y., Arinaga, N., and Umezawa, T. (2013). et al, Osmotic stress responses and plant growth controlled by potassium transporters in Arabidopsis C W. *Plant Cell* 25 (2): 609–624.

Papanatsiou, M., Petersen, J., Henderson, L. et al. (2019). Optogenetic manipulation of stomatal kinetics improves carbon assimilation,water use, and growth. *Science* 363: 1456–1459.

Pilot, G., Lacombe, B., Gaymard, F. et al. (2001). Guard cell inward K^+ channel activity in Arabidopsis involves expression of the twin channel subunits KAT1 and KAT2. *J. Biol. Chem.* 276: 3215–3221.

Pilot, G., Pratelli, R., Gaymard, F. et al. (2003). Five-group distribution of the Shaker-like K^+ channel family in higher plants. *J. Mol. Evol.* 56: 418–434.

Platten, J.D. et al. (2006). Nomenclature for HKT transporters, key determinants of plant salinity tolerance. *Trends Plant Sci.* 11: 372–374.

Ragel, P., Raddatz, N., Leidi, E.O. et al. (2019). Regulation of K^+ nutrition in plants. *Front. Plant Sci.* 10.

Ragel, P., Ródenas, R., García-Martín, E. et al. (2015). The CBL-interacting protein kinase CIPK23 regulates HAK5-mediated high-affinity K^+ uptake in Arabidopsis roots. *Plant Physiol.* 169: 2863–2873.

Reintanz, B., Szyroki, A., Ivashikina, N. et al. (2002). AtKC1, a silent Arabidopsis potassium channel α-subunit modulates root hair K^+ influx. *Proc. Natl. Acad. Sci. U. S. A.* 99: 4079–4084.

Ren, X.L., Qi, G.N., Feng, H.Q. et al. (2013). Calcineurin B-like protein CBL10 directly interacts with AKT1 and modulates K^+ homeostasis in Arabidopsis. *Plant J.* 74: 258–266.

Rengel, Z. and Damon, P.M. (2008). Crops and genotypes differ in efficiency of potassium uptake and use. *Physiol. Plant.* 133: 624–636.

Rigas, S., Debrosses, G., Haralampidis, K. et al. (2001). TRH1 encodes a potassium transporter required for tip growth in Arabidopsis root hairs. *Plant Cell* 13: 139–151.

Rigas, S., Ditengou, F.A., Ljung, K. et al. (2013). Root gravitropism and root hair development constitute coupled developmental responses regulated by auxin homeostasis in the Arabidopsis root apex. *New Phytol.* 197: 1130–1141.

Ródenas, R., Ragel, P., Nieves-Cordones, M. et al. (2021). Insights into the mechanisms of transport and regulation of the Arabidopsis high-affinity K^+ transporter HAK5. *Plant Physiol.* 1860–1874.

Rodríguez-Navarro, A. (2000). Potassium transport in fungi and plants. *Biochim. Biophys. Acta Rev. Biomembr.* 1469: 1–30.

Rodríguez-Navarro, A. and Rubio, F. (2006). High-affinity potassium and sodium transport systems in plants. *J. Exp. Bot.* 57: 1149–1160.

Rubio, F., Fon, M., Rodenas, R. et al. (2014a). A low K^+ signal is required for functional high-affinity K^+ uptake through HAK5 transporters. *Physiol. Plant.* 152: 558–570.

Rubio, F., Fon, M., Ródenas, R. et al. (2014b). A low K^+ signal is required for functional high-affinity K^+ uptake through HAK5 transporters. *Physiol. Plant.* 152: 558–570.

Rubio, F., Nieves-Cordones, M., Alemán, F., and Martínez, V. (2008). Relative contribution of AtHAK5 and AtAKT1 to K^+ uptake in the high-affinity range of concentrations. *Physiol. Plant.* 134: 598–608.

Rubio, F., Nieves-Cordones, M., Horie, T., and Shabala, S. (2020). Doing 'business as usual' comes with a cost: evaluating energy cost of maintaining plant intracellular K^+ homeostasis under saline conditions. *New Phytol.* 225: 1097–1104.

Saito, S. and Uozumi, N. (2020). Calcium-regulated phosphorylation systems controlling uptake and balance of plant nutrients. *Front. Plant Sci.* 11.

Santa-María, G.E., Moriconi, J.I., and Oliferuk, S. (2015). Internal efficiency of nutrient utilization: what is it and how to measure it during vegetative plant growth? *J. Exp. Bot.* 66: 3011–3018.

Santa-María, G.E., Oliferuk, S., and Moriconi, J.I. (2018). KT-HAK-KUP transporters in major terrestrial photosynthetic organisms: a twenty years tale. *J. Plant Physiol.* 226: 77–90.

Sanz-Fernández, M., Rodríguez-González, A., Sandalio, L.M., and Romero-Puertas, M.C. (2021). Role of potassium transporter KUP8 in plant responses to heavy metals. *Physiol. Plant.* 173: 1–11.

Schachtman, D.P., Schroeder, J.I., Lucas, W.J. et al. (1992). Expression of an inward-rectifying potassium channel by the Arabidopsis KAT1 cDNA. *Science* 258: 1654–1658.

Scherzer, S., Böhm, J., Krol, E. et al. (2015). Calcium sensor kinase activates potassium uptake systems in gland cells of Venus flytraps. *Proc. Natl. Acad. Sci. U. S. A.* 112: 7309–7314.

Sentenac, H., Bonneaud, N., Minet, M. et al. (1992). Cloning and expression in yeast of a plant potassium ion transport system. *Science* 256: 663–665.

Shabala, S. and Cuin, T.A. (2008). Potassium transport and plant salt tolerance. *Physiol. Plant.* 133: 651–669.

Shabala, S. and Pottosin, I. (2014). Regulation of potassium transport in plants under hostile conditions: implications for abiotic and biotic stress tolerance. *Physiol. Plant.* 151: 257–279.

Shen, Y., Shen, L., Shen, Z. et al. (2015). The potassium transporter OsHAK21 functions in the maintenance of ion homeostasis and tolerance to salt stress in rice. *Plant Cell Environ.* 38: 2766–2779.

Shi, X., Long, Y., He, F. et al. (2018). The fungal pathogen Magnaporthe oryzae suppresses innate immunity by modulating a host potassium channel. *PLoS Pathog.* 14: 1–21.

Su, H., Golldack, D., Katsuhara, M. et al. (2001). Expression and stress-dependent induction of potassium channel transcripts in the common ice plant. *Plant Physiol.* 125: 604–614.

Sunarpi, Horie, T., Motoda, J. et al. (2005). Enhanced salt tolerance mediated by AtHKT1 transporter-induced Na^+ unloading from xylem vessels to xylem parenchyma cells. *Plant J.* 44: 928–938.

Sze, H. and Chanroj, S. (2018). Plant endomembrane dynamics: studies of K^+/H^+ antiporters provide insights on the effects of pH and Ion homeostasis. *Plant Physiol.* 177: 875–895.

Tagliani, A., Tran, A.N., Novi, G. et al. (2020). The calcineurin β-like interacting protein kinase CIPK25 regulates potassium homeostasis under low oxygen in Arabidopsis. *J. Exp. Bot.* 71: 2678–2689.

Takahashi, R., Nishio, T., Ichizen, N., and Takano, T. (2007). Cloning and functional analysis of the K+ transporter, PhaHAK2, from salt-sensitive and salt-tolerant reed plants. *Biotechnol. Lett.* 29: 501–506.

Tester, M. and Blatt, M.R. (1989). Direct measurement of k channels in thylakoid membranes by incorporation of vesicles into planar lipid bilayers. *Plant Physiol.* 91: 249–252.

Véry, A.A., Nieves-Cordones, M., Daly, M. et al. (2014). Molecular biology of K^+ transport across the plant cell membrane: what do we learn from comparison between plant species? *J. Plant Physiol.* 171: 748–769.

Vicente-Agullo, F., Rigas, S., Desbrosses, G. et al. (2004). Potassium carrier TRH1 is required for auxin transport in Arabidopsis roots. *Plant J.* 40: 523–535.

Walker, D.J., Leigh, R.A., and Miller, A.J. (1996). Potassium homeostasis in vacuolate plant cells. *Proc. Natl. Acad. Sci. U. S. A.* 93: 10510–10514.

Wang, X.P., Chen, L.M., Liu, W.X. et al. (2016). AtKC1 and CIPK23 synergistically modulate AKT1-mediated low-potassium stress responses in Arabidopsis. *Plant Physiol.* 170: 2264–2277.

Wang, Y., He, L., Li, H.D. et al. (2010). Potassium channel α-subunit AtKC1 negatively regulates AKT1-mediated K^+ uptake in Arabidopsis roots under low-K^+ stress. *Cell Res.* 20: 826–837.

Wang, Z., Hong, Y., Zhu, G. et al. (2020). Loss of salt tolerance during tomato domestication conferred by variation in a Na^+/K^+transporter. *EMBO J.* 39: 1–14.

White, P.J. (2012a). Chapter 2 – Ion uptake mechanisms of individual cells and roots: short-distance transport. In: *Marschner's Mineral Nutrition of Higher Plants*, 3e (ed. P. Marschner), 7–47. San Diego: Academic Press.

White, P.J. (2012b). Chapter 3 – Long-distance transport in the Xylem and Phloem. In: *Marschner's Mineral Nutrition of Higher Plants*, 3e (ed. P. Marschner), 49–70. San Diego: Academic Press.

White, P.J. and Karley, A.J. (2010). Potassium. In: *Cell Biology of Metals and Nutrients* (ed. R.R. Mendel and R. Hell), 199–224. Heidelberg: Springer.

Xicluna, J., Lacombe, B., Dreyer, I. et al. (2007). Increased functional diversity of plant K^+ channels by preferential heteromerization of the Shaker-like subunits AKT2 and KAT2. *J. Biol. Chem.* 282: 486–494.

Xu, J., Li, H.D., Chen, L.Q. et al. (2006). A protein kinase, interacting with two calcineurin B-like proteins, regulates K^+ transporter AKT1 in Arabidopsis. *Cell* 125: 1347–1360.

Yang, T., Zhang, S., Hu, Y. et al. (2014). The role of a potassium transporter oshak5 in potassium acquisition and transport from roots to shoots in rice at low potassium supply levels. *Plant Physiol.* 166: 945–959.

Zhang, M., Huang, P., Ji, Y. et al. (2020). KUP 9 maintains root meristem activity by regulating K + and auxin homeostasis in response to low K. *EMBO Rep.* 21: 1–20.

Zhang, M., Liang, X., Wang, L. et al. (2019). A HAK family Na^+ transporter confers natural variation of salt tolerance in maize. *Nat. Plants* 5: 1297–1308.

Zhu, H., Li, C., and Gao, C. (2020). Applications of CRISPR-Cas in agriculture and plant biotechnology. *Nat. Rev. Mol. Cell Biol.* 21 (11): 661–677.

Zörb, C., Senbayram, M., and Peiter, E. (2014). Potassium in agriculture – status and perspectives. *J. Plant Physiol.* 171: 656–669.

8

Role of Nutrients in Modifications of Fruit Quality and Antioxidant Activity

Tomo Milošević[1] *and Nebojša Milošević*[2]

[1] *Department of Fruit Growing and Viticulture, Faculty of Agronomy, University of Kragujevac, Čačak, Republic of Serbia*
[2] *Department of Pomology and Fruit Breeding, Fruit Research Institute Čačak, Čačak, Republic of Serbia*

Introduction

Growing of fruits is one of the important and age-old effort, practiced in India, China, and other old civilizations since ancient times. Cultivation of fruit types plays an important role in overall status of the mankind (people) and the nation. The standard of living of the people of a country is depending on the production and *per capita* consumption of fruits – fresh and/or processed. Fruit growing have social and more economic advantages such as high productivity (from a unit area of soil more yield is realized from fruit crops than any of the agronomic crops), source of raw material for agro-based industries (Jemrić et al. 2017), efficient utilization of resources, utilization of devastated, waste, and barren soils for production (Glisic et al. 2008; Milošević et al. 2012), and foreign exchange (many fresh fruits and processed products are exported to many countries earning good amount of foreign exchange) (Petrovic and Milosevic 1999).

In the last 30 years, fruit production has grown rapidly. The reasons for this tendency are numerous. Firstly, fruit has become available to many people around the world so that fruit consumption has increased. On the other hand, in many countries of the world, thanks to the profitability of fruit growing, these types of cultivated plants have suppressed others agronomic crops or started to be grown on new areas.

Importance of fruits in human nutrition is well recognized. Namely, human cannot live on cereals alone. Nutritionists advocate a diversified diet that will include, in addition to cereals, vegetables, pulses, eggs, meat, and also fruits. Fruits (including vegetables) are fundamental for balanced diet, good health, and long life because they are an excellent source of primary and secondary metabolites such as carbohydrates (sugars and starch), pectin, cellulose, fats, proteins, vitamins, bioactive compounds, and minerals (Wills et al. 1983; Vashisth et al. 2017; Frías-Moreno et al. 2021). Without them, human body cannot maintain proper health and develop resistance to disease and other stresses. Recently, antioxidants

Plant Ionomics: Sensing, Signaling, and Regulation, First Edition.
Edited by Vijay Pratap Singh and Manzer H. Siddiqui.

(natural or synthetic) have played a very important diet in order to preserve and improve human health (Griffiths et al. 2016). Also, antioxidants are essential for animal and plant life, since they are involved in complex metabolic and signaling mechanisms (Wilson et al. 2017). Hence, increased consumption of fruits has been associated with potential protection against age-related diseases (Ames et al. 1993; Grusak and Della Penna 1999).

Little attention has been given to the influence of preharvest cultural practices on fruit quality. Data from relevant literature revealed that the key cultural practices of fruit trees that have contributed to a paramount increase in yield per unit area and the improvement but also the deterioration of fruit quality are fertilization (Crisosto et al. 1995, 1997; Milošević and Milošević 2015) and irrigation (Krüger et al. 2002; Fallahi et al. 2010a; Chen et al. 2018).

In this synopsis, we tried to articulate and identify hitherto known and/or unknown correlations and interactions between nutrient deficiency and the ionome, which would give new implications for plant responses in order to improve fruit quality attributes. Also, the objective of this review is to critically discuss the impacts of essential nutrients (N, P, K, Ca, Mg, S, Fe, Mn, Cu, Zn, B, Cl, Mo, Co, Ni, and Se) on main fruit quality attributes. An effort was made to highlight important synergistic interactions where effects of two or more nutrients are involved to bring out an effect on crop quality.

Short Overview About Fruit Quality

Fruit quality includes organoleptic, nutritional, and healthy aspects tightly related to their biochemical composition, especially low-molecular-weight organic compounds (metabolites, including sugars, organic acids, amino acids, phenolic compounds, isoprenoids, and alkaloids), besides minerals, starch, and cell walls.

Fruit quality is a complex sum of physical, chemical, and sensorial properties and highly depends on genetics, management, and the environment (Parent et al. 2019). Physical specifications, mechanical, electrical, thermal, visual, and acoustic traits are among attributes of useful engineering application (Safwat and Moustafa 1971). These authors also reported that fruit mass or weight, volume, and center of gravity are the most important physical parameters of agricultural products, including fruits, used in sizing systems. The fundamental morphological parameters measurable through sizing systems are dimensions (length, width, and height), surface area, and fruit weight (Khojastehpour 1996). Fruit weight denotes the yield of fruit tree and load to be transported or conveyed, while volume and surface area can be used to predict drying rates and duration in the dryer while it is processing. Also, the size, shape, volume, and surface area of the fruit have special importance not only due to being components of yield, but they also determine the time of harvesting and acceptance of fruit by the stakeholders in the field and the market (Arshad et al. 2014).

The major fruit chemicals are primary and secondary metabolites. Primary metabolites are fundamental to the growth of the cell. They are produced continuously during the growth phase and are involved in primary metabolic processes such as respiration and photosynthesis (Kumar et al. 2016). These compounds, which are identical in most fruit types, include sugars, amino acids, tricarboxylic acids, the universal blocks, and energy sources. Sugar, organic acids, and amino acids are among the key primary metabolites that determine fruit quality (Kader 2008). The fruits of certain types contain the most starch (chestnut), protein, and fat (walnut, hazelnut, and almond). Glucose, fructose, and sucrose are predominant sugars, and sucrose accumulates as a consequence of a reduction in glucose and fructose and constitutes a large portion of fruit nutritional profile.

Secondary metabolites are the compounds which are derived by pathways from primary metabolic routs and are not essential to sustain the life of cells (Kumar et al. 2016). These compounds do not have a continuous production. Very often, secondary metabolites are produced during nongrowth phase of cells. They are synthesized by the secondary metabolism taking place in the plants (Singh et al. 2018). These compounds are final products of primary metabolites such as alkaloids, phenolics, steroids, essential oils, lignins, resins, and tannins. According to the first classification, secondary metabolites are classified based on the chemical nature as isoprenoids or terpenes which include steroids, essential oils, and carotenoid pigment. Second classification includes nitrogen-containing compounds which include alkaloids and nonprotein amino acids. Lastly, the secondary metabolites are classified as phenolic compounds including lignin, tannins, aflatoxins, and flavonoids (Singh et al. 2018).

In general, the internal quality mainly is determined by aroma, flavor, taste, texture, nutritional quality (e.g. soluble sugar content, starch, organic acids, soluble solids content, carotenoids, total flavonoids, total phenolics, and antioxidant activity), flesh firmness, diseases, and chemical residues. The external quality mainly concerns the appearance, size, color, and bruises.

Fruit quality and yield are highly dependent on adequate uptake of nutrients. Fruit quality is also greatly influenced by the synergistic and antagonistic interactions in various nutrients uptake and utilization. Balanced nutrition is noted to be of paramount importance for improving tree growth and development and fruit quality (Njira and Nabwami 2015; Milošević and Milošević 2019). Hence, the mineral nutrition of higher plants is of fundamental importance to agriculture, horticulture, human health, and environment protection (Crisosto et al. 1995, 1997).

Main Role of Mineral Elements on Trees Growth, Development, and Fruit Quality

In the scientific literature, many papers address the effect of mineral elements on fruit plant growth, development, and quality. These biological-physiological, agronomic, and pomological phenomena and properties of fruit trees are influenced by a number of factors including temperature, available water, light, and cultural practices, especially available nutrients in the soil. Many studies show that there are over 100 chemical elements, but research has determined 17 nutrients that are also called essential elements (Marschner 1995), whereas other data indicate that between 16 and 22 elements are necessary for the normal fruit tree growth and development of different fruit species (Bergmann 1992). Their deficiency caused numerous physiological, morphological, and anatomical disorders that adversely affect the yield and fruit quality.

The criteria that determine the mineral elements as essential are as follows (Foth and Ellis 1988):

- it must be needed by a plant to complete its life cycle;
- its function cannot be replaced by another element;
- it is directly involved in plant growth and reproduction; and
- it must be needed by most plants.

According to the presence in plants, the necessary elements are divided into macro- and microelements or macro- and micronutrients (Table 8.1).

Between 16 and 22 the essential chemical elements, carbon (C), hydrogen (H), and oxygen (O) are not mineral elements because they are obtained from soil or water. The rest are mineral elements. They include six macronutrients [nitrogen (N), phosphorus (P), potassium (K), calcium (Ca), magnesium (Mg), and sulfur (S)] and eight trace elements [iron (Fe), manganese (Mn), copper (Cu), zinc (Zn), boron (B), chlorine (Cl), nickel (Ni), and molybdenum (Mo)] (Njira and Nabwami 2015; Milošević and Milošević 2019). However, Bergmann (1992) adds some more trace elements such as aluminum (Al), cobalt (Co), sodium (Na), silicon (Si), and vanadium (V) that are important for certain plant and fruit species (Table 8.1). For example, in 1-year-old and 2-year-old shoots, flower buds and flowers of some cultivars of European plum (*Prunus domestica* L.) important levels of sodium (Na) were detected (Milošević and Milošević 2012). Also, Karagiannis et al. (2021) noted that silicon (Si) improved primary and secondary metabolites contents and sensorial traits of apple fruits. Other nutrients that do not fall within the list of essential macro- and microelements but are needed in specific cases are referred to as beneficial elements (Njira and Nabwami 2015). Other classifications were also existed in the available literature. One of them divides the chemical elements on the basis of physicochemical characteristics into "metals" (K, Ca, Mg, Fe, Mn, Zn, Cu, and Mo) and "non-metals" (N, S, P, B, and Cl) (Marschner 1995).

Unlike C, O, and H, which rarely limit the growth and development of fruit trees because they are naturally in sufficient amounts, other macro- and microelements, especially N, P,

Table 8.1 Elements those are essential for all higher plants including fruit types.

Organic nutrients (Basic elements in organic matter)	Mineral elements		
	Macroelements (Macronutrients)	**Microelements (Micronutrients)**	
Carbon (C)	Nitrogen (N)	Iron (Fe)	Aluminum (Al)
Hydrogen (H)	Phosphorus (P)	Manganese (Mn)	Cobalt (Co)
Oxygen (O)	Potassium (K)	Copper (Cu)	Sodium (Na)
	Calcium (Ca)	Zinc (Zn)	Nickel (Ni)
	Magnesium (Mg)	Boron (B)	Silicon (Si)
	Sulfur (S)	Molybdenum (Mo)	Vanadium (V)
		Chlorine (Cl)	

Source: Bergmann (1992)/Spektrum Akademischer Verlag.

and K, are often deficient and need to be added as fertilizers. In some sources of literature, C, O, H, N, P, and K are classified as primary macronutrients, whereas Ca, Mg, and S are classified as secondary macroelements (Milošević and Milošević 2019). As a rule, agricultural soils do not lack microelements because they are in sufficient quantities. However, Fe and B are most often missing in the soil.

Generally, all types of fertilizers (organic, organo-mineral, and mineral) and some soil conditioners such natural zeolites as a source of some nutrients should be used in preplanting fertilization, fertilization during juvenile stage and, especially, fertilization for maintaining fruit orchards (Milošević and Milošević 2015).

The Ionomic Analysis of Fruit Crops

The ionome is defined as the mineral nutrient and trace element composition of an organism and represents the inorganic component of cellular and organismal systems (Salt et al. 2008). Easier, the ionome is the mineral nutrient composition of all living organisms (Huang and Salt 2016). For example, some micro- and macronutrient deficiencies (e.g. Fe, Zn, and Ca) account for almost 2 billion people in developing as well as developed countries (Tulchinsky 2010). Ionome covers all mineral nutrients and trace elements (essential and nonessential) presented in an organism including plants (Lahner et al. 2003). It is the dynamic network of elements, controlled by physiology and biochemistry of the plant, which is ultimately controlled by the genetic and environmental factors (Baxter et al. 2010). The idea of plant ionomics began with the mixing of metabolomics and mineral nutrition and was first suggested in the late 60s and early 70s (Robinson and Pauling 1974; Singh et al. 2013). Ionomics is the study of the ionome and involves the simultaneous measurement of the elemental composition of an organism and its changes in response to environmental, physiological, or genetic modifications (Fleet et al. 2011). One definition says that ionomics is a tool to target healthier food through genetics and biofortification (Parent et al. 2019).

Since the concept of ionomics was first introduced in 2000s, significant progress has been made in the identification of genes and gene networks that control the ionome (Huang and Salt 2016). Ionomics becomes essential to identify potential gene(s) responsible for the uptake, transport, and storage of ions in plants. It involves the measurement of elemental composition of an organism and change in their composition in relation to physiological, developmental, environmental, and genetic factors. In this regard, plant ionomics could be a useful technology to explore the relation of gene(s) with transport and accumulation of ions (Singh et al. 2013).

Application of the ionome is a promising approach to clarify multi-element interactions, as it has been argued that mineral biology, including the nutritional status of plants, should be examined as a system (Fleet et al. 2011). Until recently, studies of fruits nutritional disorders have focused on particular nutrients, not multiple elements. Therefore, the application of ionomic analysis is a promising way to shed light on the multi-element interactions for accurate diagnosis of nutritional disorders through simultaneous quantitative measurements. This could be especially effective as major decreases in yield and fruit quality are often caused by the interaction of several elements. Ionomics can simultaneously measure the elemental composition of organisms and their changes in response to physiological stimuli, developmental state, genetic changes, and environment (Salt et al. 2008; Singh et al. 2013). Ionomics requires the application of high-throughput elemental analysis technologies, such as inductively coupled plasma mass spectrometry (ICP-MS), quadrupole inductively coupled plasma mass spectrometry (ICP-QMS), or inductively coupled plasma optical emission spectroscopy (ICP-OES). It has the ability to detect changes in the ionome composition in relation to alterations in a plant's physiology gene function (functional genomics) and physiological status (Salt et al. 2008). Therefore, the ionomic approach is useful to reveal the role of different nutrients in fruit tree nutritional disorders, how changes in ionome composition (multi elemental composition) respond to a given agricultural practice and climate conditions, and how ionomic signatures could be used to determine better cultivation conditions for higher yield and fruit quality.

Although the use of ionomic analysis in nutritional studies and nutritional diagnosis of fruit species is still limited, some results and reports exist. Analysis was performed in different fruit species in order to characterize element concentrations in edible parts and to compare element concentrations between edible parts and vegetative organs and between species. For example, Matsuoka et al. (2018) reported that ionomes of 13 elements in blueberries were influenced by soil type and soil treatments. In the same study, N, P, K, Mn, Cu, and Zn were in significant positive correlation regarding concentrations in the soil and the content in the blueberries, while Na, Mg, Al, Ca, Fe, Rb, and Cs were not. Long-term application of N fertilizers has changed ionomic structure in fruits and leaves of apple cultivar "Jonathan" due to changes in nutrition not only with N, but also K and Mn thanks to reduced availability of these elements in soil (Matsuoka et al. 2019). Shibuya et al. (2015) reported that ionome profiles of apple and Japanese pear were similar across same organs in different species. Other authors who investigated ionomics (Nashima et al. 2013, 2014; Isuzugawa et al. 2014) detected and quantified 12 elements, i.e. Ca, K, Mg, P, S, B, Cu, Fe, Mn, Mo, Na, and Zn in European pear (*Pyrus domestica* L.) fruits. They stated that concentrations of all these elements were high, especially in the youngest fruits, and decrease significantly, although declining rates varied among the elements. Seventeen elements, i.e. B, C, Cl, Ca, Cu, Fe, H, K, Mg, Mn, Mo, N, Ni, O, P, S, and Zn, are necessary for plant growth, and excess amount of some elements, such as Al, Cd, and Na, has harmful effect on plants and human also (Shiratake and Suzuki 2016). Therefore, ionomics could be very important tool to observe plant health as well as food safety for human. Hence, such models should be developed in future.

Requirements of Fruit Trees to Chemical Elements

Fertilization also known as nutrition of fruit crops is an important tool used by the most farmers in order to boost crop yield and quality. However, excessive fertilization has been verified, especially on the horticultural enterprises, where the fertilizer costs represented less than 10% of the variable crop costs (Huett and Dirou 2000). Except for an economic point of view, excessive fertilization has been connected to ground and stream water contamination (Cuquel et al. 2011) and causes an increment of pest (Marschner 1995) and diseases occurrence (Tratch et al. 2010).

The nutrient requirement of a fruit trees is determined by the soil and tissue nutrient content which is necessary for normal vegetative growth, development, productivity (Ebert 2009), and acquirement of fruits with excellent external and internal quality, of course biologically valuable and health-safe (Milošević et al. 2013a, 2013b). Basically, requirement is closely related to the defined normal (optimum) range of nutrient amounts in soils (Hanson 1996) and also in leaves, flowers, fruits, and/or 1-year- and 2-year-old shoots (Leece 1975a, 1975b; Leece and van den Ende 1975; Bergmann and Neubert 1976; Bergmann 1992; Jiménez et al. 2004, 2007).

Since the soil testing is insufficient, in order to obtain better information about real requirements of fruit trees for nutrients, it is necessary to carry out a chemical analysis of some plant parts. It is necessary to integrate both techniques, chemical analysis of plants and soil testing, besides visual diagnosis to maximize fertilization efficiency in terms of cost and prevention of environmental damage (Milošević and Milošević 2019). Aforementioned authors also report previously defined fruit tree organs, sample numbers, and the sampling period for diverse fruit cultures. Usually, leaf is the most common organ used for chemical analysis. Optimal levels of macro- and microelements in leaf of some pome and stone fruits are presented in (Table 8.2).

Deficient, low, normal (optimal and adequate), high, and excess or toxic levels of macro- and microelements in fruit leaves were proposed by several authors and described previously in detailed overview of Milošević and Milošević (2019). On this line, surveys of nutrient levels in "deficient" and "adequate" ranges have been used to establish standard nutrient levels for different fruit types. However, leaf mineral analysis is a useful tool to diagnosis fruit tree deficiencies but often is poorly related to fruit quality, especially in apple (Fallahi et al. 2010b).

Table 8.2 Leaf composition standards for some pome and stone fruit types based on mid-shoot leaves sampled in mid-summer expressed on a dry matter basis.

	Apple[1]	Pear[1]	Plum[2]	Peach[3]	Apricot[4]	Sweet cherry[5]	Sour cherry[6]
Element	**Macroelements (% on dry matter)**						
N	2.20–2.80	2.20–2.80	2.40–3.00	3.00–3.50	2.40–3.00	2.20–2.60	2.60–3.00
P	0.18–0.30	0.15–0.30	0.14–0.25	0.14–0.25	0.14–0.25	0.14–0.25	0.16–0.22
K	1.10–1.50	1.20–2.00	1.60–3.00	2.00–3.00	2.00–3.50	1.60–3.00	1.60–2.10
Ca	1.30–2.20	1.20–1.80	1.50–3.00	1.80–2.70	2.00–4.00	1.40–2.40	1.50–2.60
Mg	0.20–0.35	0.20–0.35	0.30–0.80	0.30–0.80	0.30–0.80	0.30–0.80	0.30–0.75
	Microelements ($\mu g\, g^{-1}$ on dry matter)						
Fe	40–280	60–240	100–250	100–250	100–250	100–250	100–200
Mn	35–100	30–100	40–160	40–160	40–160	40–160	40–60
Cu	5–12	5–12	6–16	5–16	5–16	5–16	8–28
Zn	22–50	22–50	20–50	20–50	20–50	20–50	20–50
B	25–50	20–50	25–60	20–60	20–60	20–60	20–55

Source: Adapted from [1]Bergmann and Neubert (1976), [2,5]Leece (1975a, 1975b), [3]Leece et al. (1971), [4]Leece and van den Ende (1975), [6]Heckman (2004).

The Role of Elements in the Metabolism of Fruit Trees and in Improving Quality

Each of the nutrients is needed in different levels and carries specific functions in the plant. Depending on the amount that is available for plant uptake, these nutrients influence yields and fruit quality (Havlin et al. 2005). We give a brief overview of the roles of most important elements in the metabolism of fruit trees and effects on yield and fruit quality.

Macroelements

Nitrogen (N)

N plays a role in almost all metabolic processes. It has many functions including promotion of tree growth, increasing leaf size and fruit quality, and enhancing fruit and seed development. Also, it forms an integral component of many important components in plants including amino acids that are building blocks of proteins and enzymes, that are involved in catalyzing most biochemical processes (Brady and Weil 2008). As determined by its functions, N generally influences tree growth and fruit quality (Bruneto et al. 2015).

High N levels stimulate vigorous vegetative growth, causing shading out of lower fruiting wood and its death, delay stone fruit maturity, induce poor visual red color development, and inhibit ground color change from green to yellow (Crisosto et al. 1997; Fallahi et al. 2010b). These authors also reported that although high N trees may look healthy and lush, excess N does not increase fruit size, production, or soluble solids content. In addition, excess N causes susceptibility of nectarine fruits to some pathogens such as brown rot [*Monilinia fructicola* (Wint.) Honey] and insects (Michailides et al. 1993; Daane et al. 1995). Yet, a recent study suggests that lower N rates may advance fruit maturation by increasing color and soluble solids content (Rubio Ames et al. 2020). On this point of view, fertilizers with a lower N content caused a higher amount of secondary metabolites and stronger antioxidant activity (Milošević et al. 2019). With commercial fertilizer containing typical macro- and micronutrients, the lowest fertilization level, especially N, increased the contents of flavonols and ellagic acid in strawberry (*Fragaria* ×*ananassa* Duch.) (Anttonen et al. 2006).

N deficiency leads to small fruit with poor flavor, low fruit size, high firmness, early maturation, late maturation disorders (cork spot), pre- and postharvest decay, and unproductive trees (Sugar et al. 1992; Daane et al. 1995). Otherwise, foliar-applied N is absorbed rapidly with higher efficiency and therefore represents an interesting means to supplement soil N supply (Toselli et al. 2004). Fertilizers with a balanced ratio between N and PK improved phenolic compounds content, antioxidant activity, and macro- and microelements content in fruits of raspberry (Stojanov et al. 2019a). Some studies pointed out that the mode and rate of N distribution could have a major role in the quality of apricot fruit (Radi et al. 2003). These authors also reported that a balanced N : K ratio to be the most important factor to increase both sugars and phenolic compounds accumulation. Similar tendencies were observed by Milošević et al. (2013a, 2013b) for the same fruit type.

Phosphorus (P)

P is used in relative small amounts by plants; however, this element is involved in most growth processes. Shortly, it is an essential component of most organic compounds in the plant including nucleic acids, proteins, phospholipids, sugar phosphates, enzymes, and energy-rich phosphate compounds, a common example being adenosine triphosphate (ATP) (Brady and Weil 2008; Njira and Nabwami 2015). P is needed for the formation of cell membranes, carbohydrate metabolism, protein synthesis, photosynthesis, respiration sugar metabolism, energy storage, and transfer (Fregoni 1980).

Due to the little plant requirement, P deficiency is unlikely occur (Milošević et al. 2013a, 2013b; Milošević and Milošević 2015, 2019). However, its deficiency promotes high juice, low size, and juice acidity in pome fruit types, whereas excess induces late maturation, less aroma, and less sweetness (Faust 1989).

In the study of Osman et al. (2014), soil P application improved tree growth, its yield, fruit size, and content of primary metabolites such as soluble solids, sugars, fruit juice acidity, and macro- and microelements content in peach. In apple, adequate P application promoted fruit firmness, soluble solids, and control of scald and bitter pit (Raese 1987). In pear cv. "Williams," foliar P application increased the contents of glucose, sorbitol, soluble solids, malic acid, citric acid, and potassium (Hudina and Stampar 2002). However, responsiveness to P fertilization has often been made difficult by the absence of unique leaf P deficiency symptoms at low tissue P contents (Benson and Covey 1979).

Potassium (K)

K is absorbed by roots of plants in larger amounts than any other nutrient except N. Unlike N, P, and most other nutrients, K is not incorporated into structures of organic compounds; instead, K remains in ionic form (K^+) in solution in the cell and acts as an activator of many cellular enzymes (Havlin et al. 2005).

Usually, K has been associated with fruit quality in general. K is the most abundant nutrient in the fruit where it affects positively the size, firmness, skin colour, soluble solids content, i.e. sugars amount, acidity, ascorbic acid, juiciness, and aroma (Lester et al. 2010). This nutrient is important during storage (increased shelf life), since an imbalanced K : Ca ratio may promote cork spot (Curtis et al. 1990; Taiz and Zeiger 2004; Anjum et al. 2008). However, there is some controversy about the effect of K on fruit quality; namely, no correlation between leaf K content and fruit soluble solids amount has been shown in peach (Layne and Bassi 2008). In contrast, some authors provided evidence that fruit size, soluble solids content, and color can all be improved when peach trees are not deficient in K (Tagliavini and Marangoni 2002). Contrary to results of Lester et al. (2010), Fallahi et al. (2010b) reported that K fertigation in four apple cultivars decreased firmness at harvest.

The negative effect of K on quality of stored pear fruits is less frequent than in apple fruits because pear trees generally show higher absorption and transport of Ca to the fruit than apple. Also, K increased tree growth, yield, and cell strength and encourages good tolerance to insect pests and diseases (Ebert 2009; Milošević and Milošević 2019), shelf life and shipping quality of many horticultural crops (Lester et al. 2010), and content of phenolic compounds and antioxidant power (Stojanov et al. 2019b). Hence, among the many plant mineral elements, K stands out as a cation having the strongest impact on quality attributes that determine fruit marketability, consumer preference, and the amount of critically important human health-associated phytochemicals or bioactive compounds such as ascorbic acid and β-carotene (Lester et al. 2010). A condensed review on the effect of K on fruits quality had been published earlier (Lester et al. 2010).

Calcium (Ca)

After N and K, Ca is used in large amounts by plants. This element is a major component of the middle lamella (Ca-pectates) of the cell wall. On this point of view, Ca strengthens the cell walls, involved in cell elongation and division, membrane permeability, and activation of several critical enzymes (Brady and Weil 2008). Ca is important in N metabolism and protein formation by enhancing NO_3^- uptake, and it is also important in translocation of carbohydrates and other nutrients (Havlin et al. 2005). In general, Ca application, especially Ca sprays, influences crop and food quality, since it is an important mediating agent in the control of cell metabolism (Lobos et al. 2021). It is less mobile element, so its influence on fruit quality is easily noted with foliar application. Ca foliar application improves some fruit quality attributes such as firmness, storage capability, i.e. shelf life, Ca content in fruits (high quality), decreased bitter pit incidence (Crisosto et al. 1997; Launaskas and Kvikliene 2006), increased total phenolic content during storage, and decreased postharvest polygalacturonase activity (Lobos et al. 2021). In addition, multiple sprays of soluble Ca often reduce bitter pit and usually but not always increase Ca concentrations in subdermal cortical tissue (Fallahi et al. 2010b). Also, these authors reported that early-season Ca sprays often are more effective than later sprays at reducing bitter pit; however, later applications of Ca have a greater influence on fruit Ca concentration.

Ca deficiency induced cork spot in pears (Tomala and Trzak 1994; Duan et al. 2019) and bitter pit in apples (Faust et al. 1967; Faust and Shear 1968), black end internal breakdown, senescent breakdown, water core, high respiration, earlier maturation, and peel yellowing, whereas excess induced delay in ethylene evolution (Faust 1989; Curtis et al. 1990; Milošević and Milošević 2019).

Magnesium (Mg)

Similarly to P, Mg is used in small amounts by fruit trees. However, its role in plant metabolism is paramount. Mg is the central core of the chlorophyll molecule in plant tissue. It is important as a primary constituent of chlorophyll and as a structural component of ribosomes, and it helps in their configuration for protein synthesis (Havlin et al. 2005). It is also required for maximum activity of almost all phosphorylating enzymes in carbohydrate metabolism. If Mg is deficient, the shortage of chlorophyll results in poor and stunted plant growth. Magnesium also helps to activate specific enzyme systems (Milošević and Milošević 2019). Mg uptake can be strongly depressed by K^+, NH_4^+, and Ca^{2+} (Marschner 1995). Deficiencies in plant tissues are common in acid soils, particularly where the soil is high in plant-available K (Milošević and Milošević 2015; Milošević et al. 2013a, 2013b). In the study of Stojanov et al. (2019a, 2019b), soil application of fertilizer with Mg significantly improved fruit quality attributes in comparison with fertilizer without Mg.

Unlike Ca, Mg is mobile within the plant tissues, so deficiency symptoms appear first in older leaves. The Mg accumulates linearly at a slow rate throughout the growing season (Tagliavini et al. 2000; Milosevic et al. 2009).

Added Mg, especially through foliar application, increases fruit yield and enhances the total soluble sugars and anthocyanin content in fruits, but its effects on other fruit quality attributes are small (Moss and Higgins 1974). From a broader point of view, Mg fertilization improves yield of many horticultural crops, including hazelnut (Nedim and Daml 2015) and raspberry (Stojanov et al. 2019a, 2019b), suggesting that Mg fertilization is an important measure to boost crop production.

Sulfur (S)

In the earth's crust, S is the most abundant element (Brady and Weil 2008). In the tree, this element participates in the synthesis of S-containing amino acids cystine, cysteine, and methionine which are building blocks of proteins and is an important constituent of vitamins, hormones, and oils (Johnson and Uriu 1989).

S is also needed in the synthesis of coenzyme A and chlorophyll. It is responsible for the formation of the disulfide bond between cysteine residues that help to stabilize the tertiary structure of proteins (Sylvia et al. 2005). The S deficiency leads to accumulation of nonprotein N such as NO_3^- and amine (NH_2) (Havlin et al. 2005). Hence, S deficiency has been reported to lead to accumulation of NO_3^- in vegetables which is dangerous as these lead to fatal conditions such as methemoglobinemia in infants and formation of cancer-inducing nitrosamines (Sylvia et al. 2005). In general, S application has been reported to improve the yield and fruit quality attributes of some fruit types such as orange (Abdel-Nasser and El-Shazly 2000; Murovhi 2013), date palm (Dialami and Garshasbi 2018), mango (Nasreen et al. 2014), or peach (Al-Aareji and Bani 2020).

Microelements

For plant growth and development, microelements, also called micronutrients, such as iron (Fe), manganese (Mn), cooper (Cu), zinc (Zn), boron (B), chlorine (Cl), nickel (Ni), and molybdenum (Mo) are known to be essential (Faust 1989; Bergmann 1992; Marschner 1995; Milošević and Milošević 2019). Other microelements such as selenium (Se) and cobalt (Co) are needed in specific cases and commonly referred to as beneficial elements.

The requirements of plants for microelements are in small quantities, but their role in metabolism is huge and irreplaceable because they are primarily an integral part of numerous enzymes. All micro- or trace elements in excessive amounts are toxic to plants.

Iron (Fe)

Among microelements, Fe and Zn are more important elements. Fe complexes with proteins are constituents of enzyme system which brings about oxidation–reduction reactions in the plant, regulation of respiration, photosynthesis, and reduction of nitrates and sulfates and part of some pigments (Marschner 1995). In these processes, Fe accounts for a small percentage from the total Fe (Johnson and Uriu 1989).

Most of Fe is associated with chloroplasts, where it has some role in synthesizing chlorophyll (the green pigment in leaves). In this role, it is not very mobile within the plant and explains why deficiency symptoms first appear in young leaves. Among others, Fe deficiency (chlorosis) generally reduced plant growth and production. Otherwise, the occurrence of Fe chlorosis depends on several environmental and agronomic factors such as soil, climate, rootstock, and grafting type (Bruneto et al. 2015).

Manganese (Mn)

Mn in the plant participates in several important processes including photosynthesis, nitrogen, and carbohydrate metabolism (Johnson and Uriu 1989). It is generally considered to be somewhat immobile in the plant, but it is preferentially supplied to young growing tissue. The Mn deficiency can severely reduce fruit yield and shows symptoms similar to Fe chlorosis, especially on alcaline and calcareous soils, but also in soils with a high content of organic matter (Marschner 1995, 2012).

Soil application of Mn is problematic, since its types, plants, and even to various cultivars. Most efficiency depends on many soil factors, including soil micronutrients that are readily fixed in soil having alkaline pH. Several studies reported that foliar application mixture of Mn and Ca or Zn in adequate rate in the last week of May and first week of June increased yield and fruit weight in pomegranate (Hasani et al. 2012; Obaid and Al-Hadethi 2013).

Copper (Cu)

Data from literature revealed that very small amounts of Cu are needed by the fruit tree (Johnson and Uriu 1989). These authors also reported that more than half of the Cu in trees is located in the chloroplasts and participates in photosynthetic reactions. It is also found in other enzymes involved with protein and carbohydrate metabolism. Its deficiency is rarely observed, but it can occur in some orchards (Milošević et al. 2013a, 2013b). Cu spray residues can lead to inaccurate interpretation due to contamination from pesticides containing Cu.

Zinc (Zn)

Zn is a main metal component and activator of 60 enzymes involved in metabolic activities and biochemical pathways (Grotz and Guerinot 2002). These authors also reported that Zn is a functional, structural, or regulatory co-factor of a large number of enzymes. Also, Zn is required in a large number of enzymes and plays an essential role in DNA transcription and in producing the growth hormone IAA (Johnson and Uriu 1989). The shortened internodes and small leaves were observed with Zn deficiency. Also, under its deficiency, fruit are deformed and are small size, sour, and early-ripen (Wójcik and Popińska 2009). Among fruit types, pear is considered to be a Zn-sensitive (Shear and Faust 1980). Zn is an integral part of the active substances of many fungicides, and their use can partially eliminate its deficiency in cultural plants.

In sweet cherry, there was a general trend for fruit set and yield to be higher in the Zn + B treatment trees compared to the control trees (Usenik and Stampar 2002), whereas Mohamed and Ahmed (1991) reported that applying the three elements together (Cu + Zn + Fe) at the higher rate was also accompanied with an improvement in soluble solids content in apple trees.

Boron (B)

Among micronutrients, B is one of the most critical in orchards. Indeed, it is believed that pome fruit types, especially pear trees, have a high B requirement (Wójcik and Wójcik 2003). Among different fruit types, peach is much less sensitive to boron (B) deficiency, and this is possibly due to the easily translocated compound that is formed from B and sorbitol (or other sugar alcohols) that accounts for the high mobility of this element (Layne and Bassi 2008). Typical symptoms of B deficiency are the reduction of fruit set and yielding, as well as small, deformed, cracked, and corked fruits. Excess of B in fruit trees induced early maturation and lower storage capacity (Shear and Faust 1980; Wójcik and Wójcik 2003).

Fruit from B-sprayed trees may exhibit quality loss due to B excess even though leaf B appears normal (Fallahi et al. 2010b). In study of Rafiullah et al. (2020), foliar application of mixture of Zn + Cu + Fe + Mn + B in adequate rate increased yield per tree, soluble solids content, sugar content, juice acidity, and juice content in European plum.

Excess B can lead to severe toxicity, so application rate should be carefully calculated.

As noted in the literature, macro- and microelements play a specific and autonomous role in the metabolism of plants, including fruit trees. However, they do not act in isolation but together. Hence, most nutrients produce the best effects under balanced nutrition. Therefore, the balanced and so-called "smart" fertilization imposes as an imperative in modern fruit production in order to obtain regular, optimal, and sustainable yields with excellent fruit quality (Milošević and Milošević 2019).

Other Essential Microelements

According to data from relevant literature, Co is required by bacteria that fix nitrogen in legumes (Njira and Nabwami 2015). In the fruit trees, the foliar spray of Co sulfate increased fruit yield and improved fruit weight, size, soluble solids content/acid ratio (ripening index), and the total sugars in the fruit of mango (Sing and Sing 1993). Otherwise, Co is transition element and essential component of several enzymes and coenzymes which interact with other elements to form complexes (Palit et al. 1994). On the other hand, the foliar application of Se enhanced the flesh firmness, titratable acidity, and soluble solid content of the apple (Babalar et al. 2019) or peach and pear fruits (Pezzarossa et al. 2012).

As it is known, Na is an essential element for animals and humans and must be present in relatively large amounts in the diet (Subbarao et al. 2003) and is the principal electrolyte in animal and human systems and plays an important role in maintaining the ionic balance of body tissues and fluids. In contrast, the principal electrolyte for plants is K, and even in ecosystems where there is a predominance of Na, plants still exhibit a strong preference for K (Walker et al. 1996). Data from literature revealed that Na and K have almost identical chemical properties in the soil. The role and functions of K in the nutrition of plant are fully known, but those of Na are not well documented (Idowu and Aduayi 2007). The key is that Na cannot be substituted completely by K (Marschner 1995, 2012) but can detrimentally compete with the absorption of K and Ca (Labate et al. 2018). However, the new trends on the role of Na in plant nutrition are moving steadily toward its being recognized as an essential plant nutrient element. Generally, reports on the functions of Na on higher plants are inconsistent. Some authors reported that Na is not strictly an essential element, so it cannot be expected to have a specific role in the metabolic activities of higher plants, including fruit trees (Milošević and Milošević 2012). There are other opinions and results because some field crops and vegetables showed a yield and fruit quality response to Na even in the presence of adequate amounts of K (Idowu and Aduayi 2007). In the study of Milošević and Milošević (2012), the presence of Na in plum twigs and flower buds was registered. The detailed role of Na in higher plants metabolic processes remain to be elucidated in future.

Cl is an essential plant element, included with microelements. Its essentiality was discovered by Arnon and Whatley (1949) and confirmed by Broyer et al. (1954) and Izawa et al. (1969). It has important role in cell elongation, photosynthesis, and plant growth (Izawa et al. 1969). Small fruit types are characterized as a chloride-sensitive species with a toxicity threshold at a chloride concentration of 1% of leaf dry weight and an EC value of the soil solution of $2\,dS\,m^{-1}$ (Martínez Barroso and Alvarez 1997). On this line, mineral fertilizers that contain high amount of chloride are not recommended for fertilizing these fruit types. Due to its role in important functions in higher plants, several studies revealed that fertilization with adequate doses of KCl had a positive effect on yield and fruit quality of strawberry (Svenja and Sven 2009). These authors also reported that chloride fertilization can compensate insufficient Mn supply by either improving Mn availability in the soil or Mn uptake of the plant, but the mechanisms of these processes remain to be elucidated. Otherwise, horticultural practice should provide adequate chlorine fruit tree nutrition (Komosa and Górniak 2015).

Unlike several heavy, also toxic, metals, such as Ag, Cd, Cr, Hg, and Pb, that have no biological function in plant (Seregin and Kozhevnikova 2006), Ni is considered an essential microelement for higher plants, since it is involved in N metabolism as the metal component of the enzyme urease (Marschner 2012). Ni is readily taken up by plants and, being highly mobile, is easily translocated to different plant parts (Poulik 1999). The requirement of Ni in higher plants is quite low, whereas toxicity occurs at concentrations in the range of $10\,mg\,g^{-1}$ in sensitive and $50\,mg\,g^{-1}$ in moderately tolerant species (Seregin and Kozhevnikova 2006). In the recent studies, plants receiving 50 mM of Ni enhanced some fruit quality attributes of tomatoes such as firmness, brightness, and taste (Kumar et al. 2015), whereas Gad et al. (2007) observed that increasing Ni up to $30\,mg\,kg^{-1}$ soil improved the fruit dry matter, vitamin C, and soluble solids contents in tomato. However, excessive applications of Ni in the root zone lead to an increase of Ni content in fruits, which represents a serious risk for human health (Kumar et al. 2015) due to the amount of Ni required for normal growth of plants is very low. Nickel deficiency appears as leaflet tip necrosis, or "mouse-ear" leaves such as case on pecan and other horticultural crops. Applying a foliar spray at a concentration of $0.03–0.06\,mg\,kg^{-1}$ Ni is sufficient for the elimination of deficiency (Liu et al. 2020).

A detailed overview about the role of macro- and microelements in fruit trees was given earlier by Shear and Faust (1980), Bergmann (1992), Marschner (1995, 2012), Lester et al. (2010), Saini et al. (2019), and Milošević and Milošević (2019).

Conclusion and Future Prospects

In recent few decades, fruit production increases rapidly in the world. Thanks to new cultivation technologies, especially fertilization, yield increases per unit area and fruit quality also improved with application of fertilizers. On the basis data from this review, it can be said that all macro- and microelements influenced quality attributes of fruit crops. These manifestations are characterized by changes and/or differences in amounts of primary and secondary metabolites and visual appearance of fruits. However, different rates of essential elements applied to various fruit types are paramount factor that determined fruit quality and also fruit tree growth and development. As noted in this review, macro- and microelements play a specific and autonomous role in the metabolism of plants, including fruit trees. The responses of fruit types to elements (nutrients) are dependent on the metabolic capacity of the studied plant determined by the genetic background, depending on the genus, species, genotype, and cultivar investigated, environmental factors, and developmental stage. However, they do not act in isolation but together. Hence, most nutrients produce the best effects under balanced nutrition. The balanced and so-called "smart" fertilization imposes as an imperative in modern fruit production in order to obtain regular, optimal, and sustainable yields with excellent fruit quality. Findings obtained in this study reveal the utility of ionomic analysis. Its utility in fruit crops has been recently shown by a comparison of behavior of some fruit types in different environmental conditions (habitats) as well as a comparison between, for example, mutant and wild-type pear fruit. Finally, ionomics analysis could be very important tool to observe plant health as well as food safety for human and may broadly contribute to progress in the ionomic studies of fruit crops.

References

Abdel-Nasser, G. and El-Shazly, M.S. (2000). Effect of sulphur application on soil properties, nutrient status, yield and fruit quality of balady orange trees. *J Agric Res Tanta Univ* 26: 72–92.

Al-Aareji, A.M.J. and Bani, S.H.S. (2020). Rsponseof Dixired peach trees to Sulphur and Iron application. *IOP Conf Ser Earth Environ Sci* 553: 012023. https://doi.org/10.1088/1755-1315/553/1/012023.

Ames, B.N., Shigenaga, M.K., and Hagen, T.M. (1993). Oxidants, antioxidants, and the degenerative diseases of aging. *Proc Natl Acad Sci* 90 (17): 7915–7922. https://doi.org/10.1073/pnas.90.17.7915.

Anjum, R., Kirmani, N.A., Nageena, N., and Sameera, S. (2008). Quality of apple cv. Red Delicious as influenced by potassium. *Asian J Soil Sci* 3 (3): 227–229.

Anttonen, M.J., Hoppula, K.J., Nestby, R. et al. (2006). Influence of fertilization, mulch colour, early forcing, fruit order, planting date, shading, growing environment, and genotype on contents of selected phenolics in strawberry (*Fragaria* × *ananassa* Duch.) fruits. *J Agric Food Chem* 54 (7): 2614–2620. https://doi.org/10.1021/jf052947w.

Arnon, D.I. and Whatley, F.R. (1949). Is chloride a coenzyme of photosynthesis? *Science* 110 (2865): 554–556. https://doi.org/10.1126/science.110.2865.554.

Arshad, M., Shahnawaz, M., Shahkeela, S. et al. (2014). Significance of physical properties of apple fruit influenced by preharvest orchard management factors. *Eur J Exp Biol* 4 (5): 82–89.

Babalar, M., Zabihollah, S.M., Zamani, Z., and Askari, M.A. (2019). Effect of foliar application with sodiumselenate on selenium biofortification and fruitquality maintenance of "Starking Delicious" apple during storage. *J Sci Food Agric* 99 (11): 5149–5156. https://doi.org/10.1002/jsfa.9761.

Baxter, I., Brazelton, J.N., Yu, D. et al. (2010). A coastal cline in sodium accumulation in *Arabidopsis thaliana* is driven by natural variation of the sodium transporter AtHKT1;1. *PLoS Genet* 6 (11): e1001193. https://doi.org/10.1371/journal.pgen.1001193.

Benson, N.R. and Covey, R.P. Jr. (1979). Phosphorus nutrition of young "Golden Delicious" apple trees growing in gravel culture. *J Amer Soc Hort Sci* 104: 682–685.

Bergmann, W. (1992). *Nutritional Disorders of Plants: Developments, Visual and Analytical Diagnosis*, 386. Stuttgart, New York: Gustav Fischer Verlag Jena.

Bergmann, W. and Neubert, P. (1976). *Pflanzendiagnose und Pflanzenanalyse flanzendiagnose und Pflanzenanalyse zur Ermittlung von Ernährungsstörungen und des Gesundheitszustandes der Kulturpflanzen*. Jena: Verlag VEB Gustav Fischer.

Brady, N.C. and Weil, R.R. (2008). *The Nature and Properties of Soils*, 14e. New Jersey: Pearson Prentice Hall.

Broyer, T.C., Carlton, A.B., Johnson, C.M., and Stout, P.R. (1954). Chlorine - a micronutrient element for higher plants. *Plant Physiol* 29 (6): 526–532. https://doi.org/10.1104/pp.29.6.526.

Bruneto, G., Wellington, G., De Melo, B. et al. (2015). The role of mineral nutrition on yields and fruit quality in grapevine, pear and apple. *Rev Bras Frutic* 37 (4): 1089–1104. http://dx.doi.org/10.1590/0100-2945-103/15.

Chen, R., Huang, Y., Ji, X. et al. (2018). Effects of different irrigation methods on growth, fruit quality and yield of apple tree. *Asian Agric Res* 10 (7): 54–57. https://doi.org/10.19601/j.cinki.issn1943-9903.2018.7.014.

Crisosto, C.H., Mitchell, F.G., and Johnson, S. (1995). Factors in fresh market stone fruit quality. *Postharvest News Inform* 6 (2): 17N–21N.

Crisosto, C.H., Johnson, S.R., DeJong, T., and Day, K.R. (1997). Orchard factors affecting postharvest stone fruit quality. *HortScience* 32 (5): 820–823. https://doi.org/10.21273/HORTSCI.32.5.820.

Cuquel, F.L., Motta, A.C.V., Tutida, I., and May de Mio, L.L. (2011). Nitrogen and potassium fertilization affecting the plum postharvest quality. *Rev Bras Frutic* 33 (Special issue): 328–336. https://doi.org/10.1590/S0100-29452011000500041.

Curtis, D., Righetti, T.L., Mielke, E., and Facteau, T. (1990). Mineral analysis from cork spotted and normal Anjou pear fruit. *J Am Soc Hortic Sci* 115: 969–974.

Daane, K.M., Johnson, S.R., Michailides, J.T. et al. (1995). Excess nitrogen raises nectarine susceptibility to disease and insects. *Calif Agric* 49 (4): 13–18. https://doi.org/10.3733/ca.v049n04p13.

Dialami, H. and Garshasbi, M.R. (2018). Effect of sulfur and manure on yield and quality of date fruit cv. Barhee. *J Water Soil* 32 (5): 1003–1012. https://doi.org/10.22067/jsw.v32i5.72413.

Duan, Y., Xu, Y., Wang, R., and Ma, C. (2019). Investigation and prevention of cork spot disorder in "Akizuki" pear (*Pyrus pyrifolia* Nakai). *HortScience* 54 (3): 480–486. https://doi.org/10.21273/HORTSCI13775-18.

Ebert, G. (2009). *Fertilizing for High Yield and Quality. Pome and Stone Fruits of the Temperate Zone*. IPI Bulletin No 19. France: Imprimerie de Saint-Louis.

Fallahi, E., Neilsen, D., Neilsen, G.H. et al. (2010a). Efficient irrigation for optimum fruit quality and yield in apples. *HortScience* 45 (11): 1616–1619. https://doi.org/10.21273/HORTSCI.45.11.1616.

Fallahi, E., Fallahi, B., Neilsen, G.H. et al. (2010b). Effects of mineral nutrition on fruit quality and nutritional disorders in apples. *Acta Hortic* 868: 49–60. https://doi.org/10.17660/ActaHortic.2010.868.3.

Faust, M. (1989). *Physiology of Temperate Zone Fruit Trees*, 200. New York: Wiley.

Faust, M. and Shear, C.B. (1968). Corking disorders of apples: a physiological and biochemical review. *Bot Rev* 34: 441–469. https://doi.org/10.1007/BF02859134.

Faust, M., Shear, C.B., and Smith, C.B. (1967). Investigations of corking disorders of apple. I. Mineral element gradients in a "York Imperial" apples. *Proc Am Soc Hortic Sci* 91: 69–72.

Fleet, J.C., Replogle, R., and Salt, D.E. (2011). Systems genetics of mineral metabolism. *J Nutr* 141 (3): 520–525. https://doi.org/10.3945/jn.110.128736.

Foth, H.D. and Ellis, B.G. (1988). *Soil Fertility*, 7e, 212. New York, USA: John Wiley and Sons Inc.

Fregoni, M. (1980). *Nutrizione e fertilizzazione dellavite*. Bologna, Italy: Agricole.

Frías-Moreno, M.N., Parra-Quezada, R.A., González-Aguilar, G. et al. (2021). Quality, bioactive compounds, antioxidant capacity, and enzymes of raspberries at different maturity stages, effects of organic vs. conventional fertilization. *Foods* 10 (5): 953. https://doi.org/10.3390/foods10050953.

Gad, N., El-Sherif, M.H., and El-Gereedly, N.H.M. (2007). Influence of nickel on some physiological aspects of tomato plants. *Aust J Basic Appl Sci* 1 (3): 286–293.

Glisic, P.I., Milosevic, M.T., Glisic, S.I., and Milosevic, T.N. (2008). The effect of natural zeolites and organic fertilizers on the characteristics of degraded soils and yield of crops grown in Western Serbia. *Land Degrad Dev* 20 (1): 33–40. https://doi.org/10.1002/ldr.875.

Griffiths, K., Aggarwal, B.B., Singh, R.B. et al. (2016). Food Antioxidants and Their Anti-Inflammatory Properties: a potential role in cardiovascular diseases and cancer prevention. *Diseases* 4 (3): 28. https://doi.org/10.3390/diseases4030028.

Grotz, N. and Guerinot, M.L. (2002). Molecular aspects of Cu, Fe and Zn homeostasis in plants. *Biochim Biophys Acta* 1763 (7): 595–608. https://doi.org/10.1016/j.bbamcr.2006.05.014.

Grusak, M.A. and Della Penna, D. (1999). Improving nutrients composition of plants to enhance human nutrition and health. *Annu Rev Plant Physiol Plant Mol Biol* 50: 133–161. https://doi.org/10.1146/annurev.arplant.50.1.133.

Hanson, E. (1996). *Fertilizing Fruit Crops*, Extension Bulletin E-8S2. E. Lansing, MI: Michigan State University Extension.

Hasani, M., Zamani, Z., Savaghebi, G., and Fatahi, R. (2012). Effects of zinc and manganese as foliar spray on pomegranate yield, fruit quality and leaf minerals. *J Soil Sci Plant Nutr* 12 (3): 471–480. http://dx.doi.org/10.4067/S0718-95162012005000009.

Havlin, J.L., Tisdale, S.L., Nelson, W.L., and Beaton, J.D. (2005). *Soil Fertility and Fertilizers: An Introduction to Nutrient Management*, 7e. NJ: Pearson Prentice Hall.

Heckman, J. (2004). *Leaf Analysis for Fruit Trees*, 1–2. New Brunswick, NJ: Rutgers Cooperative Research & Extension NJAES, Rutgers the State University of New Jersey.

Huang, X.J. and Salt, D.E. (2016). Plant ionomics: from elemental profiling to environmental adaptation. *Mol Plant* 9 (6): 787–797. https://doi.org/10.1016/j.molp.2016.05.003.

Hudina, M. and Stampar, F. (2002). Effect of phosphorus and potassium foliar fertilization on fruit quality of pears. *Acta Hortic* 594: 487–493. https://doi.org/10.17660/ActaHortic.2002.594.63.

Huett, D.O. and Dirou, J.F. (2000). An evaluation of the rationale fertilizer management of tropical fruit crops. *Anim Prod Sci* 40 (8): 1137–1143. https://doi.org/10.1071/EA00047.

Idowu, M.K. and Aduayi, E.A. (2007). Sodium-potassium interaction on growth, yield and quality of tomato in ultisol. *J Plant Interact* 2 (4): 263–271. https://doi.org/10.1080/17429140701713803.

Isuzugawa, K., Murayama, H., and Nishio, T. (2014). Characterization of a giant-fruit mutant exhibiting fruit-limited polyploidization in pear (*Pyrus communis* L.). *Sci Hortic* 170: 196–202. https://doi.org/10.1016/j.scienta.2014.03.009.

Izawa, S., Heath, R.L., and Hind, G. (1969). The role of chloride ion in photosynthesis. III. The effect of artificial electron donors upon electron transport. *Biochim Biophys Acta* 180 (2): 388–398. https://doi.org/10.1016/0005-2728(69)90123-6.

Jemrić, T., Vuković, M., and Milošević, T. (2017). How can fruit production be made more sustainable? *CAB Reviews* 12 (024): 1–14. https://doi.org/10.1079/PAVSNNR201712024.

Jiménez, S., Garín, A., Gogorcena, Y. et al. (2004). Flower and foliar analysis for prognosis of sweet cherry nutrition: influence of different rootstocks. *J Plant Nutr* 27 (4): 701–712. https://doi.org/10.1081/PLN-120030376.

Jiménez, S., Pinochet, J., Gogorcena, Y. et al. (2007). Influence of different vigor cherry rootstocks on leaves and shoots mineral composition. *Sci Hortic* 112 (1): 73–79. https://doi.org/10.1016/j.scienta.2006.12.010.

Johnson, R.S. and Uriu, K. (1989). Mineral nutrition. In: *Peach, Plum and Nectarine: Growing and Handling for Fresh Market* (ed. J. LaRue and R.S. Johnson), 68–81. Oakland, CA: University of California Cooperative Extension, Division of Agriculture and Natural Resources.

Kader, A.A. (2008). Flavor quality of fruits and vegetables. *J Sci Food Agric* 88 (11): 1863–1868. https://doi.org/10.1002/jsfa.3293.

Karagiannis, E., Michailidis, M., Skodra, C. et al. (2021). Silicon influenced ripening metabolism and improved fruit quality traits in apples. *Plant Physiol Biochem* (in press) https://doi.org/10.1016/j.plaphy.2021.05.037.

Khojastehpour, M. (1996). Design and construction method of potato sorting machine. MSc thesis. Faculty of Bio-systems Engineering, University of Tehran. (in Persian).

Komosa, A. and Górniak, T. (2015). The effect of chloride on yield and nutrient interaction in greenhouse tomato (*Lycopersicon esculentum* Mill.) grown in rockwool. *J Plant Nutr* 38 (3): 355–370. https://doi.org/10.1080/01904167.2014.934466.

Krüger, E., Schmidt, G., and Rasim, S. (2002). Effect of irrigation on yield, fruit size and firmness of strawberry cv. Elsanta. *Acta Hortic* 567: 471–474. https://doi.org/10.17660/ActaHortic.2002.567.99.

Kumar, P., Rouphaelm, Y., Cardarelli, M., and Colla, G. (2015). Effect of nickel and grafting combination on yield, fruit quality, antioxidative enzyme activities, lipid peroxidation, and mineral composition of tomato. *J Plant Nutr Soil Sci* 178 (6): 848–860. https://doi.org/10.1002/jpln.201400651.

Kumar, S., Nagar, A., and Patidar, O.P. (2016). *Secondary Metabolites in Fruit Crops: Health Benefits*. Biotech Articles. https://www.biotecharticles.com/Agriculture-Article/Secondary-Metabolites-in-Fruit-Crops-Health-Benefits-3663.html.

Labate, J.A., Breksa, A.P., Robertson, L.D. et al. (2018). Genetic differences in macro-element mineral concentrations among 52 historically important tomato varieties. *Plant Genet Resour* 16 (4): 343–351. https://doi.org/10.1017/S1479262117000417.

Lahner, B., Gong, J., Mahmoudian, M. et al. (2003). Genomic scale profiling of nutrient and trace elements in *Arabidopsis thaliana*. *Nat Biotechnol* 21 (10): 1215–1221. https://doi.org/10.1038/nbt865.

Launaskas, J. and Kvikliene, N. (2006). Effect of calcium foliar application on some fruit quality characteristics of "Sinap Orovskij" apple. *Agron Res* 4 (1): 31–36.

Layne, D. and Bassi, D. (2008). *The Peach: Botany, Production and Uses*. Wallingford: CABI.

Leece, D.R. (1975a). Diagnostic leaf analysis for stone fruit. 4. Plum. *Aust J Exp Agric Anim Husb* 15 (72): 112–117. https://doi.org/10.1071/EA9750112.

Leece, D.R. (1975b). Diagnostic leaf analysis for stone fruit. 5. Sweet cherry. *Aust J Exp Agric Anim Husb* 15 (72): 118–122. https://doi.org/10.1071/EA9750118.

Leece, D.R. and van den Ende, B. (1975). Diagnostic leaf analysis for stone fruit. 6. Apricot. *Aust J Exp Agric Anim Husb* 15 (72): 123–128. https://doi.org/10.1071/EA9750123.

Leece, D.R., Cradock, F.W., and Carter, O.G. (1971). Development of leaf nutrient concentration standards for peach trees in New South Wales. *J Hortic Sci* 46 (2): 163–175. https://doi.org/10.1080/00221589.1971.11514395.

Lester, G.E., Jifon, J.L., and Makus, D.J. (2010). Impact of potassium nutrition on food quality of fruits and vegetables: a condensed and concise review of the literature. *Better Crops* 94 (1): 18–21.

Liu, G., Simonne, E.H., and Li, Y. (2020). *Nickel Nutrition in Plants*, Horticultural Sciences Department, UF/IFAS Extension, No HS1191. UF/IFAS Extension Service, University of Florida.

Lobos, T.E., Retamales, B.R., and Hanson, E.J. (2021). Early preharvest calcium sprays improve postharvest fruit quality in "Liberty" highbush blueberries. *Sci Hortic* 277: 109790. https://doi.org/10.1016/j.scienta.2020.109790.

Marschner, H. (1995). *Mineral Nutrition of Higher Plants*, 2e. London: Academic Press 899 p. https://doi.org/10.1016/B978-0-12-473542-2.X5000-7.

Marschner, H. (2012). *Marschner's Mineral Nutrition of Higher Plants*, 3e. London, UK: Academic Press https://doi.org/10.1016/C2009-0-63043-9.

Martínez Barroso, M.C. and Alvarez, C.E. (1997). Toxicity symptoms and tolerance of strawberry to salinity in the irrigation water. *Sci Hortic* 71 (3-4): 177–188. https://doi.org/10.1016/S0304-4238(97)00082-4.

Matsuoka, K., Moritsuka, N., Kusaba, S., and Hiraoka, K. (2018). Effects of soil type and soil treatment on solubilization of 13 elements in the root zone and their absorption by blueberry bushes. *Hortic J* 87 (2): 155–165. https://doi.org/10.2503/hortj.OKD-100.

Matsuoka, K., Moritsuka, N., Nukada, M. et al. (2019). Identification of the effects of long-term application of nitrogen fertilizer on the availability of 11 elements in soils and their supply to "Jonathan" apple trees by using simultaneous multi-element analysis. *J Plant Nutr* 42 (16): 1952–1966. https://doi.org/10.1080/01904167.2019.1648673.

Michailides, T.J., Ramirez, H.T., Morgan, D.P. et al. (1993). *Effects of nitrogen fertilization on susceptibility of stone fruits (peach and nectarine) to brown rot*. 1992 research reports for California peaches and nectarines. California Tree Fruit Agreement, Sacramento, California.

Milošević, N. and Milošević, T. (2012). Seasonal changes and content of sodium in main organs of European plum trees (*Prunus domestica* L.), fruit size and yield as affected by rootstocks on acidic soil. *Semin-Cienc Agrar* 33 (2): 605–619. https://doi.org/10.5433/1679-0359.2012v33n2p605.

Milošević, T. and Milošević, N. (2015). Apple fruit quality, yield and leaf macronutrients content as affected by fertilizer treatment. *J Soil Sci Plant Nutr* 15 (1): 76–83. http://dx.doi.org/10.4067/S0718-95162015005000007.

Milošević, T. and Milošević, N. (2019). Soil fertility: Plant nutrition vis-à-vis fruit yield and quality of stone fruits. In: *Fruit Crops: Diagnosis and Management of Nutrient Constraints* (ed. A.K. Sirvastava and H. Chengxiao), 583–605. Paperback: ©Elsevier Inc. https://doi.org/10.1016/B978-0-12-818732-6.00041-1.

Milosevic, T., Milosevic, N., Glisic, I., and Paunovic, G. (2009). Leaf nutritional status and macronutrients dynamics in European hazelnut (*Corylus avellana* L.) under Western Serbian conditions. *Pak J Bot* 41 (6): 3169–3178.

Milošević, T., Milošević, N., Glišić, I., and Mladenović, J. (2012). Fruit quality attributes of blackberry grown under limited environmental conditions. *Plant Soil Environ* 58 (7): 322–327. https://doi.org/10.17221/33/2012-PSE.

Milošević, T., Milošević, N., Glišić, I. et al. (2013a). Fertilization effect on trees and fruits characteristics and leaf nutrient status of apricots which are grown at Cacak region (Serbia). *Sci Hortic* 164 (16): 112–123. https://doi.org/10.1016/j.scienta.2013.09.028.

Milošević, T., Milošević, N., and Glišić, I. (2013b). Tree growth, yield, fruit quality attributes and leaf nutrient content of "Roxana" apricot as influenced by natural zeolite, organic and inorganic fertiliser. *Sci Hortic* 156: 131–139. https://doi.org/10.1016/j.scienta.2013.04.002.

Milošević, T., Milošević, N., and Mladenović, J. (2019). Tree vigor, yield, fruit quality, and antioxidant capacity of apple (*Malus* × *domestica* Borkh.) influenced by different fertilization regimes: preliminary results. *Turk J Agric For* 43 (1): 48–57. https://doi.org/10.3906/tar-1803-109.

Mohamed, M.A. and Ahmed, F.F. (1991). Yield and quality of anna apple cultivar fruits as affected by application of copper, zinc and iron nutrients. *Annal Agric Sci Moshtohor* 29 (1): 513–525.

Moss, G.I. and Higgins, M.L. (1974). Magnesium influences on the fruit quality of sweet orange (*Citrus sinensis* L. Osbeck). *Plant Soil* 41: 103–112. https://doi.org/10.1007/BF00017948.

Murovhi, N.R. (2013). Effect of different rates of sulphur fertilization on fruit yield and leaf mineral composition of "Valensia" oranges in the subtropical environment of South Africa. *Acta Hortic* 1007: 303–310. https://doi.org/10.17660/ActaHortic.2013.1007.33.

Nashima, K., Takahashi, H., Nakazono, M. et al. (2013). Transcriptome analysis of giant pear fruit with fruit specific DNA reduplication on a mutant branch. *J Jpn Soc Hortic Sci* 82 (4): 301–311. https://doi.org/10.2503/jjshs1.82.301.

Nashima, K., Terakami, S., Nishitani, C. et al. (2014). Transcriptome analysis of flower receptacles of the European pear (*Pyrus communis* L.) "La France" and its giant fruit sport using next-generation sequencing technology. *J Hortic Sci Biotechnol* 89 (3): 293–300. https://doi.org/10.1080/14620316.2014.11513082.

Nasreen, S., Kamal, M.A., Siddiky, A.M. et al. (2014). Effect of nitrogen, phosphorus, potassium and sulphur on the yield of mango. *Bangladesh J Agric Res* 39 (4): 631–640.

Nedim, O. and Daml, B.O. (2015). Effect of Mg fertilization on some plant nutrient interactions and nut quality properties in Turkish hazelnut (*Corylus avellana* L.). *Sci Res Essays* 10 (14): 465–470. https://doi.org/10.5897/SRE2014.5811.

Njira, K. and Nabwami, J. (2015). A review of effects of nutrient elements on crop quality. *Afr J Food Agric Nutr Dev* 15 (1): 9777–9783.

Obaid, E.A. and Al-Hadethi, M.E.A. (2013). Effect of foliar application with manganese and zinc on pomegranate growth, yield and fruit quality. *J Hortic Sci Ornam Plants* 5 (1): 41–45. https://doi.org/10.5829/idosi.jhsop.2013.5.1.273.

Osman, E.A.M., Fatma, I.I., Abou, G., and Ismaiel, A.A. (2014). Influence of phosporus fertilizer on growth, fruiting parameters and fruit quality of peach trees irrigated by acid water. *J Soil Sci Agric Eng* 5 (4): 529–542. https://doi.org/10.21608/jssae.2014.49301.

Palit, S., Sharma, A., and Talukder, G. (1994). Effects of cobalt on plants. *Bot Rev* 60: 149–181. https://doi.org/10.1007/BF02856575.

Parent, L.E., Rozane, D.E., de Deus, J.A.L., and Natale, W. (2019). Diagnosis of nutrient composition in fruit crops: major developments. In: *Fruit Crops. Diagnosis and Management of Nutrient Constraints* (ed. K.A. Srivastava and H. Chengxiao), 145–156. Paperback: © Elsevier Inc. https://doi.org/10.1016/B978-0-12-818732-6.00012-5.

Petrovic, S. and Milosevic, T. (1999). Economic importance, tendencies and problems in apricot production in Yugoslavia. *Acta Hortic* 488: 77–82. https://doi.org/10.17660/ActaHortic.1999.488.7.

Pezzarossa, B., Remorini, D., Gentile, M.L., and Massai, R. (2012). Effects of foliar and fruit addition of sodium selenate on selenium accumulation and fruit quality. *J Sci Food Agric* 92 (4): 781–786. https://doi.org/10.1002/jsfa.4644.

Poulik, Z. (1999). Influence of nickel contaminated soils on lettuce and tomatoes. *Sci Hortic* 81 (3): 243–250. https://doi.org/10.1016/S0304-4238(99)00023-0.

Radi, M., Mahrouz, M., Jaouad, A., and Amiot, M.J. (2003). Influence of mineral fertilization (NPK) on the quality of apricot fruit (cv. Canino). The effect of the mode of nitrogen supply. *Agronomie* 23 (8): 737–745. https://doi.org/10.1051/agro:2003052.

Raese, J.T. (1987). Potassium and phosphorus. *J Plant Nutr* 10 (9–16): 2007–2015. https://doi.org/10.1080/01904168709363747.

Rafiullah, T.M., Khan, F., Shah, A.H. et al. (2020). Effect of micronutrients foliar supplementation on the production and eminence of plum (*Prunus domestica* L.). *Qual Assur Saf Crop Foods* 12 (SP1): 32–40. https://doi.org/10.15586/qas.v12iSP1.793.

Robinson, A.B. and Pauling, L. (1974). Techniques of orthomolecular diagnosis. *Clin Chem* 20 (8): 961–965. https://doi.org/10.1093/clinchem/20.8.961.

Rubio Ames, Z., Brecht, J.K., and Olmstead, M.A. (2020). Nitrogen fertilization rates in a subtropical peach orchard: effects on tree vigor and fruit quality. *J Sci Food Agric* 100 (2): 527–539. https://doi.org/10.1002/jsfa.10031.

Safwat, M. and Moustafa, A. (1971). Theoretical prediction of volume, surface area, and center of gravity for agricultural products. *Trans ASAE* 14 (3): 549–553. https://doi.org/10.13031/2013.38336.

Saini, H., Vijay, S., and Saini, P. (2019). Differential responses of Fe, Zn, B, Cu and Mg on growth and quality attributes of fruit crops. *J Pharmacog Phytochem* 8 (5): 1–5.

Salt, D.E., Baxter, I., and Lahner, B. (2008). Ionomics and the study of the plant ionome. *Annu Rev Plant Biol* 59 (1): 709–733. https://doi.org/10.1146/annurev.arplant.59.032607.092942.

Seregin, I.V. and Kozhevnikova, A.D. (2006). Physiological role of nickel and its toxic effects on higher plants. *Russ J Plant Physiol* 53 (2): 257–277. https://doi.org/10.1134/S1021443706020178.

Shear, C.B. and Faust, M. (1980). Nutritional ranges in deciduous tree fruits and nuts. *Hortic Rev* 2: 142–163. https://doi.org/10.1002/9781118060759.ch3.

Shibuya, T., Watanabe, T., Ikeda, H., and Kanayama, Y. (2015). Ionomic analysis of horticultural plants reveals tissue-specific element accumulation. *Hortic J* 84 (4): 305–313. https://doi.org/10.2503/hortj.MI-058.

Shiratake, K. and Suzuki, M. (2016). Omics studies of citrus, grape and rosaceae fruit trees. *Breed Sci* 66 (1): 122–138. https://doi.org/10.1270/jsbbs.66.122.

Sing, Z. and Sing, L. (1993). Effect of cobalt ions on floral malformation, yield and fruit quality of “Dusheri” mango (*Mangifera indica* L.). *J Hortic Sci* 68 (4): 535–540. https://doi.org/10.1080/00221589.1993.11516382.

Singh, U.M., Sareen, P., Sengar, R.S., and Kumar, A. (2013). Plant ionomics: a newer approach to study mineral transport and its regulation. *Acta Physiol Plant* 35 (9): 2641–2653. https://doi.org/10.1007/s11738-013-1316-8.

Singh, S., Thakur, A., and Singh, S.K. (2018). Secondary metabolites in fruit crop - a mini review. *Int J Res Anal Rev* 5 (4): 1–10.

Stojanov, D., Milošević, T., Mašković, P., and Milošević, N. (2019a). Impact of fertilization on the antioxidant activity and mineral composition of red raspberry berries of cv. "Meeker". *Mitt Klosterneuburg* 69 (3): 184–195.

Stojanov, D., Milošević, T., Mašković, P. et al. (2019b). Influence of organic, organo-mineral and mineral fertilisers on cane traits, productivity and berry quality of red raspberry (*Rubus idaeus* L.). *Sci Hortic* 252: 370–378. https://doi.org/10.1016/j.scienta.2019.04.009.

Subbarao, G.V., Ito, O., Berry, W.L., and Wheeler, R.M. (2003). Sodium - a functional plant nutrient. *Crit Rev Plant Sci* 22 (5): 391–416. https://doi.org/10.1080/07352680390243495.

Sugar, D., Righetti, T.L., Sanchez, E.E., and Khemira, H. (1992). Management of nitrogen and calcium in pear trees for enhancement of fruit resistance to postharvest decay. *HortTechnology* 2: 382–387.

Svenja, B. and Sven, S. (2009). Chloride improves fruit yield and quality of strawberry (*Fragaria* × *ananassa* Duch.). *The Proceedings of the International Plant Nutrition Colloquium XVI*, Department of Plant Sciences, UC Davis. https://escholarship.org/uc/item/9ck8q8n2.

Sylvia, D.M., Furhmann, J.J., Hartel, P.G., and Zuberer, D.A. (2005). *Principles and Application of Soil Microbiology*, 2e. New Jersey: Pearson Prentice Hall.

Tagliavini, M. and Marangoni, B. (2002). Major nutritional issues in deciduous fruit orchards of northern Italy. *HortTechnology* 12 (1): 26–31. https://doi.org/10.21273/horttech.12.1.26.

Tagliavini, M., Zavalloni, C., Rombola, A.D. et al. (2000). Mineral nutrient partitioning to fruits of deciduous trees. *Acta Hortic* 512: 131–140. https://doi.org/10.17660/ActaHortic.2000.512.13.

Taiz, L. and Zeiger, E. (2004). *Fisiologia Vegetal*, 5e, 719. Porto Alegre: Artmed.

Tomala, K. and Trzak, M. (1994). Occurrence of cork spot (pit) in "Alexander Lucas" pears depends on fruit mineral element content. *Acta Hortic* 368: 570–577. https://doi.org/10.17660/ActaHortic.1994.368.66.

Toselli, M., Thalheimer, M., and Tagliavini, M. (2004). Leaf uptake and partitioning of urea-N as affected by the concentration and volume of spray solution and by the shoot leaf position in apple (*Malus domestica*) trees. *J Hortic Sci Biotechnol* 79 (1): 97–100. https://doi.org/10.1080/14620316.2004.11511742.

Tratch, R., May de Mio, L.L., and Serrat BM Motta, A.C.V. (2010). Nitrogen and potassium fertilization influences on intensity of peach leaf rust. *Acta Hortic* 872: 313–318. https://doi.org/10.17660/ActaHortic.2010.872.43.

Tulchinsky, T.H. (2010). Micronutrient deficiency conditions: global health issues. *Public Health Rev* 32 (1): 243–255. https://doi.org/10.1007/BF03391600.

Usenik, V. and Stampar, F. (2002). Effect of foliar application of zinc plus boron on sweet cherry fruit set and yield. *Acta Hortic* 594: 245–249. https://doi.org/10.17660/ActaHortic.2002.594.28.

Vashisth, T., Olmstead, M.A., Olmstead, J., and Colquhoun, T.A. (2017). Effects of nitrogen fertilization on subtropical peach fruit quality: organic acids, phytochemical content, and total antioxidant capacity. *J Amer Soc Hort Sci* 142 (5): 393–404. https://doi.org/10.21273/JASHS04011-16.

Walker, N.A., Sanders, D., and Maathuis, F.J.M. (1996). High-affinity potassium uptake in plants. *Science* 273 (5277): 977–979. https://doi.org/10.1126/science.273.5277.977.

Wills, R.B.H., Scriven, F.M., and Greenfield, H. (1983). Nutrient composition of stone fruit (Prunus spp.) cultivars: Apricot, cherry, nectarine, peach and plum. *J Sci Food Agric* 34 (12): 1383–1389. https://doi.org/10.1002/jsfa.2740341211.

Wilson, D.W., Nash, P., Buttar, H.S. et al. (2017). The role of food Antioxidants, benefits of functional foods, and influence of feeding habits on the health of the older person: an overview. *Antioxidants* 6 (4): 81. https://doi.org/10.3390/antiox6040081.

Wójcik, P. and Popiñska, W. (2009). Response of Lukasovka pear trees to foliar zinc sprays. *J Elem* 14 (1): 181–188. https://doi.org/10.5601/jelem.2009.14.1.19.

Wójcik, P. and Wójcik, M. (2003). Effects of boron fertilization on "Conference" pear tree vigor, nutrition and fruit yield and storability. *Plant Soil* 256 (2): 413–421. https://doi.org/10.1023/A:1026126724095.

9

Nutrients Use Efficiency in Plants

Neda Dalir

Department of Soil Science, Tarbiat Modares University, Tehran, Iran

Introduction

One of the most important challenges to achieve food security is to improve global food production. Increased food production could be achieved by expanding the land area under crops and by increasing yields per unit area through intensive farming. The intensive cultivation of agricultural soils inevitably accelerates the degradation of land and lowers its fertility and productivity (Gomiero 2016). Moreover, many agricultural soils of the world lack one or more essential nutrients for satisfactory growth and quality production. Due to the limited cultivable area, and the ever-increasing demand for food, the application rate of chemical fertilizers has been enhanced by farmers all over the world to ensure high yield. During the past decade, high quantities of inorganic fertilizer, particularly N, have been used to increase world food production (Fageria 2002). However, there has been poor development of fertilizer recommendations. Excessive fertilizer application does not guarantee constantly increasing yields, but it might even decrease the nutrient utilization efficiency and lead to a negative impact on soil quality and environmental problems in agro-ecosystems. The proper application rate of fertilizers or amendments at the right time of year, with the right method and placement, play important roles in enhancing crop nutrient efficiency (Grimme et al. 1984).

Nutrient Use Efficiency (NUE) is a critically important concept in agricultural productivity, which can be highly affected by fertilizer management as well as soil and plant water management. NUE shows the ability of crops to take up nutrients efficiently from the soil, but it also depends on the internal transport, storage, and remobilization of nutrients (Reich et al. 2014). Genetic variation within and among crops for NUE is well recognized. Therefore, the three main important processes involved in NUE are uptake, assimilation, and utilization of nutrients (Kaur et al. 2017).

The efficiency of crops in using nutrients from applied fertilizers may be between 30 to 50% depending upon the nature of the crop, climate, soil, and management practices (Panhwar et al. 2019).

Plant Ionomics: Sensing, Signaling, and Regulation, First Edition.
Edited by Vijay Pratap Singh and Manzer H. Siddiqui.

There is a great effort by the fertilizer industry to promote approaches to fertilizer nutrients to meet the global demand for food, although there is the limited fertilizer resources available. Considering the challenges ahead, it is crucial to increase the efficiency of nutrient use by plants. The improvement of NUE by plants could reduce fertilizer input costs, decrease loss of nutrients, and enhance crop yields. Effectiveness of nutrient supply will be increased by matching nutrient supply with plant needs and maintaining nutrient availability (Rajan et al. 2021). This could be achievable by integrated nutrient management considering both physiological and agronomic traits.

In this chapter we assess the concept of NUE and its effect on improving crop yield and quality and also investigate different physiological mechanisms in plant NUE.

Nutrient Use Efficiency (Concepts and Importance)

NUE is a measure of how well plants use the available mineral nutrients. It depends on the ability to take up the nutrients from the soil, but also on transport, storage, mobilization, usage within the plant, and even on the environment. In this regard, the NUE is determined based on (i) uptake efficiency (acquire from soil and short-distance translocation), and (ii) utilization efficiency including long-distance translocation of nutrient, compartmentation of nutrients, and partitioning within the plant (remobilization and redistribution) (Baligar et al. 2001).

From the operational sense, NUE is often divided into more specific categories. The first and common definition of NUE emphasizes on productivity and yield parameter in which nutrient efficiency is defined as the ability to produce a maximum yield in a soil, or other media under limiting conditions (Buso and Bliss 1988). The second category is agronomic efficiency, in which the emphasis is on the internal nutrient requirement of the plant which refers to plant production and include quantity and quality of the desired yield product, per unit input (fertilizer, nutrient content) (Hawkesford et al. 2016).

Another suggested concept is apparent nutrient recovery (ANR), a more complex form of NUE and most commonly defined as the difference in nutrient uptake in above-ground parts of the plant between the fertilized and unfertilized crop relative to the quantity of nutrient applied (Fixen et al. 2015).

Role of Nutrient-Efficient Plants for Improving Crop Yields

Currently, the genetic variation in NUE in plants is well documented. A nutrient-efficient plant have potential to produce higher yield under conditions of relatively low nutrient supply compared to a standard genotype (Sattelmacher et al. 1994).

Many researchers have since reported nutrient efficiency in different genotypes supplied with insufficient amounts of nutrients (Fageria and Stone 2006; Hacisalihoglu et al. 2004; Yan et al. 2006). In this regard, several mechanisms have been proposed to explain Zn efficiency in crop plants, e.g. efficient metabolic Zn utilization, within-plant (re)-translocation of Zn, high efficient root Zn uptake, or the ability to mobilize soil Zn by exudation of phytosiderophores into the rhizosphere (Cakmak et al. 1994; Dalir et al. 2017; Hart et al. 1998;

Rengel and Römheld 2000). Dalir et al. (2017) suggested that differences in Zn compartmentalization may play a critical role in the physiological mechanisms of Zn efficiency in wheat, since the root-to-shoot translocation differed in the two genotypes with the Zn inefficient genotype showing a higher translocated Zn fraction than the Zn efficient one at high Zn supply. Hart et al. (1998) attributed the difference in Zn efficiency between two wheat cultivars to a more efficient high-affinity uptake system in the Zn efficient cultivar, enabling the plant to take up more Zn at very low concentrations in the experimental nutrient solution (0.01–200 nM). High-affinity Zn uptake systems are thought to play an important role for Zn uptake at Zn concentrations in soil solution between 1 nm and 1 mM, which are typical for many soils (Welch and Shuman 1995). A similar explanation was proposed by Meng et al. (2014) who found significantly higher Zn influx rates for a Zn efficient genotype than for an inefficient genotype under low Zn conditions in rice. In contrast, Hacisalihoglu et al. (2001) found similar uptake rates in Zn efficient and inefficient wheat genotypes at Zn solution concentrations (0–160 nM Zn^{2+} activity) under which the high-affinity uptake system is expected to be active and concluded that Zn efficiency is not necessarily related to the ability of roots to take up more Zn under low Zn conditions. Instead, they suggested that differences in Zn compartmentalization, i.e. in vacuolar Zn storage, was the reason for the different performance of their cultivars under Zn deficient conditions.

Physiological Mechanisms in Plant Nutrient Use Efficiency

Uptake Efficiency

Uptake efficiency defined as acquire nutrient from root media and radial transport in roots. Mechanisms of nutrient acquisition include alterations to the chemical and biological properties of the rhizosphere to increase nutrient availability, increases in the volume of soil explored by increased root growth, changed root architecture, interactions with microbial populations in the rhizosphere, and changes in the expression of ion transporters in the roots to enhance uptake (Fageria and Stone 2006; Li et al. 2016).

Acquisition of Available Nutrients

Root Morphology The spatial configuration and distribution of root system in the growth medium determines the media exploration in time and space (Lunch and Brown 2001). The potential variation in root morphology parameters such as length, thickness, surface areas, volume, and root hairs plays an important role in plant NUE acquisition. Increasing root to shoot ratio is an efficient strategy to enhance nutrient acquisition especially under nutrient deficient conditions. In the case of P, for example, enhanced root growth under its low availability has been proposed as an effective adaptive strategy that enables a plant to explore a larger P pool (Wissuwa et al. 2016). The allocation of biomass to different plant organs, of course, depends on not only species but also the growth conditions such as nutritional status (Poorter and Nagel 2000). According to the optimal allocation theory, plants should allocate more heavily to organs that capture the most limiting resource and less to organs that are involved in obtaining nonlimiting resources (Niklas and Enquist 2002; Weiner 2004). On the other hand, increase the overall surface area of the root per unit mass can also improve nutrient acquisition without an increase in carbon allocation to the root. In this regard, greater numbers of smaller diameter roots, i.e. root hairs, rather than less large root can facilitate nutrient uptake primarily by increasing the root surface and enable plants to efficiently increase nutrient access and improve the volume of media that can be explored for nutrient (Misra et al. 1988). Root hairs are believed to contribute 70–90% of the total root surface (Brown et al. 2012; Raghothama 2005) and comprise a cheap, cost-effective strategy trait for improving nutrient acquisition by plants (Brown et al. 2013).

Root Physiology In addition to root morphological traits that could significantly affect ion uptake rates, the potential variation among genotypes regarding their transporter types and densities determine the physiological NUE. Nutrients can be absorbed by plant roots both passively and actively. Passive uptake involves diffusion of ions along a chemical potential (concentration) gradient across cell wall and intercellular spaces which are fully permeable (the apoplast path), without the interference of active transporters. Although, the passive transport of solutes may also occur across the membrane transport proteins when nutrients are found in high external concentrations (Griffiths and York 2020).

Active ion uptake takes place against the concentration gradient with high selectivity of ions and energy-consuming mechanism (Marschner 1995). When nutrients are scarce, ATP-mediated active transport is likely to occur, which reflects the resource acquisition capability of the plant (Griffiths and York 2020). Michaelis–Menten equation (Michaelis and Menten 1913) is an uptake model used by plant physiologists to describe root nutrient uptake from the growth medium (Gardiner et al. 1990). This mathematical model provides a basis for quantifying the rate of nutrient uptake and its dependence on the concentration in soil solution at the root surface (Claassen et al. 1986). It has been traditionally assumed that plant evolve two types of system for nutrient uptake. The low-and high-affinity uptake systems acts at high and lower nutrient concentrations, respectively (Epstein 1972). Regarding this, the up-regulation of high-affinity P transports under low P concentrations in the growth medium is a key component of P uptake by plants at the very low P supply (Dong et al. 1999; Liu et al. 1998).

Increasing Nutrient Availability

Plant roots can modify rhizosphere processes through their physiological activities, particularly the exudation low-molecular-weight organic compounds to enhance the acquisition of nutrients. Root exudates affect the physicochemical conditions and the bioavailability of nutrients whose diffusion is the main important mechanism to the root surface (e.g. Fe, Zn, and P) and also has a great influence on the uptake of limiting nutrients (Marschner 2011). Enhanced root exudation of phytosiderophores (PS) and direct uptake of the Fe(III)–PS complex by grass roots have been well demonstrated (Reichman and Parker 2005). The quantity and quality of root exudates are determined mainly by the cultivar and plant developmental stage (Gransee and Wittenmayer 2000; Xue et al. 2013). Root exudates can improve the bioavailability of nutrients in the rhizosphere by influencing soil pH, competing for mineral adsorption sites, chelating mineral nutrients, and solubilizing soil minerals (Raynaud et al. 2008). Root exudates also indirectly affect plant growth by stimulating microbial activity and population growth (Paterson et al. 2006). Among several substances identified in root exudates, sugars, amino acids (AA), and organic acids have drawn considerable interest due to their role in processes at the root–soil interface such as metal chelating (Fageria and Stone 2006; Jones et al. 2004). However, it has been shown that some amino acids had little effect in mobilizing metal micronutrients in soils, due to their rapid microbial degradation (Jones 1999; Jones et al. 1994). On the other hand, concentrations of AA in the soil solution are often much higher than those of trace elements (Schwab et al. 2008). Therefore, AA may play a significant role in forming complexes with metals and increasing their bioavailability.

Mycorrhizal Symbiosis An effective plant strategy to cope with low nutrient availability in root medium is the formation of a symbiosis with mycorrhizal fungi, which has been reported in almost all plant species (Begum et al. 2019). Mycorrhizal symbiosis may affect several aspects of plant root morphology such as changes in the root-to-shoot ratio (Henkes et al. 2018), root length, root volume and branching number of the host plant (Chen et al. 2020). Arbuscular mycorrhizal fungi (AM) could improve plant nutrition by promoting acquisition of water and nutrients, especially P and N in soil (Harrison 1999). Mycorrhizal symbiosis is reported to positively increased the concentrations of N, Fe and Zn in wheat as well as promoting plant root morphological characteristics such as root biomass, specific root length, and root density (Ingraffia et al. 2019).

Utilization Efficiency

In addition to the uptake efficiency, the utilization efficiency would determine the efficiency in which the absorbed nutrients are utilized to produce yield. In this regard, loading into the xylem and long distance transport of metal ions toward the aboveground organs is of high importance for plant nutrition (Tiffin 1966). In long-distance transport process, water and minerals are taken up from the root medium and predominantly translocated toward the upper plant parts by the vascular tissues of the xylem (Mengel and Kirkby 2012). Such transport is strongly regulated by metal–ligand complexes and by some proteins that specifically bind and transport the metal (Haydon and Cobbett 2007). The transport rate of di- or trivalent cations both in longitudinal and lateral directions has been reported to be enhanced significantly by complexation with organic acids (Perchlik and Tegeder 2017), amino acids (Dalir and Khoshgoftarmanesh 2014, 2015), or peptides (Nagajyoti et al. 2010). In contrast to reduced carbon that is translocated only by phloem, amino acids translocation occurs both in phloem and xylem. Therefore, re-translocation of amino acids helps in nitrogen recycling between roots and shoots and hasten translocation of immobile nutrient elements i.e., Zn in plant (Ortiz-Lopez et al. 2000).

Conclusion and Future Prospects

The increased productivity through NUE depends on matching nutrient supply with plant demand and also maintaining nutrient availability. Identification of precise factors related to nutritional traits such as nutrient absorption, transport and utilization in plants will allow the development of efficient plants, with improved nutrient efficiency and high yield.

Current knowledge and future research are directions for breeding programs for developing nutrient-efficient generations who have to sustain production in low-input agriculture, considering the inherent potential of crop plants.

References

Baligar, V., Fageria, N., and He, Z. (2001). Nutrient use efficiency in plants. *Communications in Soil Science and Plant Analysis* 32: 921–950.

Begum, N., Qin, C., Ahanger, M.A. et al. (2019). Role of arbuscular mycorrhizal fungi in plant growth regulation: implications in abiotic stress tolerance. *Frontiers in Plant Science* 10: 1068.

Brown, L., George, T., Thompson, J. et al. (2012). What are the implications of variation in root hair length on tolerance to phosphorus deficiency in combination with water stress in barley (Hordeum vulgare)? *Annals of Botany* 110: 319–328.

Brown, L., George, T., Dupuy, L., and White, P. (2013). A conceptual model of root hair ideotypes for future agricultural environments: what combination of traits should be targeted to cope with limited P availability? *Annals of Botany* 112: 317–330.

Buso, G. and Bliss, F. (1988). Variability among lettuce cultivars grown at two levels of available phosphorus. *Plant and Soil* 111: 67–73.

Cakmak, S., Gülüt, K.Y., Marschner, H., and Graham, R.D. (1994). Efect of zinc and iron deficiency on phytos1derophore release in wheat genotypes differing in zinc efficiency. *Journal of Plant Nutrition* 17: 1–17.

Chen, W., Meng, P., Feng, H., and Wang, C. (2020). Effects of arbuscular mycorrhizal fungi on growth and physiological performance of Catalpa bungei CA Mey. under drought stress. *Forests* 11: 1117.

Claassen, N., Syring, K., and Jungk, A. (1986). Verification of a mathematical model by simulating potassium uptake from soil. *Plant and Soil* 95: 209–220.

Dalir, N. and Khoshgoftarmanesh, A.H. (2014). Symplastic and apoplastic uptake and root to shoot translocation of nickel in wheat as affected by exogenous amino acids. *Journal of Plant Physiology* 171: 531–536.

Dalir, N. and Khoshgoftarmanesh, A.H. (2015). Root uptake and translocation of nickel in wheat as affected by histidine. *Journal of Plant Physiology* 184: 8–14.

Dalir, N., Tandy, S., Gramlich, A. et al. (2017). Effects of nickel on zinc uptake and translocation in two wheat cultivars differing in zinc efficiency. *Environmental and Experimental Botany* 134: 96–101.

Dong, B., Ryan, P., Rengel, Z., and Delhaize, E. (1999). Phosphate uptake in Arabidopsis thaliana: dependence of uptake on the expression of transporter genes and internal phosphate concentrations. *Plant, Cell and Environment* 22: 1455–1461.

Epstein, E. (1972). *Mineral Nutrition of Plants: Principles and Perspectives*, 68–82. New York: Wiley.

Fageria, N. (2002). Nutrient management for sustainable dry bean production in the tropics. *Communications in Soil Science and Plant Analysis* 33: 1537–1575.

Fageria, N. and Stone, L. (2006). Physical, chemical, and biological changes in the rhizosphere and nutrient availability. *Journal of Plant Nutrition* 29: 1327–1356.

Fixen, P., Brentrup, F., Bruulsema, T. et al. (2015). Nutrient/fertilizer use efficiency: measurement, current situation and trends. In: *Managing Water and Fertilizer for Sustainable Agricultural Intensification*, vol. 270 (ed. P. Drechsel, P. Heffer, H. Magen, et al.), 8–38. Paris: IFA.

Gardiner, D., Christensen, N., and Myrold, D. (1990). A comparison of methods for estimating phosphorus uptake kinetics under steady-state conditions. *Journal of Plant Nutrition* 13: 1079–1093.

Gomiero, T. (2016). Soil degradation, land scarcity and food security: reviewing a complex challenge. *Sustainability* 8: 281.

Gransee, A. and Wittenmayer, L. (2000). Qualitative and quantitative analysis of water-soluble root exudates in relation to plant species and development. *Journal of Plant Nutrition and Soil Science* 163: 381–385.

Griffiths, M. and York, L.M. (2020). Targeting root ion uptake kinetics to increase plant productivity and nutrient use efficiency. *Plant Physiology* 182: 1854–1868.

Grimme, H., Juo, A., Kang, B., (1984). Nutrient and Water Use Efficiency. International Institute of Tropical Agriculture, Annual Report for 1983, 154–156.

Hacisalihoglu, G., Hart, J.J., and Kochian, L.V. (2001). High-and low-affinity zinc transport systems and their possible role in zinc efficiency in bread wheat. *Plant Physiology* 125: 456–463.

Hacisalihoglu, G., Ozturk, L., Cakmak, I. et al. (2004). Genotypic variation in common bean in response to zinc deficiency in calcareous soil. *Plant and Soil* 259: 71–83.

Harrison, M.J. (1999). Molecular and cellular aspects of the arbuscular mycorrhizal symbiosis. *Annual Review of Plant Biology* 50: 361–389.

Hart, J.J., Norvell, W.A., Welch, R.M. et al. (1998). Characterization of zinc uptake, binding, and translocation in intact seedlings of bread and durum wheat cultivars. *Plant Physiology* 118: 219–226.

Hawkesford, M.J., Kopriva, S., and De Kok, L.J. (2016). *Nutrient Use Efficiency in Plants*. Springer.

Haydon, M.J. and Cobbett, C.S. (2007). Transporters of ligands for essential metal ions in plants. *New Phytologist* 174: 499–506.

Henkes, G.J., Kandeler, E., Marhan, S. et al. (2018). Interactions of mycorrhiza and protists in the rhizosphere systemically alter microbial community composition, plant shoot-to-root ratio and within-root system nitrogen allocation. *Frontiers in Environmental Science* 6: 117.

Ingraffia, R., Amato, G., Frenda, A.S., and Giambalvo, D. (2019). Impacts of arbuscular mycorrhizal fungi on nutrient uptake, N2 fixation, N transfer, and growth in a wheat/faba bean intercropping system. *PLoS One* 14: e0213672.

Jones, D. (1999). Amino acid biodegradation and its potential effects on organic nitrogen capture by plants. *Soil Biology and Biochemistry* 31: 613–622.

Jones, G.J., Bourne, D.G., Blakeley, R.L., and Doelle, H. (1994). Degradation of the cyanobacterial hepatotoxin microcystin by aquatic bacteria. *Natural Toxins* 2: 228–235.

Jones, D.L., Hodge, A., and Kuzyakov, Y. (2004). Plant and mycorrhizal regulation of rhizodeposition. *New Phytologist* 163: 459–480.

Kaur, B., Kaur, G., and Asthir, B. (2017). Biochemical aspects of nitrogen use efficiency: an overview. *Journal of Plant Nutrition* 40: 506–523.

Li, X., Zeng, R., and Liao, H. (2016). Improving crop nutrient efficiency through root architecture modifications. *Journal of Integrative Plant Biology* 58: 193–202.

Liu, C., Muchhal, U.S., Uthappa, M. et al. (1998). Tomato phosphate transporter genes are differentially regulated in plant tissues by phosphorus. *Plant Physiology* 116: 91–99.

Lunch, J. and Brown, K. (2001). Topsoil foraging: an architectural adaptation of plants to low phosphorus. *Plant and Soil* 237: 225–237.

Marschner, H. (1995). The soil-root interface (rhizosphere) in relation to mineral nutrition. In: *Mineral Nutrient of Higher Plants* (ed. H. Marschner), 537–595. London: Academic Press.

Marschner, H. (2011). *Marschner's Mineral Nutrition of Higher Plants*. Academic press.

Meng, F., Liu, D., Yang, X. et al. (2014). Zinc uptake kinetics in the low and high-affinity systems of two contrasting rice genotypes. *Journal of Plant Nutrition and Soil Science* 177: 412–420.

Mengel, K. and Kirkby, E.A. (2012). *Principles of Plant Nutrition*. Springer Science and Business Media.

Michaelis, L. and Menten, M. (1913). Die kinetik der invertinwirkung. *Biochemische Zeitschrift* 49: 333–369.

Misra, R., Alston, A., and Dexter, A. (1988). Role of root hairs in phosphorus depletion from a macrostructured soil. *Plant and Soil* 107: 11–18.

Nagajyoti, P.C., Lee, K.D., and Sreekanth, T. (2010). Heavy metals, occurrence and toxicity for plants: a review. *Environmental Chemistry Letters* 8: 199–216.

Niklas, K.J. and Enquist, B.J. (2002). Canonical rules for plant organ biomass partitioning and annual allocation. *American Journal of Botany* 89: 812–819.

Ortiz-Lopez, A., Chang, H.-C., and Bush, D. (2000). Amino acid transporters in plants. *Biochimica et Biophysica Acta (BBA) - Biomembranes* 1465: 275–280.

Panhwar, Q.A., Ali, A., Naher, U.A., and Memon, M.Y. (2019). Fertilizer management strategies for enhancing nutrient use efficiency and sustainable wheat production. In: *Organic Farming* (ed. S. Chandran, M.R. Unni, and S. Thomas), 17–39. Elsevier.

Paterson, E., Sim, A., Standing, D. et al. (2006). Root exudation from Hordeum vulgare in response to localized nitrate supply. *Journal of Experimental Botany* 57: 2413–2420.

Perchlik, M. and Tegeder, M. (2017). Improving plant nitrogen use efficiency through alteration of amino acid transport processes. *Plant Physiology* 175: 235–247.

Poorter, H. and Nagel, O. (2000). The role of biomass allocation in the growth response of plants to different levels of light, CO_2, nutrients and water: a quantitative review. *Functional Plant Biology* 27: 1191–1191.

Raghothama, K.G. (2005). Phosphorus and plant nutrition: an overview. *Phosphorus: Agriculture and the Environment* 46: 353–378.

Rajan, M., Shahena, S., Chandran, V., and Mathew, L. (2021). Controlled release of fertilizers—concept, reality, and mechanism. In: *Controlled Release Fertilizers for Sustainable Agriculture* (ed. F.B. Lewu, T.G. Volova, S. Thomas and K.R. Rakhimol), 41–56. Academic Press, Elsevier.

Raynaud, X., Jaillard, B., and Leadley, P.W. (2008). Plants may alter competition by modifying nutrient bioavailability in rhizosphere: a modeling approach. *The American Naturalist* 171: 44–58.

Reich, M., Aghajanzadeh, T., and Kok, L.J.D. (2014). Physiological basis of plant nutrient use efficiency–concepts, opportunities and challenges for its improvement. *Nutrient Use Efficiency in Plants* 1–27.

Reichman, S. and Parker, D. (2005). Metal complexation by phytosiderophores in the rhizosphere. In: *Biogeochemistry of Trace Elements in the Rhizosphere* (ed. P.M. Huang and G.R. Gobran), 129–154. Toronto, Canada: Elsevier.

Rengel, Z. and Römheld, V. (2000). Root exudation and Fe uptake and transport in wheat genotypes differing in tolerance to Zn deficiency. *Plant and Soil* 222: 25–34.

Sattelmacher, B., Horst, W.J., and Becker, H.C. (1994). Factors that contribute to genetic variation for nutrient efficiency of crop plants. *Zeitschrift für Pflanzenernährung und Bodenkunde* 157: 215–224.

Schwab, W., Davidovich-Rikanati, R., and Lewinsohn, E. (2008). Biosynthesis of plant-derived flavor compounds. *The Plant Journal* 54: 712–732.

Tiffin, L.O. (1966). Iron translocation II. Citrate/iron ratios in plant stem exudates. *Plant Physiology* 41: 515–518.

Weiner, J. (2004). Allocation, plasticity and allometry in plants. *Perspectives in Plant Ecology, Evolution and Systematics* 6: 207–215.

Welch, R.M. and Shuman, L. (1995). Micronutrient nutrition of plants. *Critical Reviews in Plant Sciences* 14: 49–82.

Wissuwa, M., Kretzschmar, T., and Rose, T.J. (2016). From promise to application: root traits for enhanced nutrient capture in rice breeding. *Journal of Experimental Botany* 67: 3605–3615.

Xue, K., Wu, L., Deng, Y. et al. (2013). Functional gene differences in soil microbial communities from conventional, low-input, and organic farmlands. *Applied and Environmental Microbiology* 79: 1284–1292.

Yan, X., Wu, P., Ling, H. et al. (2006). Plant nutriomics in China: an overview. *Annals of Botany* 98: 473–482.

10

Nutrients Uptake and Transport in Plants: An Overview

Neda Dalir

Department of Soil Science, Tarbiat Modares University, Tehran, Iran

Introduction

Uptake of nutrients by crop plants in adequate amount and proportion is very important for producing higher yields. Similarly, distribution of absorbed or accumulated nutrients in shoot and grain is associated with yield improvement (Fageria and Stone 2006).

Nutrients are primarily taken up by plants in the form of inorganic ions. However, not all available nutrients in the soil are taken up, even where there is a shortage. The large surface area of roots and their ability to absorb inorganic ions at low concentrations from the soil solution make mineral absorption by plants a very effective process. As root volume occupies only a small proportion of the soil volume, the transport of most nutrients to the roots is mainly restricted to the small soil layer surrounding the roots (the rhizosphere). The rate of uptake depends mainly on the nutrient concentration in the soil solution, if they are within the reach of roots. So, anything that affects root growth and its activity, i.e. physical conditions (soil structure), affects nutrient uptake, and plants with an extensive root system will generally be more efficient in nutrient uptake. The difference in nutrient uptake and utilization may also be associated with the ability of plants to take up sufficient nutrients from lower concentrations; plants' ability to solubilize nutrients in the rhizosphere; better transport, distribution, and utilization within plants; and balanced source–sink relationship (Luo et al. 2015; Hakim et al. 2021).

The forms of ions taken up by roots differ within plant species and with growing conditions, controlled by a combination of soil properties. The first step in uptake is nutrient transport to the root surface in both the water flux created by transpiration (called mass flow) and the diffusive flux toward the root and its passing the outer layer (Barber 1962). Root interception is another nutrient uptake mechanism in which roots absorb ionic nutrients by direct contact with the ion-exchange complex on the surface of clay particles and organic matter. Marschner (2011) reported that this process meets a small percentage of the total nutrient requirement. Other authors have neglected root interception, because

Plant Ionomics: Sensing, Signaling, and Regulation, First Edition.
Edited by Vijay Pratap Singh and Manzer H. Siddiqui.

there is no possibility of a direct exchange between soil particles and plant roots without a liquid environment (Oliveira et al. 2010).

The second step in nutrient uptake involves movement of the nutrient ion into the cytoplasm by crossing the membrane that may be passive or active (Tester and Leigh 2001). The precise point of entry of minerals into the root system has been a topic of considerable interest. Some researchers have claimed that nutrients are absorbed only at the apical regions of the root axes or branches (Bar-Yosef et al. 1972); others claim that nutrients are absorbed over the entire root surface (Nye and Tinker 1977). Experimental evidence supports both possibilities, depending on both plant species and the nutrient being investigated.

In order to maintain nutrient homeostasis, plants must regulate nutrient uptake and its transport within the plant. Transport in plants occurs at two levels: short-distance transport of substances from one cell to another and long-distance transport of sap within xylem and phloem (Marschner 2011). A better understanding of the characteristics and physiological mechanisms by which plants absorb nutrients from the root media may provide additional basic information to aid in the development of better plant nutrition purposes and the selection of adapted crop plant genotypes. Besides, insights into the understanding of the heavy metal uptake mechanisms and bioaccumulation by plants are essential for improving the safety of the food chain.

In this section, we will consider the structure and composition of cellular membranes and the solute's electrochemical gradient. We will also discuss the potential entry pathways for nutrients into the plant and their movement to the xylem.

Routes from the Soil to the Stele

A solute in the external solution may move to the xylem cells of roots through apoplastic and/or symplastic pathways: (i) It may enter the apoplast and diffuse between the epidermal cells through the cell walls, called apoplastic pathway. (ii) In the second pathway that involves the living part of the cell, solute may immediately enter the symplast by crossing the plasma membrane of an epidermal cell (Sattelmacher and Horst 2007; Nouchi et al. 2012). The plasma membrane (also known as the cell membrane or cytoplasmic membrane) is a biological membrane that separates the cell on the interior from the outside environment (Ray et al. 2016).

Apoplastic Pathway

Apoplast, or apoplasm, is the space outside the plasma membrane that allows free movement of material. The apoplastic pathway involves the movement of water and dissolved minerals extracellularly through interfibrillar and intermicellar spaces in the cell walls (Alberts et al. 2002a), which is a nonmetabolic, passive process, driven by diffusion or mass flow. The rate of diffusion is directly proportional to the concentration gradient between the external solution and apoplastic free space (Barberon and Geldner 2014). Solutes in these cell wall channels can flow freely and also exchange with an external solution, and therefore this pathway is often referred to as the free space. Uptake into this cell wall phase is rapid, reversible, and nonmetabolic, assuming that the solute concentration in the free space is the same as that in the external solution. However, the physicochemical properties of cell walls would influence the solute movement.

All plant cells contain middle lamella and primary wall, and many cells produce a third thicker layer called the secondary wall. The middle lamella consists of pectins with different degrees of methylation cements between adjacent plant cells. The primary cell wall is classified as hydroxyproline-rich glycoproteins composed of pectin, cellulose, hemicellulose, and proteins laid down by cells that are dividing and growing (Underwood 2012; Houston et al. 2016). A secondary cell wall is a thicker additional layer of cellulose, which is responsible for most of the plant's mechanical support. The cellulose/hemicellulose networks in both primary and secondary cell wall including the interfibrillar and intermicellar spaces rang in size from 3.5 to 5 nm (Gogarten 1988; Shepherd and Goodwin 1989). The movement of large molecules is limited by the diameter of the pores, while they are freely permeable to nutrients and low-molecular-weight organic solutes (Carpita et al. 1979; Sattelmacher 2001), although cell walls behave as ion exchangers with relatively high capacities. The capacity for ion exchange depends upon a composition of polymeric matrix and structure of polygalacturonic acid, originating mainly from the middle lamella (Meychik and Yermakov 2001; Marschner 2011). The cation reservoir properties of apoplast are of high importance. If the amount of negative fixed charges in the cell wall matrix is high, then it would be high capacity to control the pH or the concentration of other cations in the free space (Starrach et al. 1985). The phenomenon of free space is generally introduced as "apparent free diffusion space" for ions and includes ion wall interactions (Hope and Stevens 1952). This term is generally used as: (i) water free space (WFS), which is freely accessible to ions, and the movement of charged and uncharged molecules is not restricted by electrical charges; (ii) Donnan free space (DFS), in which cation exchange and anion repulsion take place (Figure 10.1). However, it is not a

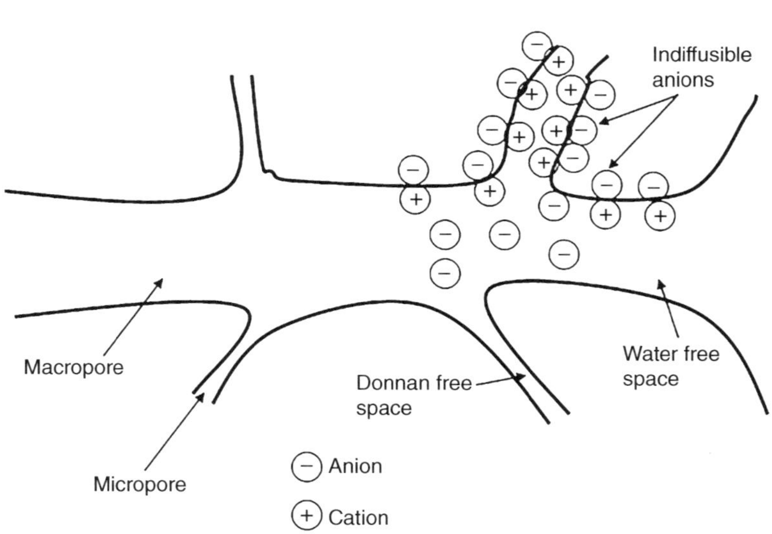

Figure 10.1 The pore system of the apparent free space of cell walls contains "Donnan free space" and "water free space." *Source:* Epstein (1972).

clear spatial differentiation between the two compartments (Starrach and Mayer 1986). Root CEC (Cation Exchange Capacity) differ among species, and dicotyledonous species generally have higher root CEC value than monocotyledonous species (grasses) (Woodward et al. 1984). Compared to monocots, dicot species have greater tendency to take up divalent ions, while monocots take up more monovalent cations.

Symplastic Pathway

The symplastic route to the vascular stele involves at first a selective uptake into a cell and then cell to cell transport by plasmodesmata. Plasmodesmata are narrow membranous structures embedded in cell walls that allow symplastic (cytoplasm-to-cytoplasm) molecular flux (Deinum et al. 2019). In addition to changes in the structure of plasmodesmata, their frequency can also determine the rate of symplastic transport (Barberon and Geldner 2014). In this pathway, the selective plasma membrane of the root cells controls the intake of ion and water. In addition, since symplast is made up of living components, the symplastic route is affected by the metabolic state of the root. Despite the primary selectivity barrier by ion binding in the negatively charged cell wall constituents of the root cell, the plasma membrane of individual cells is the principal site of selectivity in the uptake of cations and anions, as well as solutes.

Movement of Solutes Across Membranes

The complex interactions among phospholipids, proteins, and carbohydrate groups as the principal components of the plasma membrane regulate the transportation of molecules across the membrane. Cell membranes are mainly composed of two layers of phospholipid molecules. Each phospholipid molecule has a head and two tails with a hydrophobic, or water-hating, interior and a hydrophilic, or water-loving, exterior of the membrane (Figure 10.2). Given enough time, most of the molecules could diffuse across a protein-free lipid bilayer down to its concentration or electrochemical gradient. The diffusion rate is dependent partly on the size of the molecule, but mostly on its relative solubility in oil (Alberts et al. 2002b). The smaller the molecule, the more hydrophobic it is and the more rapidly it will diffuse across the membrane. Although lipid phase of the membrane provides some permeability, the membrane permeability is especially performed by proteins that may occur either actively or passively.

Figure 10.2 The structure of a phospholipid.

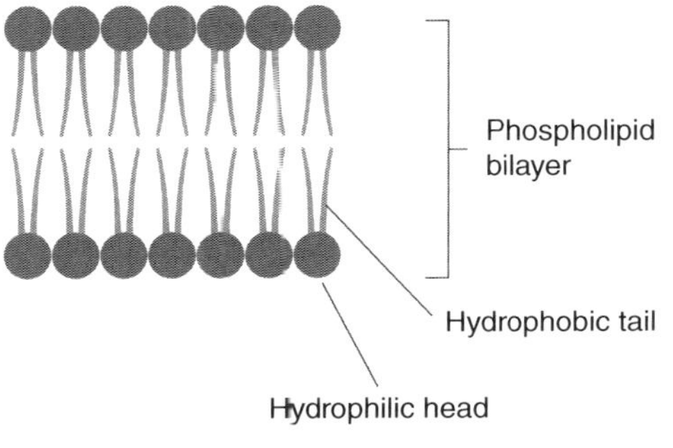

Passive Transport

During passive transport, substances move along the concentration or electrochemical gradient, from an area of high concentration to an area of low concentration and this occurs without expenditure of metabolic energy. The rate of diffusion depends on the lipid organization of the cell membrane and the characteristics of many transmembrane proteins. There are three main types of passive transport including simple diffusion, osmosis, and facilitated diffusion.

Simple Diffusion

It involves the movement of small, relatively hydrophobic molecules across a phospholipid bilayer. Gases (such as O_2 and CO_2) and small polar but uncharged molecules (such as H_2O and ethanol) are able to diffuse across the plasma membrane. In this process, no membrane proteins are involved, and the direction of transport is determined by their relative concentrations inside and outside of the cell (Figure 10.3).

Facilitated Diffusion

In facilitated diffusion, the movement of molecules across the cell membrane is mediated with the assistance of membrane proteins, such as channels and carriers. These allow polar and charged molecules, such as carbohydrates, amino acids, nucleosides, and ions, to cross the cell membrane without interacting with its hydrophobic interior (Figure 10.3).

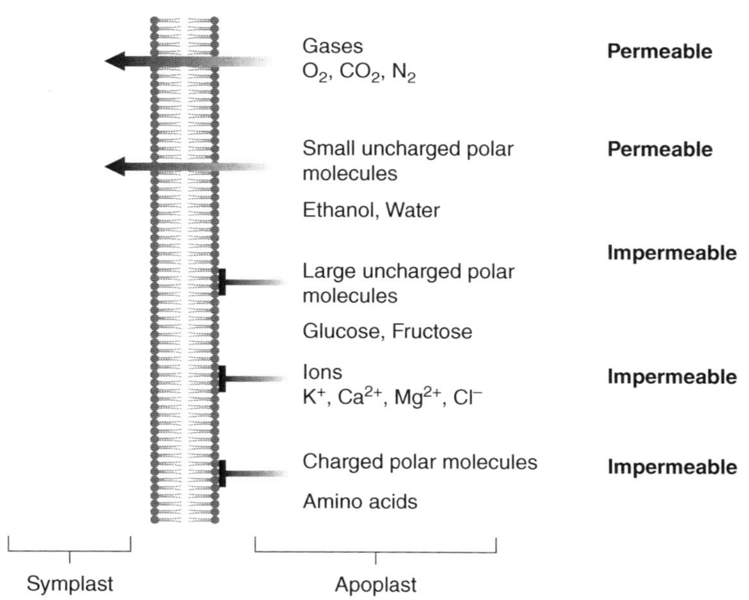

Figure 10.3 Passive transport across the plasma membrane.

Osmosis

Osmosis is the diffusion of water molecules across a selectively permeable membrane according to the concentration or electrochemical gradient and the hydrostatic pressure of the water (Figure 10.3).

Active Transport

Active transport is the movement of substances across the membrane against a concentration or electrochemical gradient with the utilization of energy often in the form of ATP. These can be classified into: (i) primary active transport (via pumps) and (ii) secondary active transport "coupled transporters."

Primary Active Transport

Electrogenic pumps are primary active transporters that are involved in the establishment and maintenance of membrane voltages (Spanswick 1981). Most of the ionic gradients across membranes of higher plants are generated and maintained by electrochemical potential gradients of H^+ (Tazawa et al. 1987). The plasma membrane (PM) H^+-ATPase, for example, is the primary electrogenic pump in plants that uses the energy of ATP hydrolysis to translocate positive charges (protons) out of the cytosol, thereby forming a membrane potential difference across the plasma membrane (negative on the inside) (Falhof et al. 2016).

Since the transport process uses ATP as an energy source, it is considered as primary active transport.

Secondary Active Transport

The other important mechanism by which solutes can be transported across membrane against the gradient of electrochemical potential is secondary active transport. The electrochemical gradients set up by primary active transport store energy are used to drive the transport of many other substances against their gradient of electrochemical potentials and thus does not directly require a chemical source of energy such as ATP.

In plants, the H^+ ions circulate across the membrane, outward through the primary active transport proteins and back into the cell through the secondary transport proteins.

There are two types of secondary transport:

a) Antiport: The mechanism is facilitated by a protein called an antiporter that carries two different ions or molecules, in the opposite direction through the membrane. For example, the sodium potassium antiporter exports three sodium ions out of the cell to transport two potassium ions in.
b) Symport: The symporter carries two different ions or molecules, both in the same direction through the membrane.

Radial Transport of Mineral Ions

To reach the xylem, ions travel via symplastic and apoplastic routes. It is generally assumed that apoplastic uptake has no role in trace element uptake. Apoplastic trace elements are considered to be blocked in the roots behind the apoplastic barriers (Casparian band) and therefore not translocated into the shoots (Pauluzzi et al. 2012). In most angiosperms, another apoplasmic barrier, the exodermis, can develop in parallel with the endodermis (Ma and Peterson 2003). However, if there is a substantial adsorption of trace elements on the root apoplast, this might act as a driving force to move the trace element from the soil toward the roots, compete with the symplastic absorption, and contribute to the element being taken up by the plant, at least into its roots (Redjala et al. 2010). Although ion transporters located in the plasma membrane of the endodermis are particularly important and should receive more attention (Bao et al. 2019).

In the case of passing from cell to cell (symplastic pathway), the density and size exclusion limit of plasmodesmata that connects the cytosolic compartments of neighboring cells is of great importance (Barberon and Geldner 2014). Moreover, the regulation of plasmodesmatal conductance represents another mechanism of cellular control of ion fluxes across the root. For example, the concentration of cytoplasmic free calcium (Ca^{2+}) induces callose deposition and plasmodesmata closure (Cheval et al. 2020).

Long Transport of Mineral Ions

The final step in the radial transfer of materials via the symplastic and/or apoplastic pathway or a combination of the two is their further transport up the stem to all parts of the plant is through the transpiration stream. Transport over longer distances through the vascular system (the xylem and the phloem) is called translocation. Releasing into the nonliving xylem vessels occurs across the plasma membrane of xylem parenchyma cells (Clarkson 1993). During long-distance transport (transport from the roots to the leaves) in the nonliving xylem vessels, important interactions take place between solutes and cell walls of both vessels and the surrounding xylem parenchyma cells. The major interactions are the adsorption and exchange processes that occur on the negatively charged cell walls and retrieval (uptake) and release of elements and of organic solutes by surrounding living cells (xylem parenchyma and phloem) (Marschner 2011).

The factors that influence the rate of translocation in xylem are the valence of the cation, its own activity, as well as surface charge (Wolterbeek 1987). The differences between monocots and dicots are related to the cell wall structure, composition, and pH (Albenne et al. 2014; Calderan-Rodrigues et al. 2019; Su et al. 2019).

Over the long distance, elements and organic solutes are exchanged between the xylem and phloem (VAN BEL 1990). The same as xylem, all nutrients could be identified in high concentrations in the phloem sap (Dinant et al. 2010).

An important aspect is the direction of transport in these two tissues. Transport in xylem (of water and minerals) is essentially unidirectional, from roots to the stems. While translocation in phloem (organic and mineral nutrients) is bidirectional (Ludewig and

Frommer 2002). Shoot–root phloem transport is an important component in cycling of nutrients in vascular plants and for systemic signals reflecting the nutritional status of the shoot to the roots (Xu et al. 2021).

Plants produce a number of chelating agents including organic acids and amino acids (Callahan et al. 2006) that may influence metal uptake by roots and its translocation to the shoots. Several authors have also reported the involvement of organic or amino acid chelation in enhancing the rate of root uptake and root-to-shoot transport of transition metal ions (Krämer et al. 1996; Richau et al. 2009). Amino acids histidine and glycine, for example, have been reported to greatly affect the movement of Ni and Zn in the xylem, both in longitudinal and lateral directions (Dalir and Khoshgoftarmanesh 2014, 2015; Khodamoradi et al. 2015).

Conclusion and Future Prospects

The optimum nutrition in plants is a key determinant in crop productivity that could be achieved by optimizing nutrient uptake and translocation in plants. Nutrients are taken up as inorganic ions by roots, and the rate of uptake depends directly on the soil solution concentration. The mechanism of nutrient uptake by higher plants is characterized by transport selectivity and accumulation into the intracellular compartments, such as the vacuole. Long-distance transport from roots to the aboveground parts of plants is also of great importance for improving crop nutrient efficiency. A better understanding of the characteristics and physiological mechanisms by which plants absorb nutrients from the root media may provide additional basic information to aid in the development of better plant nutrition purposes and may contribute to genetic modification of plants for improving plant nutrition.

References

Albenne, C., Canut, H., Hoffmann, L., and Jamet, E. (2014). Plant cell wall proteins: a large body of data, but what about runaways? *Proteomes* 2: 224–242.

Alberts, B., Johnson, A., Lewis, J. et al. (2002a). *The Plant Cell Wall, Molecular Biology of the Cell*, 4e. Garland Science.

Alberts, B., Johnson, A., Lewis, J. et al. (2002b). *Principles of Membrane Transport, Molecular Biology of the Cell*, 4e. Garland Science.

Bao, Z., Bai, J., Cui, H., and Gong, C. (2019). A missing link in radial ion transport: ion transporters in the endodermis. *Frontiers in Plant Science* 10: 713.

Barber, S.A. (1962). A diffusion and mass-flow concept of soil nutrient availability. *Soil Science* 93: 39–49.

Barberon, M. and Geldner, N. (2014). Radial transport of nutrients: the plant root as a polarized epithelium. *Plant Physiology* 166: 528–537.

Bar-Yosef, B., Kafkafi, U., and Bresler, E. (1972). Uptake of phosphorus by plants growing under field conditions: I. Theoretical model and experimental determination of its parameters. *Soil Science Society of America Journal* 36: 783–788

Calderan-Rodrigues, M.J., Guimarães Fonseca, J., de Moraes, F.E. et al. (2019). Plant cell wall proteomics: a focus on monocot species, Brachypodium distachyon, *Saccharum* spp. and *Oryza sativa*. *International Journal of Molecular Sciences* 20: 1975.

Callahan, D.L., Baker, A.J., Kolev, S.D., and Wedd, A.G. (2006). Metal ion ligands in hyperaccumulating plants. *JBIC, Journal of Biological Inorganic Chemistry* 11: 2–12.

Carpita, N., Sabularse, D., Montezinos, D., and Delmer, D.P. (1979). Determination of the pore size of cell walls of living plant cells. *Science* 205: 1144–1147.

Cheval, C., Samwald, S., Johnston, M.G. et al. (2020). Chitin perception in plasmodesmata characterizes submembrane immune-signaling specificity in plants. *Proceedings of the National Academy of Sciences* 117: 9621–9629.

Clarkson, D.T. (1993). Roots and the delivery of solutes to the xylem. *Philosophical Transactions of the Royal Society of London. Series B: Biological Sciences* 341: 5–17.

Dalir, N. and Khoshgoftarmanesh, A.H. (2014). Symplastic and apoplastic uptake and root to shoot translocation of nickel in wheat as affected by exogenous amino acids. *Journal of Plant Physiology* 171: 531–536.

Dalir, N. and Khoshgoftarmanesh, A.H. (2015). Root uptake and translocation of nickel in wheat as affected by histidine. *Journal of Plant Physiology* 184: 8–14.

Deinum, E.E., Mulder, B.M., and Benitez-Alfonso, Y. (2019). From plasmodesma geometry to effective symplasmic permeability through biophysical modelling. *eLife* 8: e49000.

Dinant, S., Bonnemain, J.-L., Girousse, C., and Kehr, J. (2010). Phloem sap intricacy and interplay with aphid feeding. *Comptes Rendus Biologies* 333: 504–515.

Epstein, E. (1972). *Mineral Nutrition of Plants: Principles and Perspectives*, 68–82. New York: Wiley.

Fageria, N. and Stone, L. (2006). Physical, chemical, and biological changes in the rhizosphere and nutrient availability. *Journal of Plant Nutrition* 29: 1327–1356.

Falhof, J., Pedersen, J.T., Fuglsang, A.T., and Palmgren, M. (2016). Plasma membrane H^+-ATPase regulation in the center of plant physiology. *Molecular Plant* 9: 323–337.

Gogarten, J.P. (1988). Physical properties of the cell wall of photoautotrophic suspension cells from *Chenopodium rubrum* L. *Planta* 174: 333–339.

Hakim, S., Naqqash, T., Nawaz, M.S. et al. (2021). Rhizosphere engineering with plant growth-promoting microorganisms for agriculture and ecological sustainability. *Frontiers in Sustainable Food Systems* 5: 16.

Hope, A. and Stevens, P. (1952). Electric potential differences in bean roots and their relation to salt uptake. *Australian Journal of Biological Sciences* 5: 335–343.

Houston, K., Tucker, M.R., Chowdhury, J. et al. (2016). The plant cell wall: a complex and dynamic structure as revealed by the responses of genes under stress conditions. *Frontiers in Plant Science* 7: 984.

Khodamoradi, K., Khoshgoftarmanesh, A.H., Dalir, N. et al. (2015). How do glycine and histidine in nutrient solution affect zinc uptake and root-to-shoot translocation by wheat and triticale? *Crop & Pasture Science* 66: 1105–1110.

Krämer, U., Cotter-Howells, J.D., Charnock, J.M. et al. (1996). Free histidine as a metal chelator in plants that accumulate nickel. *Nature* 379: 635–638.

Ludewig, U. and Frommer, W.B. (2002). Genes and proteins for solute transport and sensing. *The Arabidopsis Book/American Society of Plant Biologists* 1: e0092.

Luo, X., Ma, C., Yue, Y. et al. (2015). Unravelling the complex trait of harvest index in rapeseed (*Brassica napus* L.) with association mapping. *BMC Genomics* 16: 1–10.

Ma, F. and Peterson, C.A. (2003). Current insights into the development, structure, and chemistry of the endodermis and exodermis of roots. *Canadian Journal of Botany* 81: 405–421.

Marschner, H. (2011). *Marschner's Mineral Nutrition of Higher Plants*. Academic press.

Meychik, N. and Yermakov, I. (2001). Ion exchange properties of plant root cell walls. *Plant and Soil* 234: 181–193.

Nouchi, I., Hayashi, K., Hiradate, S. et al. (2012). Overcoming the difficulties in collecting apoplastic fluid from rice leaves by the infiltration–centrifugation method. *Plant and Cell Physiology* 53: 1659–1668.

Nye, P.H. and Tinker, P.B. (1977). *Solute Movement in the Soil-Root System*. Univ of California Press.

Oliveira, E.M.M., Ruiz, H.A., Alvarez, V.H. V et al. (2010). Nutrient supply by mass flow and diffusion to maize plants in response to soil aggregate size and water potential. *Revista Brasileira de Ciência do Solo* 34: 317–328.

Pauluzzi, G., Divol, F., Puig, J. et al. (2012). Surfing along the root ground tissue gene network. *Developmental Biology* 365: 14–22.

Ray, S., Kassan, A., Busija, A.R. et al. (2016). The plasma membrane as a capacitor for energy and metabolism. *American Journal of Physiology-Cell Physiology* 310: C181–C192.

Redjala, T., Sterckeman, T., Skiker, S., and Echevarria, G. (2010). Contribution of apoplast and symplast to short term nickel uptake by maize and *Leptoplax emarginata* roots. *Environmental and Experimental Botany* 68: 99–106.

Richau, K.H., Kozhevnikova, A.D., Seregin, I.V. et al. (2009). Chelation by histidine inhibits the vacuolar sequestration of nickel in roots of the hyperaccumulator *Thlaspi caerulescens*. *New Phytologist* 183: 106–116.

Sattelmacher, B. (2001). The apoplast and its significance for plant mineral nutrition. *New Phytologist* 149: 167–192.

Sattelmacher, B. and Horst, W.J. (2007). *The Apoplast of Higher Plants: Compartment of Storage, Transport and Reactions: The Significance of the Apoplast for the Mineral Nutrition of Higher Plants*. Springer Science and Business Media.

Shepherd, V. and Goodwin, P. (1989). The porosity of permeabilised Chara cells. *Functional Plant Biology* 16: 231–239.

Spanswick, R.M. (1981). Electrogenic ion pumps. *Annual Review of Plant Physiology* 32: 267–289.

Starrach, N. and Mayer, W.-E. (1986). Unequal distribution of fixed negative charges in isolated cell walls of various tissues in primary leaves of Phaseolus. *Journal of Plant Physiology* 126: 213–222.

Starrach, N., Flach, D., and Mayer, W.-E. (1985). Activity of fixed negative charges of isolated extensor cell walls of the laminar pulvinus of primary leaves of Phaseolus. *Journal of Plant Physiology* 120: 441–455.

Su, Y., Ashworth, V., Kim, C. et al. (2019). Delivery, uptake, fate, and transport of engineered nanoparticles in plants: a critical review and data analysis. *Environmental Science: Nano* 6: 2311–2331.

Tazawa, M., Shimmen, T., and Mimura, T. (1987). Membrane control in the Characeae. *Annual Review of Plant Physiology* 38: 95–117.

Tester, M. and Leigh, R.A. (2001). Partitioning of nutrient transport processes in roots. *Journal of Experimental Botany* 52: 445–457.

Underwood, W. (2012). The plant cell wall: a dynamic barrier against pathogen invasion. *Frontiers in Plant Science* 3: 85.

VAN BEL, A.J.E. (1990). Xylem-phloem exchange via the rays: the undervalued route of transport. *Journal of Experimental Botany* 41: 631–644.

Wolterbeek, H.T. (1987). Cation exchange in isolated xylem cell walls of tomato. I. Cd^{2+} and Rb^{+} exchange in adsorption experiments. *Plant, Cell and Environment* 10: 39–44.

Woodward, R.A., Harper, K., and Tiedemann, A. (1984). An ecological consideration of the significance of cation-exchange capacity of roots of some Utah range plants. *Plant and Soil* 79: 169–180.

Xu, J., Guo, Z., Jiang, X. et al. (2021). Light regulation of horticultural crop nutrient uptake and utilization. *Horticultural Plant Journal* 7: 367–379.

11

Regulation of Phytohormonal Signaling by Nutrients in Plant

Harshita Joshi, Nikita Bisht, and Puneet Singh Chauhan

Microbial Technologies Division, Council of Scientific and Industrial Research-National Botanical Research Institute (CSIR-NBRI), Lucknow, Utter Pradesh, India

Introduction

Plants obtain chemical elements from the soils that are grouped or referred to as "essential mineral nutrients" and are required by plants to complete their life cycle (Toor et al. 2021). Besides oxygen, hydrogen, and carbon, which can be fundamentally acquired from water and carbon dioxide, plants actively take up at least 20 elements. The essential nutrients play basic functions in plants physiology and metabolism as a component of enzyme cofactors (e.g. Mo, Mg, Zn, Fe, Cu, Mo, Ni, etc.), macromolecules, or metabolite (e.g. Bo, S, N, and P), secondary messengers in signaling cascade (e.g. Ca), or electrolytes (e.g. P, Cl, Na, etc.) (Rubio et al. 2009). Despite limited nutrient availability, plants maintain constant levels of mineral nutrients for their optimal growth and development. The nutrient limitation may be due to their low solubility in soils and low concentration or accessibility. The plants trigger various developmental and physiological responses to survive reduced nutrient availability, which may alter the plant's metabolism and its morphology (López-Bucio et al. 2003; Schachtman and Shin 2007). The responses include accumulation/production of enzymes, compounds, and induced expression of specific high-affinity transporters that aid in nutrient remobilization (Schachtman and Shin 2007).

Moreover, nutrient supply or availability is known to modulate the synthesis and action of phytohormones. Several effects of nutrient unavailability on plant growth and development are due to the phytohormonal imbalance they cause in the plants (Engels et al. 2012). Phytohormones are signal molecules or chemical messengers that play a crucial role in the modulation of the development and growth of higher plants. They perform a plethora of other functions like stress management including nutrient homeostasis. These are transported either from organ to organ or cell to cell as their site of synthesis and site of action are generally physically separated. (Engels et al. 2012). Various evidence has shown how mineral nutrient variation influences signaling molecules like phytohormone and their biosynthesis, therefore supporting a close association between nutritional homeostasis and

Plant Ionomics: Sensing, Signaling, and Regulation, First Edition.
Edited by Vijay Pratap Singh and Manzer H. Siddiqui.

hormonal stimuli (Krouk et al. 2011). Several studies suggest nutrients can modulate biosynthesis, degradation, de-conjugation, signaling, and transport of various phytohormones, adapting nutrient availability and plant development and growth.

Furthermore, numerous hormones are required in the mineral nutrient status of plants, and cross talk between a nutrient and hormone has been described for the following pairs. Phosphorous (P) and cytokinins, sulfur (S) and cytokinins, P and auxins, potassium (K) and auxin, iron (Fe), K and jasmonic acid (JA), and cytokinins, K and ethylene, Ca and ethylene, N and ethylene, (Iqbal et al. 2013), S and polyamines (Hasanuzzaman et al. 2018), and many more.

Phytohormones: Structure, Sites of Biosynthesis, and its Effects

Phytohormones participate in various processes throughout a plant's lifecycle. Additionally to the five classical phytohormones such as abscisic acid, auxins, cytokinins, ethylene, and gibberellins, phytohormones like brassinosteroids, jasmonic acid, strigolactones salicylic acid (SA), and polyamine also play important roles in plant growth and stress responses (Devireddy et al. 2021). The basic structures of the different phytohormone classes are shown in Figure 11.1. Some of the major characteristics of the phytohormones are given in Table 11.1.

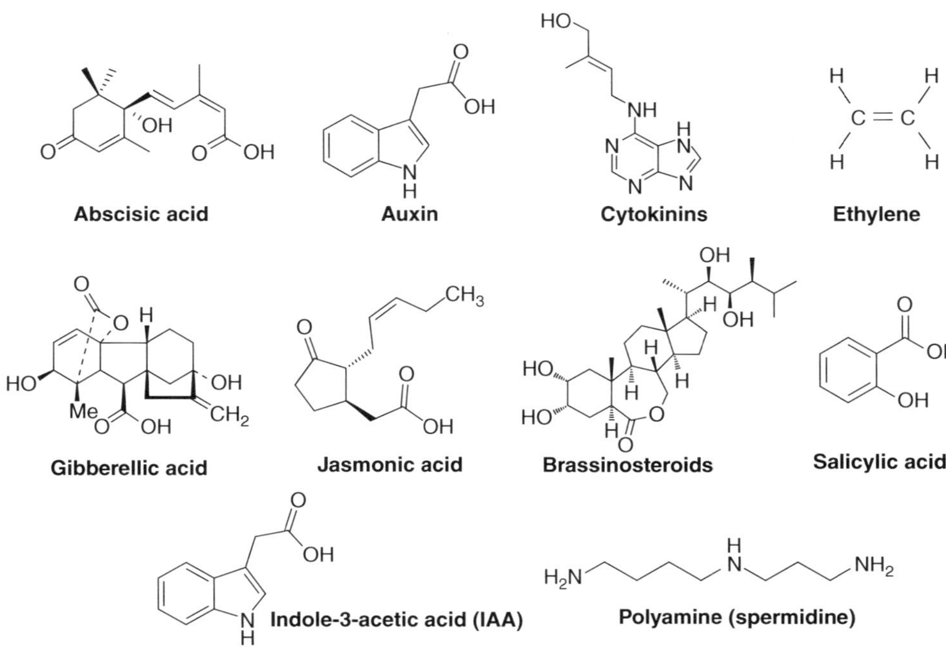

Figure 11.1 Molecular structure of phytohormones.

Table 11.1 Phytohormone: pathway, site of biosynthesis and its effect.

S. No.	Phytohormone	Biosynthetic precursor	Site of biosynthesis	Effects
1.	Cytokinins	Purine derivatives	Root and shoot meristem and embryo. Long-distance transport through the xylem.	RNA and protein synthesis, delay senescence, cell division, and expansion, suppress auxin-regulated apical dominance
2.	Auxin	Indole derivatives of tryptophan	The apical meristem, young leaves. Basipetal transport: cell to cell, long-distance in the proximity of phloem.	Apical dominance, tropism, cell growth, and division, adventitious root development
3.	Abscisic Acid	Carotenoids neoxanthin and violaxanthin.	Root and shoot tissues	Regulate dormancy of seeds and buds, induces stomatal closure, inhibits cell extension, induce abscission of fruits and leaves
4.	Ethylene (ET)	Methionine	Numerous plant organs	Formation of root hairs, arenchymya, flowering snd epinastic curvature of leaves, defense responses to pests and pathogens, and promotes seed germination, ripening, and senescence.
5.	Gibberellins (GA)	Gibbane C skeleton by terpenoid pathway	Developing tissues and seeds	Delay fruit and leaf senescence induce cell expansion, promote enzymatic activities (e.g. hydrolases), break seed and bud dormancy, stimulate shoot elongation, promote flowering
6.	Jasmonic acid (JA)	Linolenic acid	Fruits, root, and shoot	Promotes leaf senescence, fruit ripening, regulate defense responses to pathogens and pests inhibits seed germination, root, and shoot growth, promote tendril coiling.
7.	Brassinosteroids (BRs)	Isopentyl diphosphate.	Vegetative tissues, pollen, and seeds	Regulate cell elongation and division, prevent leaf abscission induce stem elongation and apical dominance, enhance stress resistance.
8.	Salicylic acid (SA)	Phenylalanine	All tissues.	Inhibits leaf senescence and promote flowering and defense responses to pathogen and pests
9.	Polyamines (PA)	Arginine and ornithine.	All tissues	Stimulate the synthesis of DNA, RNA proteins, and cell division, delay leaf senescence, induce root initiation, tuber formation embryogenesis, fruit ripening, and flower development

Auxins are the indole derivatives of tryptophan (amino acid), the most eminent being is indole-3-acetic acid (IAA). They are produced in young expanding tissues or meristems. They can be redistributed locally from cell to cell and are transported in the phloem (Gallei et al. 2020). The direction of distribution is determined by the polar locations of the AUX/LAX auxin influx carriers: of the multidrug-resistant/P-glycoprotein (MDR/PGP) subfamily of ATP-binding cassette (ABC) proteins and the PIN auxin efflux carriers in the plasma membrane (Robert and Friml 2009). Auxin is suggested in apical dominance, xylogenesis, adventitious root development, shoot elongation, sand plant tropism, and promotes cell expansion and cell division (Wang et al. 2020).

Cytokinin is produced from purine derivatives and is known to suppress auxin-induced dominance and delay leaf senescence. They are known to promote cell differentiation and division, delay protein degradation, and stimulate protein synthesis and transcription (Argueso et al. 2009). The major sites of their biosynthesis are in roots, and they are readily mobile within plants. Although the root to shoot xylem transport rules, the cytokinins are also vigorous in the phloem and carried to developing seeds and inflorescences from the source leaves. The cytokinins are reversibly inactivated by glucosylation and are degraded by cytokinin oxidases, which are induced in response to high concentrations of tissue cytokinins (Engels et al. 2012; Li et al. 2020).

Abscisic acid (ABA) is a stress hormone, a type of metabolite known as isoprenoids or terpenoids, and is derived from a five-carbon precursor molecule, isopentenyl (IDP) (Vishwakarma et al. 2017). The important site of biosynthesis is the roots and shoots, and it can circulate within the plant and is highly mobile in both phloem and xylem. ABA can be transformed into numerous biologically inactive metabolites comprising glucose conjugates, dihydrophaseic acid, and phasseic acid (Engels et al. 2012). It is known to control various growth and developmental characteristics of plants like inhibition of fruit ripening, leaf abscission, promotes desiccation tolerance of seeds, prevents water loss from leaves, and induces dormancy of seeds and buds (Chen et al. 2020).

Gibberellin includes a large group of diterpenoid carboxylic acids that are classified according to their structure (Peter Hedden 2020). The gibbane carbon skeleton is produced by the terpenoid pathway, which generates over 100 GAs (gibberellin structures) in plants. GA_1, GA_3, and GA_4 are the three most common biologically active GAs, and the rest serve either as active gas or their degradation products. Plants also contain inactive GA conjugates such as β-glucosyl esters and GA-O-β-glucosides (He et al. 2020). GA accumulation is higher in emerging seeds than in vegetative tissues. Gibberellins are known to stimulate shoot elongation, break seed and bud dormancy, delay leaf and fruit senescence, induce flowering, and promote seed germination (Kalra and Bhatla 2018).

Ethylene, contrary to other phytohormones, is a gas and is a site of activity and synthesis that are in the same tissues. It is synthesized from methionine via the consecutive effort of S-adenosyl-L-methionine synthase, 1-aminocyclopropane-1-carboxylic acid (ACC) synthase, and ACC oxidase (Dubois et al. 2018). Ethylene is known to perform various functions mentioned later (Engels et al. 2012)

- Root aerenchyma formation in response to flooding
- Root, stem, leaf growth repression, or enhancement
- Climacteric fruits ripening

- Leaves and flower senescence induction and accelerates germination
- Plant tropism
- Defense response to pathogens and pest

SA is present in all plant tissues and is synthesized from both phenylalanine ammonia-lyase (PAL) and an isochorismate synthase (ICS) pathway, both initiate from chorismate (Lefevere et al. 2020. It is a crucial phytohormone involved in defense that encourages immunity against both bio and semibiotrophic pathogens. It is known to induce flowering and leaf senescence possibly due to a reduction in the rate of ET synthesis. SA also plays an important role in local immune response and the establishment of systemic resistance to the spread of several bacterial, fungal, and viral diseases (Huang et al. 2020).

Jasmonates are synthesized from unsaturated fatty acid α-linolenic acid, although some plants use a 16 carbon fatty acid. Jasmonates are believed to act as systemic signals inducing defense responses and are mobile in the phloem. The concentration of jasmonates is found to be higher in plants that are asserted by specific pathogen and pests (Han 2017). Jasomnyl-isoleucine, an active form of JA, is found to be highly synthesized in response to various abiotic stress. It is known to inhibit root and shoot germination, and seed germination accelerates seed and fruit-induced leaf senescence and promotes tuber induction and fruit ripening (Engels et al. 2012).

Brassinosteroids are plant steroid hormones that are synthesized from isopentyl diphosphate via campesterol, which is a crucial intermediate (Fujiyama et al. 2019). Their structure is similar to plasma membrane sterols like sitosterol, campesterol, and stigmasterol. This class of phytohormones has over 60 compounds in various plant species, the first isolation was from pollen from oilseed rape. At very low concentration, they are known for promoting apical dominance, grass leave bending, and stem elongation. Brassinosteroids can be locally transported between cells and at very low concentrations as 10^{-10} M stimulate growth elongation. BR concentration in tissues is affected by various pathways for its catabolism (Engels et al. 2012).

Polyamines (PA) are observed as another class of phytohormones. The main polyamines in plants are spermidine (Spd) and Putrescine (Put), which are known for regulating various physiological processes (Chen et al. 2019). They are synthesized from ornithine and arginine and are omnipresent in plant cells and mobile in both phloem and xylem. PA is known for the synthesis of proteins, DNA, and RNA, and proteins stimulate cell division, embryogenesis, tuber formation, fruit ripening, and flower development. They act as potential osmotica to protect cells under abiotic stress and as antioxidants to protect cells from oxidative damage. PA is the potential inhibitor of ET biosynthesis as decreased PA content is correlated with increased ET production during fruit ripening (Engels et al. 2012).

Interaction between Nutrient Availability and Phytohormone Signaling

The plant life cycle notably depends on 17 essential nutrients such as N, P, K (primary elements), C, H, O (basic elements) Mg, S, Ca (secondary elements), and Cu, Cl, B, Fe, Mo, Mn, Ni, Zn as micronutrients. And it is established that nutrient availability modulates plant growth and development. Additionally, plant development is remarkably modulated by various hormonal signals. There is a tremendous increase in the evidence that exhibits coordination between nutritional and hormonal signaling. Therefore, this chapter provides evidence supporting the involvement of nutrients in phytohormone signalings. A list of summarized evidence of how nutrient stress affects the phytohormone is provided in Table 11.2.

Table 11.2 Effects of nutrients on phytohormones.

S. No.	Nutrients	Plants	Phytohormones	References
1.	Low N stress	Wheat	Concentrations of cytokinins (CKs), indole-3-acetic acid (IAA), gibberellin (GA_3), jasmonic acid (JA) enhanced, and salicylic acid (SA) concentration decreased	Lv et al. 2021
2.	Zn, Fe and Zn-Fe stress	Maize	Alteration in genes related to phytohormone metabolism, viz, gibberellins, cytokinins, auxin, ethylene	Mallikarjuna et al. 2020
3.	Exogenous Si	*Solanum Lycopersicum*	Decreased SA and ABA concentration	Khan et al. 2020
4.	Fe deficiency; K deficiency	*Pyrus betulifolia*	alterations of the gene expression mediated by auxin; increased concentration of ABA, ETH, JA, gibberellins	Yang et al. 2020
5.	B deficiency	*B. napus; Plantago major L.*	Decreased IAA and increased Jasmonate and ABA concentration induced expression of Auxin and ABA biosynthesis gene; decreased abscisic acid and salicylic acid and increased cytokinins and brassinosteroids content.	Zhou et al. 2016; Pommerrenig et al. 2019
6.	Contrasting Pi availability	*Arabidopsis*	Increased level of auxin, gibberellins, and the stress-related hormones, ABA, SA, and JA.	(Munné-Bosch et al. 2018)
7.	Pi defeciency	Tibetian wild barley accession; *Arabidopsis*	Relative expression of indole acetic acid (IAA) and ethylene synthesis responsive genes; stmulation of strigolactone pathway and increased ABA, JA, and SA level	Nadira et al. 2016
8.	Fe, S, K defeciency	*Arabidopsis*	Metabolites of stress-related phytohormones, jasmonic, salicylic and abscisic acid	Forieri et al. 2017
9.	Fe defeciency	*Arabidopsis*	Increased SA contents that elevates auxin and ethylene signaling	Shen et al. 2016
10.	S defeciency	Oilseed rape	Induction of ACC synthase and 12-oxophitodienoate reductase observed in ethylene and jasmonate biosynthesis	D'Hooghe et al. 2013

Nutrients in Cytokinin (CK) Signaling

Cytokinin is a functional phytohormone that plays crucial roles not only in development processes and multiple plant growth but also in stress responses. Cytokinin signaling cascades are evolutionarily linked to the bicomponent network in unicellular organisms that engage in the conduction of signals that are activated by numerous environmental stimuli like nutrient concentrations. The function of its metabolism, transport, and signaling in responses to variation in levels of both macronutrients (N, P, K, and S) and micronutrients (B, Se, Si, and Fe) is summarized later.

Nitrogen is observed as among the most important growth-restricting nutrients for plants. In *Arabidopsis*, cytokinin biosynthesis is regulated by the availability of nitrate via controlled expression of the enzymes that catalyze the first rate-limiting step, isopentyl transferase (IPT3, 5) and the following synthesis of cytochrome P450 (CYP735A2) and trans-Zeatin (tZ) (Pavlů et al. 2018). The nitrate-upregulated cytokinin biosynthesis involves signaling components like the NLP-NIGT1 transcriptional cascade controlling CYP735A2, nitrate transporter-receptor NRT1 working upstream of IPT3 and IPT3 (Maeda et al. 2018). Recent research shows nutrient availability mediated adaptation of organogenesis, and shoot apical meristem size is accounted for the long-distance transport of cytokinin precursor via expression modulation of WUSCHEL (Landrein et al. 2018). A diagrammatic representation of how nitrate regulates the phytohormonal signaling and pathway is given in Figure 11.2.

Similarly, an intricate signaling system is essential to sustain Pi (inorganic phosphate) homeostasis, and the suboptimal Pi conditions are required for various hormones. The Pi-limiting condition causes decreased expression of IPT3 and cytokinin signaling components like AHK4, a cytokinin receptor (Pavlů et al. 2018). In *Arabidopsis* roots, CK is responsible for negatively regulating Pi-starvation responses (PSR), like an expression of Pi-deprivation-inducible genes (AtPT1 and AtPS1) and proliferation of late1ral roots. Moreover, the mechanism whereby CK signaling applies its effect on PSR is unclear. Initially, it was suggested that CK could serve as long-distance signals during the amelioration of Pi deficiency (Rubio et al. 2009).

Potassium is one of the primary macronutrients and the most abundant inorganic cation in plants. Potassium deprivation in *Arabidopsis* plants reduces cytokinin contents, and CK signaling controls potassium uptake and root growth inhibition (Nam et al. 2012). Similarly, sulfur deficiency is known to induce several adaptive responses. *Arabidopsis* plants grown in the sulfur-deficient media, a link between cytokinin status and sulfur deficiency is indicated by the downregulation of IPT3 in roots (Ham et al. 2018). A sulfur deficiency marker gene, GGCT2;1 encodes a highly cytokinin-responsive gene, a key enzyme involved in glutathione degradation. This suggests the role of cytokinin in glutathione homeostasis and its probable role in nutrient mobilization (Pavlů et al. 2018).

Boron, an essential micronutrient, when present in suboptimal concentration is known to induce decreased expression of cytokinin signaling genes (Eggert and von Wirén 2017). A meta-analysis observed the expression of BOR4, codifying a B transporter identical to ARR1 and LOG7, a cytokinin metabolism gene (Pavlů et al. 2018). Similarly, in oilseed rape, boron content correlates to cytokinin content and is known to enhance cytokinin synthesis and convert the feebly active form to high active cytokinin. Conversely, a current

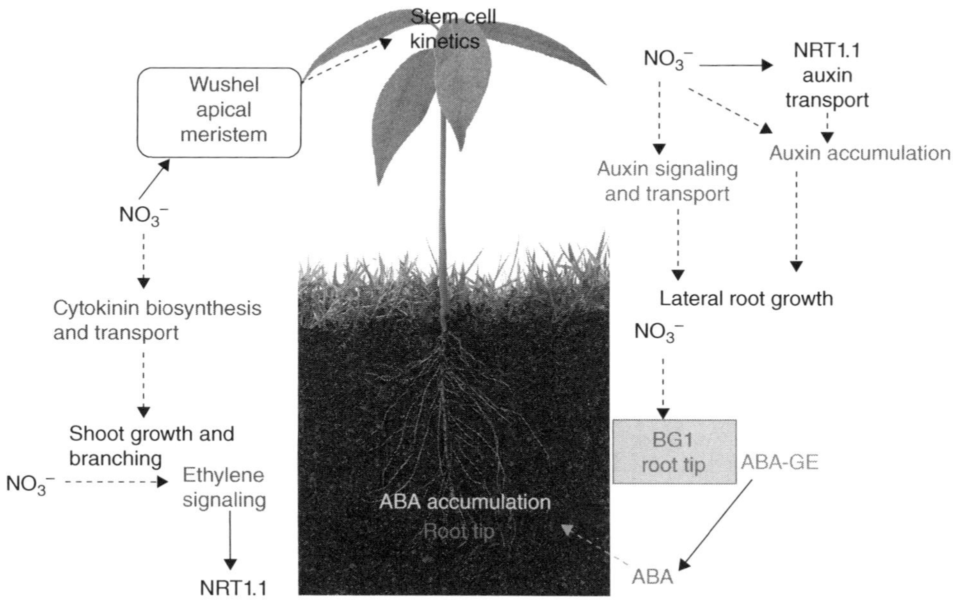

Figure 11.2 NO_3^- regulates phytohormonal signaling and pathways and impacts root and shoot morphology. Here, we emphasize a model with the link between NO_3^- and different hormone signaling pathways. NO_3^- alters auxin- transport, signaling, and biosynthesis, modulating both primary and lateral root development. In root tips, NO_3^- causes ABA synthesis by the expression of the gene coding enzyme BG1 (B-glucosidase 1). NO_3^- also induces ethylene signaling that controls the expression of basic nitrate transporters, NRT1.1. In shoot apical meristem, cytokinin, and WUSHEL function in modulating cell cycle activity that lies in NO_3^- supply. *Source:* Modified from Vega et al. (2019).

finding suggests that boron deficiency suppresses the growth of root meristem by cytokinin-mediated downregulation of cyclin CYCD3 (Poza-Viejo et al. 2018). Likewise, silicon accumulation in *Sorghum* and *Arabidopsis* is known to induce IPT7, a cytokinin synthesis gene, and hinder dark-induced senescence of leaf via activation of cytokinin pathways (Markovich et al. 2017).

Selenium has two major forms, selenite and selenite, that are easily absorbed by plants through phosphate and sulfate transporters, respectively (Pavlů et al. 2018). In contrary, Pi and sulfate pathways regulated by cytokinins may signify a cross-talk between cytokinin signaling and selenium. In a recent study, a loss of function mutation in TPS22, a terpenoid synthase gene, in selenium-tolerant *Arabidopsis* mutant exhibited reductions in the expression of cytokinin receptors AHK3 and AHK4 and cytokinin levels (Jiang et al. 2018).

Nutrients in Ethylene (ETH) Signaling

In plants, ethylene plays a crucial function in response or adaptation under biotic and abiotic stress conditions. Ethylene production often improves the tolerance to suboptimal environmental conditions. Some mineral nutrients mostly influence ethylene biosynthesis and perception with a strong impact on plant physiology. The effects of suboptimal or optimal nutrient conditions on the ethylene pathway and plant responses are summarized later.

In *Arabidopsis*, ethylene production is advanced in its roots when transferred from low nitrate (0.1 mM) to high nitrate (10 mM) concentration. The enhanced ethylene generation and associated reduction in lateral root number and length were actively influenced by mutations in the ETHYLENE INSENSITIVE RECEPTOR and ETHYLENE INSENSITIVE 2 (EIN2) genes (Vega et al. 2019). Conversely, shifting seedlings from high nitrate (10 mM) to low (0.2 mM) nitrate concentration, the low nitrate supply causes the speedy production of ethylene. However, in this study, expression of the NRT2.1 gene was highly expressed, accompanied by a strong induction of critical genes like EIN3 and EIN3-LIKE 1, EIL1, concerned in ethylene signaling (Zheng et al. 2013).

The function of ethylene signaling by Pi deprivation has been studied. P deficiency-induced ethylene production is known to be involved in the mobilization of Pi by the modulation of Phosphate transporters and ATPase activity (Iqbal et al. 2013). Likewise, other research also reported a relation between ethylene and P. Transcriptome studies exhibited that transcript levels of genes responsive for ethylene biosynthesis are enhanced in *Arabidopsis* and white lupin under P deficient conditions (Thibaud et al. 2010). Studies have shown that Pi limitation elevates the accumulation of EIN3 protein, a vital element of the ethylene signaling pathway. EIN3 directly binds to the promoter of genes targeted by *RSL4*, thereby increasing root hair production (Song et al. 2016). Similarly, low potassium supply regulates the expression of numerous genes associated with ethylene biosynthesis and is associated with increased ethylene production (Iqbal et al. 2013).

Calcium is interrelated with ethylene signaling and modulates plant responses. Studies carried out on *A. thaliana* utilizing intact seedlings demonstrated that Ca transduction through phosphoinositide cycle is essential for ethylene modulation of 1-aminocyclopropane-1-carboxylic acid synthase (ACS) (Iqbal et al. 2013). Ca is also a crucial cofactor in organ abscission. Calcium increased the petiole abscission in tomatoes by enhancing the ethylene evolution. At the molecular level, increased expression of the subsequent genes: ACS2, ETR1, ACO1, and ETR4 involved in ethylene synthesis and perception was observed under the Ca treatment (Xu et al. 2010). Likewise, an enhanced ethylene biosynthesis under sub-optimal Mg levels has been suggested. It was observed that genes codifying for ACS iso-forms were highly induced by Mg deficiency; ACS11 was highly expressed in both roots and leaves, while ACS, ACS7, and ACS were only expressed in leaves (Iqbal et al. 2013).

Fe deficiency is known to enhance the gene expression essential in ethylene synthesis (ACO, SAM synthetase, and ACS genes) and signaling (AtETR1, AtCTR1, AtEIL1, AtEIN2, AtEIN3, and AtEIL3). In response to Fe deficiency, initiation of the root hairs are observed and are differentially affected by antagonists of ethylene and error in the ethylene signaling (Divte et al. 2021). Similar to the nutrients discussed, copper, zinc, and selenium also show some relationships with ethylene.

Nutrients in Auxin Signaling

Auxin is among the most important plant growth regulators and a signaling molecule with the ability to induce plant growth and development. Auxin is essential in regulating root development and plays a pivotal role in the regulation of root architecture and growth during P-starvation. Analysis on P-limited roots of *Arabidopsis* showed enhanced expression of auxin receptor T1R1 that cause degradation of indole acetic acid/AUX auxin-responsive repressors and activate ARF TRs to repress/activate genes associated with lateral root emergence and formation (Scheible and Rojas-Triana 2018). Under Pi starvation, the SUMO E3 ligase SIZ1 is known to modulate root structure architecture (RSA) via regulation of auxin modulation (Miura et al. 2011). Another E3 ligase, AtPUB9, and the linked ARK2 receptor kinase are also involved in the root development and spatial auxin accumulation during P-limitation (Deb et al. 2014).

Similarly, changes in N availability and sources can activate changes in auxin signaling and transport, which is directly linked with the regulation of root development (Vega et al. 2019). Auxin and nitrogen exhibit a well-established connection in *A. thaliana*, *Oryza sativa*. Under low N conditions, a key gene in auxin biosynthesis, TRYPTOPHAN AMINOTRANSFERASE RELATED 2 (TAR2), was known to be involved in the modulation of root system architecture. This suggests the alteration in levels of auxin in response to N (Ma et al. 2014). In *O. sativa*, under suboptimal nitrate nutrition (75/25 ammonium/nitrate ratio), auxin accumulation (IAA) in roots was observed as compared to plants in sole N source (with ammonium) (Bai et al. 2014). NRT1.1 is a nitrate transceptor and is known to transport both nitrate and auxin. Under nitrate deficiency, auxin accumulation is negatively regulated by NRT1.1 benefiting the basipetal auxin transport which stops lateral root outgrowth and accumulation. Therefore, it is known that auxins play a key role in nitrate responses to regulate root system architecture at various levels, involving signaling, biosynthesis, transport, and auxin distribution (Vega et al. 2019).

Various reports suggesting a correlation between K and auxin signaling/transport are available. Under K deficiency, evidence for the role of auxin-dependent processes was contributed by illustrating downregulation of CYP79B2 and CYP79B3 genes associated with tryptophan-dependent auxin biosynthesis. A current study suggested that AtKUP9 also works as an auxin efflux facilitator; it imparts both auxin and K^+ transport from the ER lumen to the cytoplasm to sustain root meristem action and root growth under K^+ limited conditions (Wang et al. 2021).

Likewise, Nikiforova et al. (2005), proposed that auxin signaling under S limitation is highly modulated. Sulfur limitation causes a high increase of auxin and the activation of auxin-induced genes. For example, under S deficiency, analysis of the transcriptional responses of plants induced the expression of the NITRILASE 3 gene (NIT3) that codifies a critical enzyme in the biosynthesis of auxin. Many auxin-relevant transcription factors like ARF-2, IAA13, and IAA28 were observed to serve as coordinators of the metabolic shifts driving S homeostasis (Pandey et al. 2019).

Various recent evidence supports the correlation between auxin signaling pathways and their synthesis under abiotic stresses like Fe deficiency, for their crucial function in biologically significant modification (Divte et al. 2021). Normally, auxin synthesis and distribution are controlled by transporter families like the PIN family, LAX or AUX1 family, and ABCB subfamily. Moreover, OsABCB14 plays an important function in auxin transport, uptake, and Fe homeostasis in *O. sativa*. Fe deficiency induces stress responses that are linked to an increased accumulation of auxin in root tips (Xu et al. 2014). In wheat, the shoot Fe status is known to directly regulate the root Fe uptake by modulating PS release into the rhizosphere via auxin signaling. Possibly, high levels of auxin elevate ethylene production by producing the synthesis of ACC synthase. For example, Fe limitation often leads to the development of branched root hairs via a signaling cascade that involves auxin and ethylene (Pandey et al. 2019).

Nutrients in Gibberellic Acid (GA) and Abscisic Acid (ABA) Signaling

Gibberellic acid (GA) and its correlation with plant adaptive responses to nutrient starvation are currently limited to the connection between PSR and GA (Jiang and Fu 2007). Pi starvation is known to induce the accumulation of the DELLA proteins (core components of GA-signaling) and reduce bioactive GA and transcript encoding enzymes of GA metabolism. Additionally, DELLAs are known to participate in K uptake under conditions of K limitation. Current advances show that GAs significantly regulate the responses to other nutrient accessibility. In *Arabidopsis*, for example, the temporal distribution of GA-modulated DELLA growth repressors in roots controls the root architecture and the Fe-uptake mechanism to the plant's Fe requirement (Gao and Chu 2020). Similarly, P-scarcity induces transcription factor MYB62 which in turn influences the expression of numerous GA biosynthesis genes (Scheible and Rojas-Triana 2018). Nitrate and GA have been known to be linked with the modulation of flowering time. Increased nitrate availability hinders floral induction through GA signaling and the age-related pathway (Vega et al. 2019). Similarly, under low N, NGR5 (NITROGEN MEDIATED TILLER GROWTH RESPONSE), an APETALA2-domain transcription factor, is a key gene controlling tiller number. NGR5 is responsive to N and is a target of GA receptor, G1D1 protein, and is negatively regulated by GA and its receptor (Wu et al. 2020).

Abscisic acid (ABA) is a principal phytohormone for stress signals, vital in the tolerance response to salinity and drought stress. Correlation between nitrate homeostasis and ABA has been studied by analyzing the effect of plant response of altered ABA signaling to nitrate resupply and, conversely, by observing regulation of ABA biosynthesis by nitrate (Ondzighi-Assoume et al. 2016). Nitrate regulates the production of bioactive ABA form, ABA-glucose ester (ABA-GE) utilizing the enzyme of BG1, B-glucosidase 1(BG1). Nitrate is known to transcriptionally upregulate BG1. The nitrate-induced production of ABA is independent of de novo biosynthesis (Vega et al. 2019). Under the low Pi condition, *Arabidopsis* mutants display reduced ABA biosynthesis or low sensitivity, aba1 and abi2-1, exhibit decrease PSR accumulation and gene expression of anthocyanins. Moreover, alteration in ABA signaling in these mutants did not affect other responses such as root to shoot activity or low Pi-increased phosphatase activity (Rubio et al. 2009). Conversely, ABA synthesis is not altered in cotton under suboptimal P levels, and neither the analyzed developmental and biochemical responses are altered in aba1 or abi2-1 mutants to P-starvation (Scheible and Rojas-Triana 2018). Sulfate assimilation is known to be induced by oxidative stress and so is ATPS, APS reductase (APR), SAT, sulfite reductase, and sulfate transporter. Moreover, sulfate influence ABA signaling to root to shoot and affects stomatal closure under drought stress. Likewise, S-containing H_2S (gaseous molecule) works at the downstream pathway of ABA and aid in ABA-modulated stomatal closure in response to drought (Hasanuzzaman et al. 2018).

Nutrient Availability and Signaling of other Phytohormones

Present knowledge of nutrients availability on signaling and biosynthesis of other phytohormones like jasmonic acid (JA), brassinosteroids, SA, and polyamines are more limited.

Jasmonic Acid (JA)

JA and its derivatives often called jasmonates (JAs) are a class of oxylipin-derived phytohormones, which are produced in response to a vast number of abiotic and biotic stresses in plants. JA signaling is known to be activated under nutrient deficiency conditions. Under low S conditions, a large number of JA biosynthesis genes are found to be induced (Rubio et al. 2009). The primary and secondary S-metabolism are coordinately influenced by jasmonates and through sulfate deficiency, JAs synthesis is induced. S has some role in JA signaling through cyclophilin and CYP20-3 JA precursors. The OPDA-CYP20-3 complex, a precursor of Cyt, communicates with serine acetyltransferase (SAT), to achieve this mechanism S metabolites are necessary (Hasanuzzaman et al. 2018). Similarly, under P-deprived plants, genes modulating the first three steps of JA biosynthesis were found to be highly expressed. Conversely, K availability to the same plants reduced the expression of about 19 JA biosynthesis-associated genes. These transcriptional activities were mostly found in the shoot, suggesting organ specificity in the K-mediated regulation of JA accumulation (Rubio et al. 2009).

Brassinosteroids (BR)

Brassinosteroids are found to play a crucial role in plant developmental responses under nutrient availability. Recently, BRs under mild N deficiency in *Arabidopsis* have been known to modulate the root N foraging. Under mild N starvation screening of genetic adaptation among *A. thaliana* accessions leads to the identification of a gene, BRASSINOSTEROID SIGNALING KINASE 3 (BSK3), related to PR elongation (Jia et al. 2019). N-starvation is also found to promote BR activity and increase the BZR1 (BR signaling transcription factor) levels in *Solanum lycopersicum* (Wang et al. 2019). Likewise, a system genomics perspective throughout combined nutrient limitation exhibited the prohibitory effect of Zn limitation on the root growth is suppressed by Pi deficiency. Plants under low Pi exhibited a decrease in the bioactive BR with a reduction in the cytoplasmic/nucleus ratio of BZR1/BES1, thereby, limiting BR-mediated root growth. In a recent study, the transcriptional response of B starved roots accompany the recognition of genes responsive for BR and both B and BRs controlled the expression of numerous genes. The B limitations decreased the active BR (brassinolide-BL), suppressed the BR biosynthesis genes, and restricted PR growth (Pandey et al. 2019).

Salicylic Acid (SA)

SA is known to participate in the regulation of redox status of plants, potentially by modulating the production of the antioxidant glutathione, that defend plants against the oxidative stress which accompany various nutritional deficiency (Rubio et al. 2009). A relationship between nitrate and SA has been demonstrated at different levels and in different studies. In a study, N limitation modifies poly (A) usage in many transcripts, some mediated by FIP1, a part of the polyadenylation mechanism. A hormone profile reveals that the levels of SA, a phytohormone that reduces nitrate accumulation and root growth, enhance substantially upon N limitation. However, the meta-analyses of APA-influenced and *fip1-2*-deregulated genes exhibit a relation between SA signaling and N limited response (Conesa et al. 2020). In *Arabidopsis* mutant E3SUMO ligase S1Z1, induction in the expression of pathogenesis-related gene PR1 and PR2 and enhanced contents of SA was found (Vega et al. 2019).

Polyamines and Strigolactones

Spermidine (Spd), spermine (Spm), and Putrescine (Put) are the three major PAs that develop in plants and are produced consecutively in a similar biosynthetic pathway. The cellular accumulation of PAs usually differs depending on the plant growth or stage of development, nutritional status, especially depends on the kind of N form supplied to the plant (Hasanuzzaman et al. 2018). Under N availability, the level of PAs is found to increase in plants. As PAs usually result from N-induced enhanced levels of their precursor amino acids, such as Arg and Orn, that are converted to Put. Likewise, S deficiency is known to hinder the biosynthesis of PA, therefore, the concentration of PA declines (Hasanuzzaman et al. 2018).

Strigolactones (SLs) are a recently recognized category of carotenoid-derived phytohormones. Various studies have shown that the level of nutrients in the soil is essential in the modulation of the exudation and biosynthesis of SLs (Marzec et al. 2013). For example, the effect of nutrient limitation on the exudation and biosynthesis was studied in *Trifolium pretense* L., a host for AMF and parasitic plants. The level of orobanchol, an SL that was studied under low K, N, P, Mg, and Ca conditions, was found 20 times higher under P-limited conditions. Similarly, in the roots *Sorghum bicolor* L., under N and P limitation, levels of 5-deoxystrigol, a major SL, were found to be enhanced (30 times than control) in response to the nutrient stress (Marzec et al. 2013; Kalia et al. 2021)).

Transcriptional Interrelation between Nutrient Deprivation and Phytohormones

Recently, transcriptional data is generated on the effects of nutrient starvation and phytohormone treatments. The comparison among the genes responsive to hormonal signaling with those modified by nutrient application shows a substantial association for either upregulated or down-regulated genes for nutrient limitation and phytohormone. In *Arabidopsis,* correlation among the transcriptional response of limitation stress for different nutrients like N, K, P, and S and different hormone treatments was analyzed. Significant upregulation was observed under all nutrient starvation stresses in CK and downregulation in ABA-responsive gene, except in the case of Pi limitation. Similarly, except S starvation, all other nutrient stresses modulated the expression of metabolism/signaling genes corresponding to all analyzed hormone-like ABA, CK, and SA activity linked genes. Moreover, it was found that in all four nutrient-deprived stresses there is an overrepresentation of metabolism/signaling genes representing one or more phytohormones (Rubio et al. 2009). Similarly, a transcriptional study on B efficient/inefficient genotypes of *B. napus* showed differential expression of phytohormone-responsive genes. The study showed limited B affected the expression of efflux gene *BnPIN1* and auxin biosynthesis gene *BnNIT1* and reduced the IAA concentration in both root and shoot of *B. napus*. The B deficiency also increased the abscisic acid (ABA), jasmonates (Jas) concentrations and induced the expression of ABA sensor gene *BnPYL4* and ABA biosynthesis genes in the shoot. The study suggested that B starvation interrupted the phytohormone homeostasis in *B. napus* that originated from the alteration in the phytohormone-associated genes (Zhou et al. 2016). In a recent study, in rice, the transcriptomic study in response to low ammonium supply showed involvement of phytohormonal signaling and *OsJAZ9*, a transcription factor in a rearrangement of N metabolism and its uptake (Sun et al. 2020). Similarly, a transcriptomic study on maize in response to N, K, P deficiency showed alteration in genes like *RTCS-LIKE, RTCL; ROOTLESS CONCERNING CROWN AND SMEINAL ROOTS, RTCS; ROOTLESS WITHUNDETECTABLEMERISTEM 1, RUM1* that are involved in auxin signaling (Ma et al. 2020).

Conclusions and Prospects

Nutrient and phytohormone signaling are intimately interrelated to coordinate plant function and form. As the conception of the phytohormonal signaling improves, more relatedness becomes evident between phytohormones and nutrients and nutrient responsiveness in various plant species. The fluctuation in the availability of nutrients is known to modulate the plant's developmental and growth processes. This modulation is observed to be dependent on various phytohormonal pathways. Furthermore, the transcriptomic comparability discloses a potentially immense degree of interconnection between nutrient deficiency and hormonal signaling, which is induced by overrepresentation between nutrient limitation induced alteration in the expression of genes related to metabolism/signaling of phytohormones. The recent studies show the interconnection between nutrients and phytohormone extends beyond the physiological aspect to include biotic and abiotic stress responses. Besides, hormonal signaling is also known to apply feedback control over nutrient metabolism and its regulatory networks. Although the interaction between nutrients and important phytohormones is evident and crucial. The understanding of the mechanism of the interconnections between the nutrients and phytohormones at the systemic and local level following the alteration in environmental state lingers an area of active proceeding research. The understanding of the regulatory networks behind the nutrient signaling and its association with the phytohormonal pathways is principal to engineer an advanced biotechnological approach to develop new crops which are nutrient-use efficient in real-life circumstances for agriculture sustainability.

Acknowledgments

We acknowledge Director, CSIR-National Botanical Research Institute, Lucknow, India, for providing all the necessary facilities. Harshita Joshi acknowledges Department of Science and Technology, New Delhi, India, for her INSPIRE fellowship awarded to her. This research was supported by in-house project OLP109 and CSIR, New Delhi, funded project MLP049.

References

Argueso, C.T., Ferreira, F.J., and Kieber, J.J. (2009). Environmental perception avenues: the interaction of cytokinin and environmental response pathways. *Plant Cell Environ.* 32: 1147–1160.

Bai, L., Ma, X., Zhang, G. et al. (2014). A receptor-like kinase mediates ammonium homeostasis and is important for the polar growth of root hairs in Arabidopsis. *Plant Cell* 26 (4): 1497–1511.

Chen, D., Shao, Q., Yin, L. et al. (2019). Polyamine function in plants: metabolism, regulation on development, and roles in abiotic stress responses. *Front. Plant Sci.* 9: 1945.

Chen, K., Li, G.J., Bressan, R.A. et al. (2020). Abscisic acid dynamics, signaling, and functions in plants. *J. Integr. Plant Biol.* 62: 25–54.

Conesa, C.M., Saez, A., Navarro-Neila, S. et al. (2020). Alternative polyadenylation and salicylic acid modulate root responses to low nitrogen availability. *Plan. Theory* 9 (2): 251.

D'Hooghe, P., Escamez, S., Trouverie, J., and Avice, J.C. (2013). Sulphur limitation provokes physiological and leaf proteome changes in oilseed rape that lead to perturbation of sulphur, carbon and oxidative metabolisms. *BMC Plant Biol.* 13: 1–15.

Deb, S., Sankaranarayanan, S., Wewala, G. et al. (2014). The S-domain receptor kinase AtARK2 and the U-box/ARM-repeat-containing E3 ubiquitin ligase AtPUB9 module mediates lateral root development under phosphate starvation in Arabidopsis. *Plant Physiol.* 165: 1647–1656.

Devireddy, A.R., Zandalinas, S.I., Fichman, Y., and Mittler, R. (2021). Integration of reactive oxygen species nd hormone signaling during abiotic stress. *Plant J.* 105: 459–476.

Divte, P.R., Yadav, P., Pawar, A.B. et al. (2021). Crop response to iron deficiency is guided by cross-talk between phytohormones and their regulation of the root system architecture. *Agric. Res.* 10: 347–360.

Dubois, M., Van den Broeck, L., and Inzé, D. (2018). The pivotal role of ethylene in plant growth. *Trends Plant Sci.* 23: 311–323.

Eggert, K. and von Wirén, N. (2017). Response of the plant hormone network to boron deficiency. *New Phytol.* 216: 868–881.

Engels, C., Kirkby, E., and White, P. (2012). Mineral nutrition, yield and source–sink relationships. In: *Marschner's Mineral Nutrition of Higher Plants*, 85–133. Academic Press.

Forieri, I., Sticht, C., Reichelt, M. et al. (2017). System analysis of metabolism and the transcriptome in *Arabidopsis thaliana* roots reveals differential co-regulation upon iron, sulfur and potassium deficiency. *Plant Cell Environ.* 40: 95–107.

Fujiyama, K., Hino, T., Kanadani, M. et al. (2019). Structural insights into a key step of brassinosteroid biosynthesis and its inhibition. *Nat. Plants.* 5: 589–594.

Gallei, M., Luschnig, C., and Friml, J. (2020). Auxin signalling in growth: Schrödinger's cat out of the bag. *Curr. Opin. Plant Biol.* 53: 43–49.

Gao, S. and Chu, C. (2020). Gibberellin metabolism and signaling: targets for improving agronomic performance of crops. *Plant Cell Physiol.* 61 (11): 1902–1911.

Ham, B.K., Chen, J., Yan, Y., and Lucas, W.J. (2018). Insights into plant phosphate sensing and signaling. *Curr. Opin. Biotechnol.* 49: 1–9.

Han, G.Z. (2017). Evolution of jasmonate biosynthesis and signaling mechanisms. *J. Exp. Bot.* 68: 1323–1331.

Hasanuzzaman, M., Bhuyan, M.H.M.B., Mahmud, J.A. et al. (2018). Interaction of sulfur with phytohormones and signaling molecules in conferring abiotic stress tolerance to plants. *Plant Signal. Behav.* 13: e1477905.

He, J., Xin, P., Ma, X. et al. (2020). Gibberellin metabolism in flowering plants: an update and perspectives. *Front. Plant Sci.* 11: 532. https://doi.org/10.3389/fpls.2020.00532.

Hedden, P. (2020). The current status of research on gibberellin biosynthesis. *Plant Cell Physiol.* 61 (11): 1832–1849.

Huang, W., Wang, Y., Li, X., and Zhang, Y. (2020). Biosynthesis and regulation of salicylic acid and N-hydroxypipecolic acid in plant immunity. *Mol. Plant* 13 (1): 31–41.

Iqbal, N., Trivellini, A., Masood, A. et al. (2013). Current understanding on ethylene signaling in plants: the influence of nutrient availability. *Plant Physiol. Biochem.* 73: 128–138.

Jia, Z., Giehl, R.F., Meyer, R.C. et al. (2019). Natural variation of BSK3 tunes brassinosteroid signaling to regulate root foraging under low nitrogen. *Nat. Commun.* 10: 1–13.

Jiang, C. and Fu, X. (2007). GA action: turning on de-DELLA repressing signaling. *Curr. Opin. Plant Biol.* 10: 461–465.

Jiang, L., Cao, H., Chen, Z. et al. (2018). Cytokinin is involved in TPS22-mediated selenium tolerance in *Arabidopsis thaliana*. *Ann. Bot.* 122: 501–512.

Kalia, V.C., Gong, C., Patel, S.K., and Lee, J.K. (2021). Regulation of plant mineral nutrition by signal molecules. *Microorganisms* 9: 774.

Kalra, G. and Bhatla, S.C. (2018). Gibberellins. In: *Plant Physiology, Development and Metabolism*, 617–628. Berlin/Heidelberg: Springer.

Khan, A., Khan, A.L., Imran, M. et al. (2020). Silicon-induced thermotolerance in *Solanum lycopersicum* L. via activation of antioxidant system, heat shock proteins, and endogenous phytohormones. *BMC Plant Biol.* 20: 1–8.

Krouk, G., Ruffel, S., Gutiérrez, R.A. et al. (2011). A framework integrating plant growth with hormones and nutrients. *Trends Plant Sci.* 16: 178–182.

Landrein, B., Formosa-Jordan, P., Malivert, A. et al. (2018). Nitrate modulates stem cell dynamics in *Arabidopsis* shoot meristems through cytokinins. *Proc. Natl. Acad. Sci. U. S. A.* 115: 1382–1387.

Lefevere, H., Bauters, L., and Gheysen, G. (2020). Salicylic acid biosynthesis in plants. *Front. Plant Sci.* 11: 338.

Li, S.M., Zheng, H.X., Zhang, X.S., and Sui, N. (2020). Cytokinins as central regulators during plant growth and stress response. *Plant Cell Rep.* 40 (2): 1–12.

López-Bucio, J., Cruz-Ramírez, A., and Herrera-Estrella, L. (2003). The role of nutrient availability in regulating root architecture. *Curr. Opin. Plant Biol.* 6: 280–287.

Lv, X., Zhang, Y., Hu, L. et al. (2021). Low-nitrogen stress stimulates lateral root initiation and nitrogen assimilation in wheat: roles of phytohormone signaling. *J. Plant Growth Regul.* 40: 436–450.

Ma, W., Li, J., Qu, B. et al. (2014). Auxin biosynthetic gene TAR2 is involved in low nitrogen-mediated reprogramming of root architecture in Arabidopsis. *Plant J.* 78: 70–79.

Ma, N., Dong, L., Lü, W. et al. (2020). Transcriptome analysis of maize seedling roots in response to nitrogen-, phosphorus-, and potassium deficiency. *Plant Soil* 447: 637–658.

Maeda, Y., Konishi, M., Kiba, T. et al. (2018). A NIGT1-centred transcriptional cascade regulates nitrate signalling and incorporates phosphorus starvation signals in Arabidopsis. *Nat. Commun.* 9: 1376.

Mallikarjuna, M.G., Thirunavukkarasu, N., Sharma, R. et al. (2020). Comparative transcriptome analysis of iron and zinc deficiency in maize (*Zea mays* L.). *Plan. Theory* 9: 1812.

Markovich, O., Steiner, E., Kouˇril, Š. et al. (2017). Silicon promotes cytokinin biosynthesis and delays senescence in Arabidopsis and Sorghum. *Plant Cell Environ.* 40: 1189–1196.

Marzec, M., Muszynska, A., and Gruszka, D. (2013). The role of strigolactones in nutrient-stress responses in plants. *Int. J. Mol. Sci.* 14: 9286–9304.

Miura, K., Lee, J., Gong, Q. et al. (2011). SIZ1 regulation of phosphate starvation-induced root architecture remodeling involves the control of auxin accumulation. *Plant Physiol.* 155: 1000–1012.

Munné-Bosch, S., Simancas, B., and Müller, M. (2018). Ethylene signaling cross-talk with other hormones in Arabidopsis thaliana exposed to contrasting phosphate availability: differential effects in roots, leaves and fruits. *J. Plant Physiol.* 226: 114–122.

Nadira, U.A., Ahmed, I.M., Wu, F., and Zhang, G. (2016). The regulation of root growth in response to phosphorus deficiency mediated by phytohormones in a Tibetan wild barley accession. *Acta Physio. Planta* 38: 105.

Nam, Y.J., Tran, L.S.P., Kojima, M. et al. (2012). Regulatory roles of cytokinins and cytokinin signaling in response to potassium deficiency in Arabidopsis. *PLoS One* 7: e47797.

Nikiforova, V.J., Kopka, J., Tolstikov, V. et al. (2005). Systems rebalancing of metabolism in response to sulfur deprivation, as revealed by metabolome analysis of Arabidopsis plants. *Plant Physiol.* 138 (1): 304–318.

Ondzighi-Assoume, C.A., Chakraborty, S., and Harris, J.M. (2016). Environmental nitrate stimulates abscisic acid accumulation in Arabidopsis root tips by releasing it from inactive stores. *Plant Cell* 28: 729–745.

Pandey, V., Bhatt, I.D., and Nandi, S.K. (2019). Role and regulation of auxin signaling in abiotic stress tolerance. In: *Plant Signaling Molecules* (ed. M.I.R. Khan, P.S. Reddy, A. Ferrante and N.A. Khan), 319–331. Woodhead Publishing.

Pavlů, J., Novák, J., Koukalová, V. et al. (2018). Cytokinin at the crossroads of abiotic stress signalling pathways. *Int. J. Mol. Sci.* 19: 2450.

Pommerrenig, B., Eggert, K., and Bienert, G.P. (2019). Boron deficiency effects on sugar, ionome, and phytohormone profiles of vascular and non-vascular leaf tissues of common plantain (*Plantago major* L.). *Int. J. Mol. Sci.* 20: 3882.

Poza-Viejo, L., Abreu, I., González-García, M.P. et al. (2018). Boron deficiency inhibits root growth by controlling meristem activity under cytokinin regulation. *Plant Sci. J.* 270: 176–189.

Robert, H.S. and Friml, J. (2009). Auxin and other signals on the move in plants. *Nat. Chem. Biol.* 5 (5): 325–332.

Rubio, V., Bustos, R., Irigoyen, M.L. et al. (2009). Plant hormones and nutrient signaling. *Plant Mol. Biol.* 69: 361–373.

Schachtman, D.P. and Shin, R. (2007). Nutrient sensing and signaling: NPKS. *Annu. Rev. Plant Biol.* 58: 47–69.

Scheible, W.R. and Rojas-Triana, M. (2018). Sensing, signalling, and control of phosphate starvation in plants: molecular players and applications. *Annu. Rev. Plant Biol.* 23–63.

Shen, C., Yang, Y., Liu, K. et al. (2016). Involvement of endogenous salicylic acid in iron-deficiency responses in Arabidopsis. *J. Exp. Bot.* 67 (14): 4179–4193.

Song, L., Yu, H., Dong, J. et al. (2016). The molecular mechanism of ethylene-mediated root hair development induced by phosphate starvation. *PLoS Genet.* 12: e1006194.

Sun, L., Di, D.W., Li, G. et al. (2020). Transcriptome analysis of rice (*Oryza sativa* L.) in response to ammonium resupply reveals the involvement of phytohormone signaling and the transcription factor OsJAZ9 in reprogramming of nitrogen uptake and metabolism. *J. Plant Physiol.* 246: 153137.

Thibaud, M.C., Arrighi, J.F., Bayle, V. et al. (2010). Dissection of local and systemic transcriptional responses to phosphate starvation in Arabidopsis. *Plant J.* 64 (5): 775–789.

Toor, M.D., Adnan, M., Rehman, F.U. et al. (2021). Nutrients and their importance in agriculture crop production; a review. *Indian J. Pure Appl. Biosci.* 9: 1–6.

Vega, A., O'Brien, J.A., and Gutiérrez, R.A. (2019). Nitrate and hormonal signaling crosstalk for plant growth and development. *Curr. Opin. Plant Biol.* 52: 155–163.

Vishwakarma, K., Upadhyay, N., Kumar, N. et al. (2017). Abscisic acid signaling and abiotic stress tolerance in plants: a review on current knowledge and future prospects. *Front. Plant Sci.* 8: 161.

Wang, Y., Cao, J.J., Wang, K.X. et al. (2019). BZR1 mediates brassinosteroid-induced autophagy and nitrogen starvation in tomato. *Plant Physiol.* 179: 671–685.

Wang, Y., Chen, Y.F., and Wu, W.H. (2021). Potassium and phosphorus transport and signaling in plants. *J. Integr. Plant Biol.* 63 (1): 34–52.

Wang, Q., Qin, G., Cao, M. et al. (2020). A phosphorylation-based switch controls TAA1-mediated auxin biosynthesis in plants. *Nat. Commun.* 11: 1–10.

Xu, T., Qi, M., Shuangshuang, L.V. et al. (2010). The effects of calcium on the expression of genes involved in ethylene biosynthesis and signal perception during tomato flower abscission. *Afr. J. Biotechnol.* 9: 3569–3576.

Xu, Y., Zhang, S., Guo, H. et al. (2014). Os ABCB 14 functions in auxin transport and iron homeostasis in rice (*Oryza sativa* L.). *The Plant J.* 79: 106–117.

Yang, H., Li, Y., Jin, Y. et al. (2020). Transcriptome analysis of *Pyrus betulaefolia* seedling root responses to short-term potassium deficiency. *Int. J. Mol. Sci.* 21: 8857.

Zheng, D., Han, X., An, Y. et al. (2013). The nitrate transporter NRT2.1 functions in the ethylene response to nitrate deficiency in Arabidopsis. *Plant Cell Environ.* 36: 1328–1337.

Zhou, T., Hua, Y., Huang, Y. et al. (2016). Physiological and transcriptional analyses reveal differential phytohormone responses to boron deficiency in Brassica napus genotypes. *Front. Plant Sci.* 7: 221.

12

Nutrients Regulation and Abiotic Stress Tolerance in Plants

Nikita Bisht, Harshita Joshi, and Puneet Singh Chauhan

Microbial Technologies Division, Council of Scientific and Industrial Research-National Botanical Research Institute (CSIR-NBRI), Lucknow, India

Introduction

Food production worldwide is haunted by rapid population growth and extreme fluctuations in climate (Bisht and Chauhan 2020b). Therefore, agricultural production must be increased to feed the world's growing population and maintain the well-being of humankind. Since the scope for increasing the area under agriculture is very small, the planned food production must be accomplished on already cultivated land. However, recent trends show that soil productivity and fertility are decreasing globally as a result of soil depletion and intensive usage without adequate soil management practises (Kopittke et al. 2019). As a consequence of population increase all over the world and excessive and unappropriate use of natural resources, issues such as water scarcity and salinity are becoming more prevalent. These environmental stresses play a major role in lowering crop yields well below their full output. It has been stated that abiotic stress factors are linked to a decrease in maximum crop yields as compared to yields that are obtained under ideal conditions (Waqas et al. 2019). Moreover, it is reported that as a result of abiotic and biotic stresses, farmers lose around 20.0–70.0% of their prospective crop production (Seleiman 2019). Abiotic stresses caused by unfavorable climatic conditions are seen as the main limiting factors for the decline in crop production (Meena et al. 2017; Bisht et al. 2019). Drought, salinity, low/high temperature, acidic conditions, light intensity, anaerobiosis, excess of toxic metals like Al, As, and Cd in the soil, submergence, and nutrient starvation are the dominant abiotic stresses (Wang et al. 2003; Hirel et al. 2007; Zhu 2016; Meena et al. 2017; Bisht et al. 2019). Salinity, drought, and temperature stresses have the greatest impact on plant geographical distribution, and plant production, thus risking both food and energy security (Zhu 2016; Bisht et al. 2019). Globally, drought has afflicted 64%, cold has afflicted 57%, while acidic soils, mineral deficiency, and salinity have influenced 15%, 9%, and 6%, respectively, of the total land area (Mittler 2006; Cramer et al. 2011). While an exact calculation of agricultural loss (reduction in crop output and soil health) that occur because of

Plant Ionomics: Sensing, Signaling, and Regulation, First Edition.
Edited by Vijay Pratap Singh and Manzer H. Siddiqui.

abiotic stresses cannot be made, it is obvious that such stresses have an effect on vast areas of land and have a major effect on qualitative and quantitative crop production loss.

Neverethless, in the environment, plants often cope with the rapid fluctuations, and these variations place the plant system out of homeostasis, and thus necessitate the presence of advanced genetic and metabolic pathways within the plant's system (Foyer 2018). Plants have a number of defensive mechanisms that have developed over time to help them deal with harsh environmental conditions (Yolcu et al. 2016) and trigger metabolic reprogramming in the cells to aid regular bio-physicochemical processes regardless of external circumstances (Mickelbart et al. 2015). Despite the plants' internal resilience to stress, its negative effects can be mitigated by providing a sufficient and balanced supply of mineral nutrients. For instance, plant's mineral-nutrient status appears to play a critical role in rising plant drought resistance (Waraich et al. 2011b). Furthermore, plants are often assisted in reducing the burden of environmental stresses by the microbiome in their rhizosphere (Ke et al. 2020). Therefore, for better plant growth and production management, it is imperative to recognize and consider the physiological, ecological, and biochemical interferences associated with these abiotic stresses. Also, improving plant species with increased resistance to those harmful abiotic stresses is needed to satisfy future population's food demand on a global scale.

How Abiotic Stresses Affect Plants

Plant growth and production are hampered by extreme conditions (below or above optimum levels). Plants can detect and respond to stress in a variety of ways that benefit their survival (Hilker and Schmülling 2019). The most visible consequence of unfavorable conditions appears first at the cellular level, followed by physiological symptoms (Meena et al. 2017). However, plants have developed various strategies to conserve water and control their development before they confront adverse conditions (Renau-Morata et al. 2020). When exposed to high intensity of light photooxidation is induced, which increases the development of highly reactive oxygen intermediates that manipulate biomolecules and enzymes in plants. Moreover, under extreme conditions plant production is also hampered (Li et al. 2009). Similarly, temperature (Zandalinas et al. 2018), various edaphic factors, such as soil salinity, acidity, and alkalinity (Machado and Serralheiro 2017; Dai et al. 2020), pollutants, and anthropogenic disturbances, all negatively influence plant growth and crop productivity (Meena et al. 2017).

Plants react to stress stimuli quickly and effectively, activating a complex stress-specific signalling cascade (Nejat and Mantri 2017). Accumulation of phenolic acids and flavonoids, synthesis of phytohormones (Sharma et al. 2019), various antioxidants, and osmolytes, and stress-specific transcription factors are initiated to establish an efficient defense system in plants (Zandalinas et al. 2018). Understanding the molecular machinery and its networks that operate under stress conditions is the most important element of stress mitigation in plants. This provides a detailed analysis of the abundance of metabolic pathways and their regulatory genes. Further, the use of a variety of plant and microbial biomolecules has been proposed as another strategy for alleviating abiotic stresses in plants. These methods are paving the gateways for scientists to discover new ways to mitigate abiotic stresses in plants.

Plant's Response to Abiotic Stress

Plants internal resilience system help them in sensing, managing, maintaining, or avoiding any alterations in environmental conditions. Abiotic stimuli in the soil are thought to be more sensitive to root system architecture, which responds accordingly (Khan et al. 2016). It is a complex process with dynamic and real-time alterations at the genetic, transcriptomic, cellular, metabolic, and physiological levels. Plant hormones are known to play vital role in growth and development, but they are also documented to be crucial for developing tolerance against abiotic stresses (Wani et al. 2016). Hormones control the prioritization of signals by protein switches such as kinases, transcription factors, and G-proteins, according to gene expression profiling (Depuyd and Hardtke 2011; Yao et al. 2011). Abiotic stresses such as drought, frost, salinity, and heat cause water deficiency within cells, which is accompanied by the production of biochemical, molecular, and phenotypic responses to stress. The stresses that plants face in the environment can be numerous, as can be the complexity of their responses. The response lies in the activation of specific gene(s) accompanied by metabolic reprogramming in the cells to the stress encountered (Meena et al. 2017). Figure 12.1 illustrates the effect of abiotic stress on plant's health and its defence mechanisms.

Water, the most essential component of life, is increasingly becoming a scarce resource for humans and agricultural production. One of the major abiotic factors affecting agricultural crop production around the world is a lack of water. Drought stress has a variety of effects on plant physiology and development. It reduces plant development by interfering with a variety of physiological and biochemical functions, including chlorophyll synthesis (thus, photosynthesis), ion absorption and translocation, nutrient metabolism, carbohydrates metabolism, and respiration. Drought and chilling stress create oxidative stress in plants by inducing reactive oxygen species (ROS) production which in turn disrupts cell

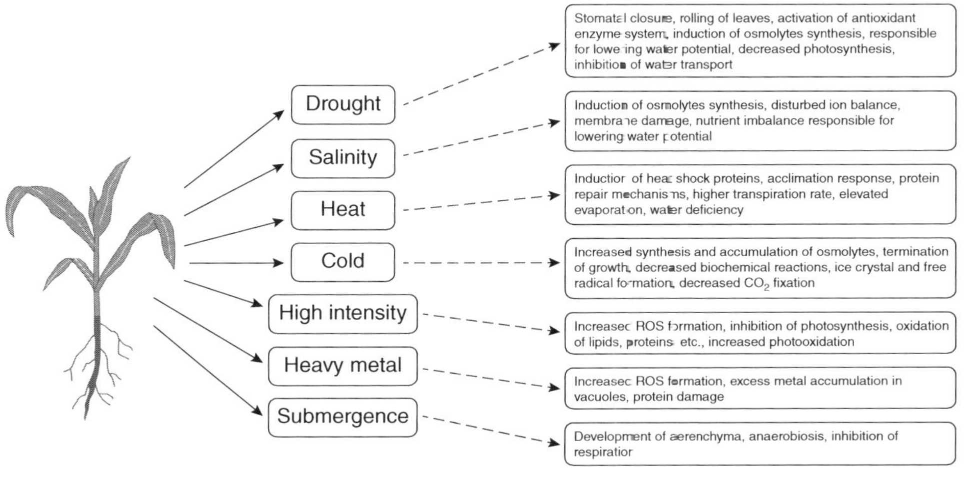

Figure 12.1 Different abiotic stresses, their effect on plant, and plant's strategic defence mechanisms.

membranes, cause enzyme inactivation, ionic imbalance, and protein degradation. Besides, ROS cause harm to the cellular macromolecules such as DNA leading to various biochemical and physiological disorders (Hussain et al. 2018). An excessive amount of salt in the soil is harmful for plant cells, and different cells in a tissue respond in a different way to the stresses induced by salinity (Voesenek and Pierik 2008). Because of the elevated salt levels in the soil, the osmotic potential of the soil is drastically reduced, resulting in ion toxicity and water stress in the plants. This condition can reduce plant fertility by inhibiting seed germination and seedling development, delaying plant senescence, and eventually causing death (McCue and Hanson 1990).

As global temperatures continue to increase, heat stress is becoming an important agricultural issue that is negatively impacting crop production. Increasing temperature creates adverse alterations in the morpho-anatomical, physiological, biochemical, and genetic level in plants. Heat influences plants at various developmental stages of growth, and high temperatures cause seed germination to be reduced, photosynthesis and respiration to be lost, and membrane permeability to be reduced (Xu et al. 2014). Plant's responses to heat stress include changes in phytohormone levels, metabolites, heat shock and associated proteins, and the development of ROS (Iba 2002). In addition to scavenging ROS, accumulating antioxidants and compatible solutes, chaperone signaling, and transcriptional modulation are some activities that aid cells in surviving heat stress (Wahid et al. 2007).

Heavy metal stress has emerged as a major issue in a variety of terrestrial habitats around the world. These days extensive industrialization has led to detrimental effects on soil as well as on crop productivity by accumulating heavy metals. In the soil, the accumulation of toxic heavy metals interferes with various physiological and molecular activities of plants and thus hampers plant growth directly and/or indirectly (Hassan et al. 2017; Tiwari and Lata 2018). These abnormalities cause the formation of ROS, such as superoxide anion radical, hydrogen peroxide, and hydroxyl radical, which disrupts cell redox homeostasis (Gill and Tuteja 2010; Ibrahim et al. 2015; Shahid et al. 2015). Heavy metal toxicity in plants is believed to be caused by a redox status imbalance. Heavy metal in soil not only affect plants negatively but also their accumulation in food crops cause harm to human health severely (Nabulo et al. 2011; Uzu et al. 2011; Shahid et al. 2015). Plants have developed a variety of protective mechanisms to cope with heavy metal stress and toxicity, including decreased heavy metal absorption, vacuolar sequestration, binding to phytochelatins/metallothioneins, and activation of various antioxidants (Shahid et al. 2015; Tiwari and Lata 2018).

Submergence/flooding is harmful to all terrestrial plants, causing stunted growth and, in some cases, death in many plant species (Voesenek et al. 2006). During submergence, the slow diffusion rates of gases in water as compared to air, as well as the relatively low solubility of O_2 in water, have negative effect on terrestrial plant life (Jackson 1985; Armstrong and Drew 2002). O_2 concentrations in non-photosynthetic plant tissues, such as roots, drop dramatically when submerged (Armstrong et al. 1994); however, CO_2 emitted during fermentation and respiration accumulates (Crawford 1992). The gas composition of submerged green tissues, on the other hand, varies significantly during the day (Raskin and Kende 1984a).

Submergence is often linked to the production of hypoxia, which can result in morphological changes and cellular acclimation responses. Ethylene, abscisic acid, gibberellic acid, and other hormones are essential in the genetically programmed survival process. ATP control, starch metabolism, elemental toxicity, the function of transporters, and redox status have all been clarified at the cellular level. The interplay of transcripts and hormones during this stress provides some key insights into submergence tolerance (Phukan et al. 2015).

Under field conditions, multiple stresses occur at the same time, so plants have several mechanisms in place to deal with rapidly changing adverse circumstances. Even though a lot of research work is being carried out to assess plant responses to single stress conditions, but there have been a few attempts to determine the impact of combined stress conditions on crop plants in simulated laboratory trials. (Meena et al. 2017). This has a big impact on our awareness and understanding of plant responses to multiple stresses, as well as our ability to predict cumulative stress tolerance mechanisms in the lab as well as in the field.

Mineral Nutrients in the Alleviation of Abiotic Stress in Plants

Throughout their life span, most cultivable crops are exposed to one or more abiotic stress(es). As a result, increasing plant resilience in response to abiotic stress is a major challenge in the effort to increase food production by 70% by the year 2050 to feed the world's growing population (Wani and Sah 2014). Plant growth is slowed by changes in plant metabolism, which are also caused by disruptions in the absorption and translocation of essential mineral nutrients. Abiotic stresses such as high salinity, heavy metals, and other environmental factors also initiate direct competition for plant nutrients at membrane transporters, thus worsening the conditions for plants. Mineral nutrients are important not only for proper growth and development but also for defending against a variety of stresses. An increased nutrient concentration in plant tissues often modifies plant responses to stress while also increasing their nutritional value. Similarly, certain beneficial elements used in small amounts may have a positive impact on plant metabolism and lead to increased resistance to harmful environmental changes. A large number of elements in their various chemical forms are studied for this purpose as enhancers of plant abiotic stress tolerance, including essential and beneficial elements. Therefore, optimizing the mineral nutrient level under such conditions will greatly reduce the toxic effects of stresses (Kathpalia and Bhatla 2018). The ability of plants to establish adaptive mechanisms to escape or withstand stress is critical for their survival and crop productivity. Several studies have documented that plant nutritional status play a vital role in their adaptation to different abiotic stresses (Cakmak 2008; Waraich et al. 2011a; Hasanuzzaman et al. 2018; Ahanger and ahmad 2019).

Macronutrients

In plant cells, nitrogen (N) is an important component of many structural, genetic, and metabolic compounds. It is also a component of many essential organic compounds, such as amino acids, proteins, nucleic acids, enzymes, and the chlorophyll molecule. It is the most often limiting nutrient for crop growth among all the necessary nutrients. N is the nutrient that causes the greatest yield response in crop plants, encouraging rapid vegetative growth and a healthy green color. Sometimes, owing to the limited availability of water, agriculture production and crop growth are hampered. However, under such conditions, efficient N application can help (Shangguan et al. 2000; Mahpara et al. 2019; Asghar and Bashir 2020). Drought-stressed plants are also more vulnerable to heat tremors. Drought stress results in crop biomass loss due to nitrogen deficiency (Gong et al. 2020). It has been documented that under drought-cumulative N stress, shoot biomass is more affected, while root biomass is not affected as much (Song et al. 2010). Plants, on the other hand, become drought resistant when there is enough N in the soil. Drought-stressed crops performed significantly better when nitrogen levels were increased (Saneoka et al. 2004; Mahpara et al. 2019). Drought-stressed crops performed significantly better when N levels were increased. N is also essential for preventing plasma membrane damage and adjusting osmotic pressure. The use of N in areas where there is a lack of water improves the absorption of other important nutrients such as potassium and calcium (Asghar and Bashir 2020) components, such as amino acids and nucleic acids. It has been reported that in many plants, NO^{3-} uptake is severely affected by salinity, thereby decreasing NO^{3-} content (Khan and Srivastava 1998). Thus, salinity-induced NO^{3-} deficiency causes reduction in yield and plant growth. Use of NO^{3-} fertilizer decreases the concentration of Cl^- in leaves due to NO^{3-}/Cl^- antagonism (Deane-Drummond 1986; Hu and Schmidhalter 1997). In another study, it was observed that application of N improved rice shoot biomass under salinity condition (Abdelgadir et al. 2005).

Phosphorus (P) is a critical component for plant growth and productivity. Since P is fixed in the soil, its availability in the soil is rarely sufficient for plant growth and development (Malhotra et al. 2018). Optimal P in crops is reported to increasee root growth and stomatal function (Naeem and Khan 2009). The availability of P also improves leaf area, plasma membrane stability, and water quality. It is reported that P levels in leaves were higher under drought conditions than when water was abundant, implying that phosphorus plays a role in drought tolerance (Asghar and Bashir 2020). In plant tissues, salinity stress reduces P uptake and concentration. As a result, plants show reduced and stunted growth, dark green leaf coloration, slender stem development, and the death of older leaves (Taiz and Zeiger 2006). Under salt stress, uptake of P into plants can be needed to maintain vacuolar membrane integrity, allowing for easier compartmentalization of Na^+ ions within vacuoles. This compartmentalization is essential to prevent Na^+ ions from interfering with metabolic processes in the cytosol (Cantrell and Linderman 2001). However, salinity and P nutrition interactions are heavily influenced by plant organisms, physiological developmental stage, climate, salt concentration, and P availability (Grattan and Grieve 1992).

Among the essential plant nutrients, potassium (K) is an essential mineral element that plays a critical role in plant growth and development. ROS production in plants is reduced when the K status is increased. K inhibits the activity of nicotinamide adenine dinucleotide phosphate (NADPH) oxidases while preserving photosynthetic electron transport, reducing ROS. K deficiency may reduce photosynthetic CO_2 fixation as well as assimilate transport and utilization (Waraich et al. 2012). Moreover, K is also involved in cell turgor maintenance, osmotic modification, and aquaporin work under drought conditions, in addition to growth and productivity. There has been evidence of a connection between K nutritional status and plant drought resistance. It has been observed that under drought conditions, a good supply of K will increase plant dry matter more than a lower concentration of K in the soil. Application of K increases root growth by increasing root surface area, which in turn improves water uptake by plant cells (Hasanuzzaman et al. 2018). Recently, Pandey and Mahiwal (2020) have reported that improving K nutrition status will help plant cope better with abiotic stress by lowering ROS levels. Cakmak (2005) reviewed the role of K in ameliorating abiotic stress in plants and suggested that crop plants grown under high light intensity for an extended period of time have a higher internal K requirement than plants grown under lower light intensities. K-deficient plants have an increased capacity to oxidize NADPH. K deficiency increased NADPH-dependent ROS generation, which is consistent with increased NADPH oxidase activity. He also documented that different environmental stresses, such as chilling, drought, and salinity, cause ROS to be released by NADPH oxidase or during photosynthesis, which cause cell damage. Therefore, plants may become extremely sensitive to environmental stresses when K levels are low. The inhibitory effect of K against ROS development during photosynthesis and NADPH oxidase appears to be linked to plant's higher K requirements under various abiotic stresses.

Magnesium (Mg) is involved in a variety of biochemical and physiological processes; it is a necessary component of plant development, as well as a vital component of plant defence mechanisms during abiotic stress (Huber and Jones 2013; Mengutay et al. 2013). It is the central atom of the chlorophyll molecule in the light-absorbing complex of chloroplasts and its contribution to carbon dioxide fixation photosynthetically is possibly the most well-known feature. Mg deficiency affects phloem packing, resulting in sucrose accumulation in photosynthetically active tissues and a deficiency in root energy supply. As a result, it can be hypothesized that plants with Mg deficiency would be more vulnerable to soil–water deficits (Senbayram et al. 2015). Mg is translocated to growing parts of the plant because of its high phloem mobility, where it is required for chlorophyll formation, phloem export of photosynthates to ensure growth and development, and enzyme activation for protein biosynthesis. Hence, adequate amount of Mg is needed to improve tolerance to various stresses and to increase crop yield and quality parameters (Senbayram et al. 2015; Yang et al. 2017).

Micronutrients

Plant's functions are controlled and changed by essential micronutrients. Plants exhibit resistance in a variety of ways under abiotic stress conditions such as drought, mineral deficiency, elevated salt concentrations in soil, and so on, thanks in part to cellular biochemical reactions aided by adequate micronutrient availability. Zinc (Zn), manganese (Mn), copper (Cu), iron (Fe), molybdenum (Mo), boron (B), and chloride (Cl) are essential micronutrients that have been studied for their role in plant survival during abiotic stresses (Pandey 2018). Plant micronutrients, as constituents of biomolecules, function as activators of several enzymes, electron carriers, and other biomolecules in response to abiotic stresses. In addition, their roles in metabolism control, reproduction, and defence from abiotic and biotic stress are critical. A large number of reactive oxygen species (ROS) are generated during various stress conditions, and they harm plant cellular metabolism. The presence of adequate micronutrients protects plants from ROS by causing enzymes and biomolecules to become constituents or activators, thus promoting growth and cellular metabolism (Pandey 2020). Tavanti et al. (2021) have suggested that antioxidative enzymes, nonoxidizing metabolism, and sugar metabolism all can be elicited and activated by foliar or soil application of fertilizers containing micronutrients in optimum concentrations to mitigate oxidative stress harm. They further documented that plants given micronutrients have a better nutritional status and are more resistant to abiotic stress due to reduced ROS formation and increased plant photosynthetic potential. Noreen et al. (2018) have also advocated that foliar application of micronutrients can be used for alleviating the negative effects of abiotic stress in plants.

Iron (Fe) homeostasis is critical for maintaining metabolism during abiotic stresses. At the time of oxidative stress, Fe improves the tolerance by inducing antioxidant system of plants. Choline monooxygenase and cytochrome P450 monooxygenase overexpression aids in the establishment of osmoprotection against high temperatures and salinity stress. Fe^{3+} and Fe^{2+} reduction at the plasmalemma promotes a variety of mechanisms that protect cell metabolism from abiotic stresses (Pandey 2018). In plants, manganese (Mn) is required in only small quantities, but it is critical for the growth of plants. It is an important component of metalloenzyme cluster of the oxygen-evolving complex (OEC) in photosystem II (Alejandro et al. 2020). The Mn–enzyme complex catalyzes cellular metabolism and also detoxifies oxygen free radicals. More than 30 enzymes have been identified for containing Mn (Burnell 1988). The main function of Mn in certain enzymes in plants is to counteract the negative effects of abiotic stress such as Mn superoxide dismutase, phosphoenol pyruvate carboxy kinase, isocitrate dehydrogenase, NAD^{+}malic enzyme, allantoate amidohydrolase, and phosphoenol pyruvate carboxylase. Another essential micronutrient that plays role in abiotic stress tolerance is zinc (Zn). In plants, Zn plays functional and structural role. Zn is reported to prevent water stress and to protect cells against damage by ROS. Zn inhibits ROS either by their detoxification or by the activity of membrane-bound NADPH oxidase (Pandey 2018). Under abiotic stresses such as hot weather and salinity, the wilting of leaf margins occurs (Sharma 2006), due to osmoregulation impairment. In plants, chlorine (Cl) is involved in the osmoregulatory process during abiotic stress. Cl acts as an osmotically active substance in roots (Flowers 1988). Higher concentrations of chloride accumulate in the extension zone of the growing root and the shoot apex, promoting turgor-induced root and shoot apical extension development. Cl itself does not act as a catalyst, but it activates the enzyme asparagine synthetase, which catalyzes the glutamate-dependent asparagine synthesis method (Pandey 2018).

Plant Growth-Promoting Rhizobacteria (PGPR), Mineral Nutrients, and Abiotic Stress

Plant tolerance to abiotic stress has an important effect of increasing crop yield. This can be accomplished by searching for, selecting, and engineering plant species that can withstand stress conditions (Enebe and Babalola 2018). However, this procedure might be time cosuming and laborious. Therefore, to address the global threat posed by climate change-related inconsistencies in rainfall, temperature rise, salinization, etc. of agricultural land, plant scientists are now focussing on the use of eco-friendly and sustainable technology, viz.

PGPR. PGPR are the rhizobacteria that directly or indirectly promote plant health, either by releasing phytohormones or any other biologically active substance, modulating endogenous phytohormone level, increasing the availability and uptake of nutrients through fixation and mobilization, or showing anatagonistic effect against phytopathogens by using the multiple mechanisms they possess (Bisht et al. 2020; Bisht and Chauhan 2020c). While specialized microbes are reported to be responsible for the majority of mineral cycling, their behavior is influenced by the biotic population (microbe–microbe, microbe–plant, or microbe–animal/human) and abiotic constituents (Naik et al. 2019). PGPRs belonging to the genera *Bacillus, Pseudomonas, Paenibacillus, Flavobacterium, Chryseobacterium, Azospirillum,* and *Achromobacter* have been reported to efficiently improve the agricultural productivity in stress affected soils (Etesami and Maheshwari 2018). Moreover, PGPR such as *Paenibacillus lentimorbus* and *Bacillus amyloliquefaciens* are reported to improve the growth of crops such as chickpea and rice grown under suboptimal nutrient conditions (Bisht et al. 2019; Bisht and Chauhan 2020a). The use of PGPRs is expected to become a viable strategy and a growing trend in long-term plant growth enhancement under abiotic stress. Because of their multiple potentials in alleviating stresses, these bacteria may be a good choice for optimal crop production and bio-fertilizer production in the future. The combined use of stress-tolerant PGPRs with proper nutrients under abiotic stresses can be used as an alternative to laborious, time-cosuming processes such as breeding and transgenic technologies. Figure 12.2 demonstrates how abiotic stress induces competition for nutrients, their uptake, and transport thereby creating nutrient imbalance and negatively influencing plant's physiological status, and how synergistic effect of microbes and nutrients negates these effects and improves plant health.

Conclusion

Plant growth and production are negatively impacted by abiotic stresses. Plant growth is slowed by changes in plant metabolism, which is also caused by disruptions in the absorption and translocation of nutrients. Abiotic stresses such as drought, heavy metals, chilling, salinity, and other environmental factors cause competition for plant nutrients at membrane transporters, exacerbating negative effects of nutrient shortage. By having nutrients accessible during stress, optimize the nutrient status and thus greatly reduce the harmful effects of stresses, affecting the physiological and biochemical performance of plants. Certain complex signaling events are initiated to control plant cellular functioning when mineral nutrients occur in suboptimal and supraoptimal amounts. A systematic understanding of plant's responses to stress and tolerance mechanisms may contribute to the strategic creation of better crop production management approaches.

Acknowledgments

We acknowledge Director, CSIR-National Botanical Research Institute, Lucknow, India, for providing all the necessary facilities. Nikita Bisht acknowledges CSIR for the fellowship awarded to her. This research was supported by in-house project OLP109 and CSIR, New Delhi funded project MLP049.

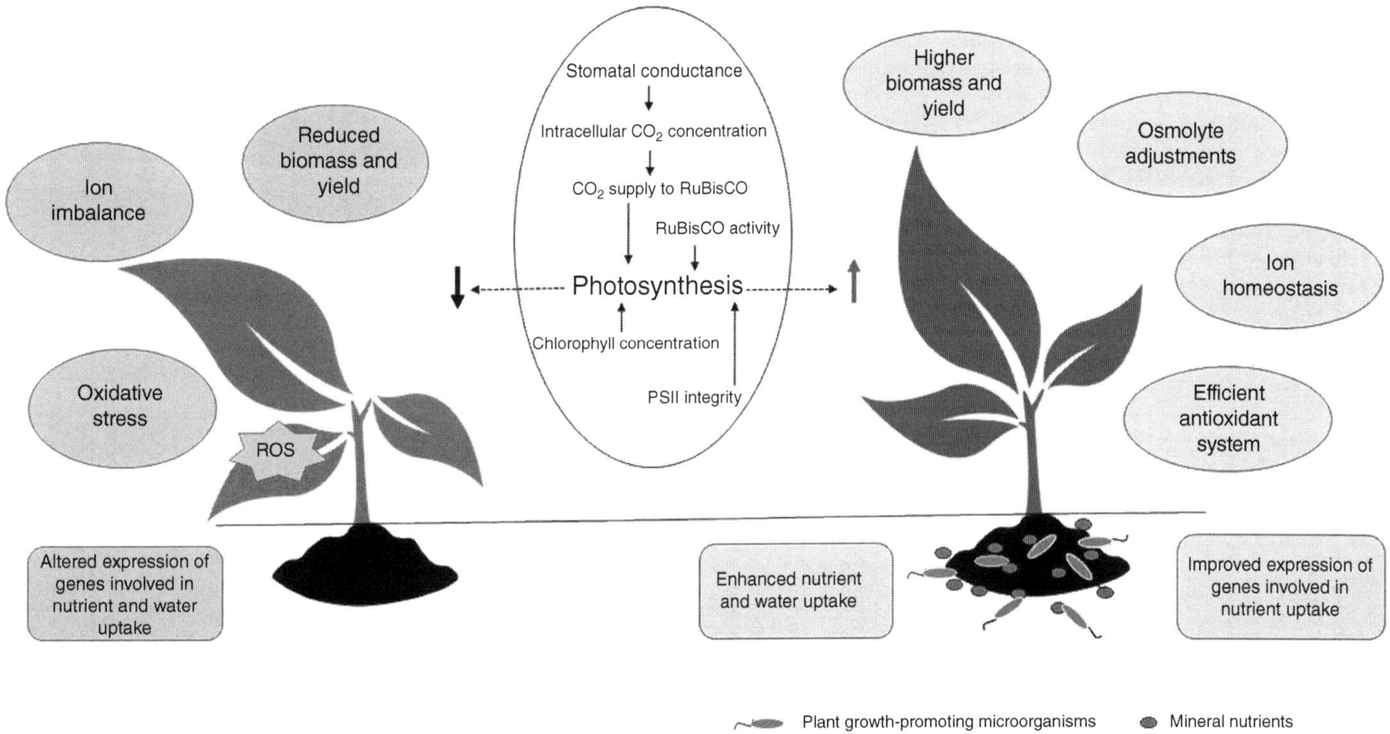

Figure 12.2 Differential response of plants under abiotic stress in absence and presence of nutrients and microbes.

References

Abdelgadir, E.M., Oka, M., and Fujiyama, H. (2005). Nitrogen nutrition of rice plants under salinity. *Biol. Plant.* 49: 99–104.

Ahanger, M.A. and Ahmad, P. (2019). Role of mineral nutrients in abiotic stress tolerance: revisiting the associated signaling mechanisms. In: *Plant Signaling Molecules* (ed. M. Iqbal, R. Khan, P.S. Reddy, et al.), 269–285. Woodhead Publishing ISBN 9780128164518.

Alejandro, S., Höller, S., Meier, B., and Peiter, E. (2020). Manganese in plants: from acquisition to subcellular allocation. *Front. Plant Sci.* 11: 300.

Armstrong, W. and Drew, M.C. (2002). Root growth and metabolism under oxygen deficiency. In: *Plant Roots: The Hidden Half* (ed. Y. Waisel, A. Eshel and U. Kafkafi), 729–761. New York, NY, USA: Marcel Dekker.

Armstrong, W., Brändle, R., and Jackson, M.B. (1994). Mechanisms of flood tolerance in plants. *Acta Bot. Neerl.* 43: 307–358.

Asghar, M.G. and Bashir, A. (2020). Protagonist of mineral nutrients in drought stress tolerance of field crops. In S. Fahad, S. Saud, Y. Chen, C. Wu, & D. Wang (Eds.), *Abiotic Stress in Plants*. IntechOpen. https://doi.org/10.5772/intechopen.94135

Bisht, N. and Chauhan, P.S. (2020a). Comparing the growth-promoting potential of Paenibacillus lentimorbus and Bacillus amyloliquefaciens in Oryza sativa L. var. Sarju-52 under suboptimal nutrient conditions. *Plant Physiol. Biochem.* 146: 187–197.

Bisht, N. and Chauhan, P.S. (2020b). Microorganisms in maintaining food and energy security in a world of shifting climatic conditions. In: *Contemporary Environmental Issues and Challenges in Era of Climate Change* (ed. P. Singh, R. Singh and V. Srivastava). Singapore: Springer https://doi.org/10.1007/978-981-32-9595-7_9.

Bisht, N. and Chauhan, P.S. (2020c). Excessive and disproportionate use of chemicals cause soil contamination and nutritional stress. In: *Soil Contamination* (ed. M.L. Larramendy and S. Soloneski). London: IntechOpen https://doi.org/10.5772/intechopen.94593.

Bisht, N., Tiwari, S., Singh, P.C. et al. (2019). A multifaceted rhizobacterium *Paenibacillus lentimorbus* alleviates nutrient deficiency-induced stress in *Cicer arietinum* L. *Microbiol. Res.* 223: 110–119.

Bisht, N., Mishra, S.K., and Chauhan, P.S. (2020). *Bacillus amyloliquefaciens* inoculation alters physiology of rice (*Oryza sativa* L. var. IR-36) through modulating carbohydrate metabolism to mitigate stress induced by nutrient starvation. *Int. J. Biol. Macromol.* S0141-8130 (19): 35586–35582.

Burnell, J.N. (1988). The biochemistry of manganese in plants. In: *Manganese in Soil and Plants (Graham RD)* (ed. R.J. Hannam and N.C. Uren), 125–137. Dordrecht: Kluwer Academic.

Cakmak, I. (2005). The role of potassium in alleviating detrimental effects of abiotic stresses in plants. *J. Plant Nutr. Soil Sci.* 168: 521–530.

Cakmak, I. (2008). Role of mineral nutrients in tolerance of crop plants to environmental stress factors. *Fertigation: Optimizing the Utilization of Water and Nutrients; International Symposium on Fertigation: Beijing*, China (20–24 September 2005), pp. 35–48. International Potash Institute.

Cantrell, I.C. and Linderman, R.G. (2001). Preinoculation of lettuce and onion with VA mycorrhizal fungi reduces deleterious effects of soil salinity. *Plant Soil* 233: 269–281.

Cramer, G.R., Urano, K., Delrot, S. et al. (2011). Effects of abiotic stress on plants: a systems biology perspective. *BMC Plant Biol.* 11: 163.

Crawford, R.M.M. (1992). Oxygen availability as an ecological limit to plant distribution. *Adv. Ecol. Res.* 23: 93–185.

Dai, Y., Zheng, H., Jiang, Z., and Xing, B. (2020). Combined effects of biochar properties and soil conditions on plant growth: a meta-analysis. *Sci. Total Environ.* 713: 136635.

Deane-Drummond, C.E. (1986). A comparison of regulatory effects of chloride on nitrate uptake, and of nitrate on chloride uptake into Pisum sativum seedlings. *Physiol. Plant.* 66: 115–121.

Depuyd, S. and Hardtke, C.S. (2011). Hormone signalling crosstalk in plant growth regulation. *Curr. Biol.* 21: 365–373.

Enebe, M.C. and Babalola, O.O. (2018). The influence of plant growth-promoting rhizobacteria in plant tolerance to abiotic stress: a survival strategy. *Appl. Microbiol. Biotechnol.* 102: 7821–7835.

Etesami, H. and Maheshwari, D.K. (2018). Use of plant growth promoting rhizobacteria (PGPRs) with multiple plant growth promoting traits in stress agriculture: action mechanisms and future prospects. *Ecotoxicol. Environ. Saf.* 156: 225–246.

Foyer, C.H. (2018). Reactive oxygen species, oxidative signaling and the regulation of photosynthesis. *Environ. Exp. Bot.* 154: 134–142.

Gill, S.S. and Tuteja, N. (2010). Reactive oxygen species and antioxidant machinery in abiotic stress tolerance in crop plants. *Plant Physiol. Biochem.* 48: 909–930.

Gong, X., Li, J., Ma, H. et al. (2020). Nitrogen deficiency induced a decrease in grain yield related to photosynthetic characteristics, carbon–nitrogen balance and nitrogen use efficiency in proso millet (*Panicum miliaceum* L.). *Arch. Agron. Soil Sci.* 66: 398–413.

Grattan, S.R. and Grieve, C.M. (1992). Mineral element acquisition and growth response of plants grown in saline environments. *Agric. Ecosyst. Environ.* 38: 275–300.

Hasanuzzaman, M., Bhuyan, M.H.M.B., Nahar, K. et al. (2018). Potassium: a vital regulator of plant responses and tolerance to abiotic stresses. *Agronomy* 8: 31.

Hassan, T.U., Bano, A., and Naz, I. (2017). Alleviation of heavy metals toxicity by the application of plant growth promoting rhizobacteria and effects on wheat grown in saline sodic field. *Int. J. Phytoremediation* 19: 522–529.

Hilker, M. and Schmülling, T. (2019). Stress priming, memory, and signalling in plants. *Plant Cell Environ.* 42: 753–761.

Hirel, B., Le Gouis, J., Ney, B., and Gallais, A. (2007). The challenge of improving nitrogen use efficiency in crop plants: towards a more central role for genetic variability and quantitative genetics within integrated approaches. *J. Exp. Bot.* 58: 2369–2387.

Hu, Y. and Schmidhalter, U. (1997). Interactive effects of salinity and macronutrient level on wheat. II. Composition. *J. Plant Nutr.* 20: 1169–1182.

Huber, D.M. and Jones, J.B. (2013). The role of magnesium in plant disease. *Plant Soil* 368: 73–85.

Hussain, H.A., Hussain, S., Khaliq, A. et al. (2018). Chilling and drought stresses in crop plants: implications, cross talk, and potential management opportunities. *Front. Plant Sci.* 9: 393.

Iba, K. (2002). Acclimative response to temperature stress in higher plants: approaches of gene engineering for temperature tolerance. *Annu. Rev. Plant Biol.* 53: 225–245.

Ibrahim, M.A., Khan, P.R., Hegazy, S.S. et al. (2015). Improving the phytoextraction capacity of plants to scavenge heavy-metal infested sites. *Environ. Rev.* 23: 1–22.

Jackson, M.B. (1985). Ethylene and the responses of plants to soil waterlogging and submergence. *Annu. Rev. Plant Physiol.* 36: 145–174.

Kathpalia, R. and Bhatla, S.C. (2018). *Plant Mineral Nutrition. Inplant Physiology, Development and Metabolism*, 37–81. Singapore: Springer.

Ke, J., Wang, B., and Yoshikuni, Y. (2020). Microbiome engineering: synthetic biology of plant-associated microbiomes in sustainable agriculture. *Trends Biotechnol.* 39 (3): 244–261. https://doi.org/10.1016/j.tibtech.2020.07.008.

Khan, M.G. and Srivastava, H.S. (1998). Changes in growth and nitrogen assimilation in maize plant induced by NaCl and growth regulators. *Biol. Plant.* 41: 93–99.

Khan, M.A., Gemenet, D.C., and Villordon, A. (2016). Root system architecture and abiotic stress tolerance: current knowledge in root and tuber crops. *Front. Plant Sci.* 7: 1584.

Kopittke, P.M., Menzies, N.W., Wang, P. et al. (2019). Soil and the intensification of agriculture for global food security. *Environ. Int.* 132: 105078.

Li, Z., Setsuko, W., Fischer, B.B., and Niyogi, K.K. (2009). Sensing and responding to excess light. *Annu. Rev. Plant Biol.* 60: 239–260.

Machado, R.M.A. and Serralheiro, R.P. (2017). Soil salinity: effect on vegetable crop growth. Management practices to prevent and mitigate soil salinization. *Horticulturae* 3: 30.

Mahpara, S., Shahnawaz, M., Rehman, K. et al. (2019). Nitrogen fertilization induced drought tolerance in sunflower: a review. *Pure Appl. Biol.* 8: 1675–1683.

Malhotra, H., Vandana, S.S., and Pandey, R. (2018). Phosphorus nutrition: plant growth in response to deficiency and excess. In: *Plant Nutrients and Abiotic Stress Tolerance* (ed. M. Hasanuzzaman, M. Fujita, H. Oku, et al.). Singapore: Springer.

McCue, K.F. and Hanson, A.D. (1990). Salt-inducible betaine aldehyde dehydrogenase from sugar beet: cDNA cloning and expression. *Trends Biotechnol.* 8: 358–362.

Meena, K.K., Sorty, A.M., Bitla, U.M. et al. (2017). Abiotic stress responses and microbe-mediated mitigation in plants: the omics strategies. *Front. Plant Sci.* 8: 172.

Mengutay, M., Ceylan, Y., Kutman, U.B., and Cakmak, I. (2013). Adequate magnesium nutrition mitigates adverse effects of heat stress on maize and wheat. *Plant Soil* 368: 57–72.

Mickelbart, M.V., Paul, M., Hasegawa, P.M., and Bailey-Serres, J. (2015). Genetic mechanisms of abiotic stress tolerance that translate to crop yield stability. *Nat. Rev. Genet.* 16: 237–251.

Mittler, R. (2006). Abiotic stress, the field environment and stress combination. *Trends Plant Sci.* 11: 15–19.

Nabulo, G., Black, C.R., and Young, S. (2011). Trace metal uptake by tropical vegetables grown on soil amended with urban sewage sludge. *Environ. Pollut.* 159: 368–376.

Naeem, M. and Khan, M.M. (2009). Phosphorus ameliorates crop productivity, photosynthesis, nitrate reductase activity and nutrient accumulation in coffee senna (*Senna occidentalis* L.) under phosphorus-deficient soil. *J. Plant Interact.* 4: 145–153.

Naik, K., Mishra, S., Srichandan, H. et al. (2019). Plant growth promoting microbes: Potential link to sustainable agriculture and environment. *Biocatal. Agric. Biotechnol.* 21: 10132.

Nejat, N. and Mantri, N. (2017). Plant immune system: crosstalk between responses to biotic and abiotic stresses the missing link in understanding plant defence. *Curr. Issues Mol. Biol.* 23: 1–16.

Noreen, S., Fatima, Z., Ahmad, S. et al. (2018). Foliar application of micronutrients in mitigating abiotic stress in crop plants. In: *Plant Nutrients and Abiotic Stress Tolerance* (ed. M. Hasanuzzaman, M. Fujita, H. Oku, et al.). Singapore: Springer.

Pandey, S.N. (2018). Biomolecular functions of micronutrients toward abiotic stress tolerance in plants. In: *Plant Nutrients and Abiotic Stress Tolerance* (ed. M. Hasanuzzaman, M. Fujita, H. Oku, et al.). Singapore: Springer.

Pandey, S.N. (2020). Role of micronutrients in biochemical responses of crops under abiotic stresses. In: *Sustainable Agriculture in the Era of Climate Change* (ed. R. Roychowdhury, S. Choudhury, M. Hasanuzzaman and S. Srivastava). Cham: Springer.

Pandey, G.K. and Mahiwal, S. (2020). Potassium in abiotic stress. In: *Role of potassium in plants. SpringerBriefs in Plant Science* (ed. K. Teng). Cham: Springer.

Phukan, U.J., Mishra, S., and Shukla, R.K. (2015). Waterlogging and submergence stress: affects and acclimation. *Crit. Rev. Biotechnol.* 16: 1–11.

Raskin, I. and Kende, H. (1984a). Regulation of growth in stem sections of deep-water rice. *Planta* 160: 66–72.

Renau-Morata, B., Carrillo, L., Dominguez-Figueroa, J. et al. (2020). CDF transcription factors: plant regulators to deal with extreme environmental conditions. *J. Exp. Bot.* 71: 803–3815.

Saneoka, H., Moghaieb, R.E., Premachandra, G.S., and Fujita, K. (2004). Nitrogen nutrition and water stress effects on cell membrane stability and leaf water relations in Agrostis palustris Huds. *Environ. Exp. Bot.* 52: 131–138.

Seleiman, M. (2019). Use of plant nutrients in improving abiotic stress tolerance in wheat. In: *Wheat Production in Changing Environments* (ed. M. Hasanuzzaman, K. Nahar and M.A. Hossain), 481–495. Springer: Singapore.

Senbayram, M., Gransee, A., Wahle, V., and Thiel, H. (2015). Role of magnesium fertilisers in agriculture: plant-soil continuum. *Crop. Pasture Sci.* 66: 1219–1229.

Shahid, M., Khalid, S., Abbas, G. et al. (2015). Heavy metal stress and crop productivity. In: *Crop Production and Global Environmental Issues* (ed. K.R. Hakeem), 1–25. Cham: Springer.

Shangguan, Z.P., Shao, M.A., and Dyckmans, J. (2000). Nitrogen nutrition and water stress effects on leaf photosynthetic gas exchange and water use efficiency in winter wheat. *Environ. Exp. Bot.* 44: 141–149.

Sharma, A., Shahzad, B., Kumar, V. et al. (2019). Phytohormones regulate accumulation of osmolytes under abiotic stress. *Biomol. Ther.* 9 (7): 285.

Song, C.J., Ma, K.M., Qu, L.Y. et al. (2010). Interactive effects of water, nitrogen and phosphorus on the growth, biomass partitioning and water-use efficiency of Bauhinia faberi seedlings. *J. Arid Environ.* 74: 1003–1012.

Taiz, L. and Zeiger, E. (2006). *Plant Physiology*, 4e. Sunderland: Sinauer Associates, Inc.

Tavanti, T.R., de Melo, A.A.R., Moreira, L.D.K. et al. (2021). Micronutrient fertilization enhances ROS scavenging system for alleviation of abiotic stresses in plants. *Plant Physiol. Biochem.* 160: 386–396.

Tiwari, S. and Lata, C. (2018). Heavy metal stress signaling andtolerance due to plant-associated microbes: anoverview. *Front. Plant Sci.* 9: 452.

Uzu, G., Sauvain, J.J., Baeza-Squiban, A. et al. (2011). in vitro assessment of the pulmonary toxicity and gastric availability of lead rich particles from a lead recycling plant. *Environ. Sci. Technol.* 45: 7888–7895.

Voesenek, L.A.C.J. and Pierik, R. (2008). Plant stress profiles. *Science* 320: 880–881.

Voesenek, L.A.C.J., Colmer, T.D., Pierik, R. et al. (2006). How plants cope with complete submergence. *New Phytol.* 170: 213–226.

Wahid, A., Gelani, S., Ashraf, M., and Foolad, M.R. (2007). Heat tolerance in plants: an overview. *Environ. Exp. Bot.* 61: 199–223.

Wang, W., Vinocur, B., and Altman, A. (2003). Plant responses to drought, salinity and extreme temperatures: towards genetic engineering for stress tolerance. *Planta* 218: 1–14.

Wani, S.H. and Sah, S.K. (2014). Biotechnology and abiotic stress tolerance in rice. *J. Rice. Res.* 2: 1000–1105.

Wani, S.H., Kumar, V., Shriram, V., and Sah, S.K. (2016). Phytohormones and their metabolic engineering for abiotic stress tolerance in crop plants. *Crop. J.* 4: 162–176.

Waqas, M.A., Kaya, C., Riaz, A. et al. (2019). Potential mechanisms of abiotic stress tolerance in crop plants induced by thiourea. *Front. Plant Sci.* 10: 1336.

Waraich, E.A., Ahmad, R., Ashraf, M.Y., and Saifullah, A.M. (2011a). Improving agricultural water use efficiency by nutrient management in crop plants. *Acta Agric. Scand B Soil Plant Sci.* 61 (4): 291–304.

Waraich, E.A., Ahmad, R., and Saifullah, A.M.Y. (2011b). Role of mineral nutrition in alleviation of drought stress in plants. *Aust. J. Crop. Sci.* 5: 764–777.

Waraich, E.A., Ahmad, R., Halim, A., and Aziz, T. (2012). Alleviation of temperature stress by nutrient management in crop plants: a review. *J. Soil Sci. Plant Nutr.* 12: 221–244.

Xu, Z., Shimizu, H., Ito, S. et al. (2014). Effects of elevated CO_2, warming and precipitation change on plant growth, photosynthesis and peroxidation in dominant species from North China grassland. *Planta* 239: 421–435.

Yang, N., Jiang, J., Xie, H. et al. (2017). Metabolomics reveals distinct carbon and nitrogen metabolic responses to magnesium deficiency in leaves and roots of Soybean [*Glycine max* (Linn.) Merr.]. *Front. Plant Sci.* 8: 1–12.

Yao, D., Zhang, X., Zhao, X. et al. (2011). Transcriptome analysis reveals salt-stressregulated biological processes and key pathways in roots of cotton (*Gossypium hirsutum* L.). *Genomics* 98: 47–55.

Yolcu, S., Ozdemir, F., Güler, A., and Bor, M. (2016). Histone acetylation influences the transcriptional activation of POX in Beta vulgaris L. and Beta maritima L. under salt stress. *Plant Physiol. Biochem.* 100: 37–46.

Zandalinas, S.I., Mittler, R., Balfagón, D. et al. (2018). Plant adaptations to the combination of drought and high temperatures. *Physiol. Plant.* 162: 2–12.

Zhu, J.K. (2016). Abiotic stress signaling and responses in plants. *Cell* 167: 313–324.

13

Nutrient Management and Stress Tolerance in Crops

Saghir Abbas[1], *Kashif Tanwir*[1], *Amna*[2], *Muhammad Tariq Javed*[2], *and Muhammad Sohail Akram*[1]

[1] *Department of Botany, Faculty of Life Sciences, Government College University, Faisalabad, Pakistan*
[2] *Department of Plant Sciences, Faculty of Biological Sciences, Quaid-i-Azam University, Islamabad, Pakistan*

Over the years, rapid increase in global population and industrialization has magnified the human's impact toward environment and pose serious threats to our environment. The rising food demand is ought to be paralleled with increasing world population, but different abiotic stresses have severely impeded the yield of food crops, curtailed the food supply, and deteriorated goals for achieving global food safety. Plants are exposed to a wide array of abiotic stresses and suffers from oxidative injury catalyzed by the production of reactive oxygen species, which impairs cellular and metabolic processes in plants thus deteriorating plant growth and development. Plants require a wide range of essential nutrients for normal growth and development. Plants fed with optimum levels of nutrients showed higher tolerance and serves important factor for increasing crops quality and yield under various abiotic stress conditions. This chapter summarizes the role of different macro- and micronutrients in plant defense systems, depicts their response in higher tolerance capacity of plants, and shows some insights into their importance for improving crop quality and production under adverse abiotic stress condition.

Introduction

Mineral nutrients are important for natural plant growth and yield. For useful crops, ample allocation of nutrients is necessary from soil (Thakur et al. 2010; Parvin et al. 2019). Plant nutrition studies the effect of elements or nutrients present in topsoil essential for plant growth and metabolism. Plants cannot complete their normal life cycle in the deficiency of these essential nutrients (McCauley et al. 2009). There are about 92 elements in diverse tissues, while only 17 elements are described as essential for plant reproductive life cycle and growth Marschner (2011). These seventeen elements are classified into macro- and micro-nutrients depending upon their concentration required in plants. Mengel and

Plant Ionomics: Sensing, Signaling, and Regulation, First Edition.
Edited by Vijay Pratap Singh and Manzer H. Siddiqui.

Kirkby (1987) further divided nutrients based on their biological composition and different metabolic functions. The first group contains H, C, O, N, and S covalently bonded in reduced form and constitute plant organic material. The second cluster includes P along with B that appears as oxyanions, silicate, and borate. The next third group contains K, Na, Mg, Ca, Mn, and Cl that are designated for their ionic and ionic balance functions. The following fourth growth includes Fe, Cu, Zn, and Mo, and these nutrients appear as mechanical chelates or metalloproteins. Each nutrient has a specific adequacy range in plants, an disparity in nutrient range can negatively impact plant growth and any imbalance might cause nutrient toxicity of deficiency, which may induced due to deficient range of nutrient phytoavailability from surrounding soil due to insufficient fertilizer application or nutrient source (Taiz and Zeiger 2002; Karthika et al. 2018). An element is counted as essential if it accomplishes the criteria purposed by Arnon and Stout (1939), (i) a deficiency of nutrient renders it unavailable for plant to achieve its life cycle, (ii) the deficiency of particular element can only be apprehended or corrected by providing the relevant element, and (iii) the element is directly involved in the plant nutrition. Based on the following measures, the elements are believed as vital for plants and are given in Table 13.1.

Plant dry material comprises the 10–20% of the fresh weight, whereas approximately 10% of the dry weight comprises of nutrients. Inadequate level of nutrient adversely disturbs

Table 13.1 Information of nutrients available form, concentration in soil and their content in plant.

Nutrient	Available form	Content in soil	Average con in plants	Year of discovery
Carbon (C)	CO_2		45%	1800
Hydrogen (H)	H_2O		6%	
Oxygen (O)	H_2O and O_2		45%	
Nitrogen (N)	NO^{3-}, NH^{4+}		1.5%	1804
Phosphorus (P)	H_2PO^{4-}, HPO_4^{2-}	0.1%	0.2%	1839
Potassium (K)	K^+	2.0%	1.0%	1839
Calcium (Ca)	Ca^{2+}	0.5%	0.5%	1839
Magnesium (Mg)	Mg^{2+}	0.05%	0.2%	1839
Sulphur (S)	SO_4^{2-}	0.04%	0.1%	1860
Iron (Fe)	Fe^{2+}	25 000 ppm	100 mg kg^{-1}	1843
Manganese (Mn)	Mn^{2+}	200–300 ppm	20 mg kg^{-1}	1922
Zinc (Zn)	Zn^{2+}	10–30 ppm	20 mg kg^{-1}	1926
Copper (Cu)	Cu^{2+}	5–50 ppm	6 mg kg^{-1}	1931
Boron (B)	H_3BO_3, $H_2BO_3^-$, HBO_3^{2-}, BO_3^{2-}	20–200 ppm	20 mg kg^{-1}	1923
Molybdenum (Mo)	MoO_4^{2-}	0.2–2.0 ppm	0.1 mg kg^{-1}	1939
Chlorine (Cl)	Cl^-		100 mg kg^{-1}	1954
Nickel (Ni)	Ni^{2+}	<100 ppm	0.1 mg kg^{-1}	1987

Source: Adapted from Tisdale et al. (1997).

the crop growth resulting in unusual visible symptoms thus causing reduced crop yield (Karthika et al. 2018; Barzana et al. 2021). It is believed that the role and movement of each nutrient is necessary to be identified to establish which particular nutrient is accountable for certain symptom in plants weather its deficiency or toxicity (McCauley et al. 2009). Therefore, a proper understanding of the role of nutrient in plant expansion is important to enhance crop productivity (Ahmed et al. 2020). Plants are subjected to various stresses that are detrimental for their growth and development (Ahanger et al. 2014). Different abiotic stresses that include high temperature, drought, heavy metals, salinity, nutrient stress, waterlogging, and environmental pollution influence plants development in large extent (Hasanuzzamn et al. 2012, 2018). The overall effects of these stresses may trigger decrease in productivity from 51 to 82% and cause significant loss to global food production (Hasanuzzamn et al. 2018a; Kumar and Mallick 2019). Plant faces dehydration and nutritional and hormonal imbalance owing to drought, salinity, and extreme temperature. In plants, different abiotic stresses trigger excessive biosynthesis of reactive oxygen species (ROS), i.e. superoxide, singlet oxygen, hydrogen peroxide (H_2O_2), as well as hydroxyl radicals at cellular level following oxidative stress (Nahar et al. 2016; Abbas et al. 2020). ROS overproduction causes lipid peroxidation, degradation of enzymatic activities protein oxidation, and electrolyte leakage (EL), which damage the intracellular machinery thereby causing cell mortality (Ali et al. 2020; Abbas et al. 2021; Tanwir et al. 2021). Naturally, plants could detoxify harmful environmental stress up to some extent for their existence (Javed et al. 2017). Plants trigger different enzymatic and nonenzymatic compounds as a first defense line to prevent ROS-induced oxidative injury thereby helping plants to improve stress resistance (Abbas et al. 2020).

Implications of Abiotic Stress in Plants

Sustainable agriculture production is predominantly a function several abiotic stress factors, which may link with mineral status of soil (Hasanuzzamn et al. 2018a). In future, considerable increase in magnitude of different stresses is expected to cause significant damage to fertile land, altered mineral uptake, and assimilation (Javed et al. 2017; Gull et al. 2019). Since different environmental stresses are possible threat to crop yield worldwide, current chapter focuses the current understandings about impacts of various abiotic stresses on overall plant development. Furthermore, role of mineral nutrients induced stress resistance in aspect of growth, physiology, and yield enhancement will be discussed in this chapter. Hence, a better understanding of nutrient management would assist to increase crop production and nutritive value for global food production.

Salinity Stress

Salinity (salt stress) is one of the severe environmental stresses that is linked with augmented levels of Na^+ and Cl^- ions that causes imbalance and decline in growth and crop output in arid and semi-arid region crops (Kumar et al. 2017; Sunita et al. 2020). Salinity stress degrades membrane integrity, hampers functions of growth regulators, impairs enzymatic activities, photosynthesis, and physio biochemical traits of plants leading to plant death (Nxele et al. 2017; Gupta et al. 2020). Salinity stress decreases seed germination, disrupts seedling vigor, retards dry matter accumulation, and obstructs plant growth and confers lower crop yield (Shu et al. 2017; Kanwal et al. 2018). The excess salinity caused increased salt content that enhances Na^+ concentration in medium contributing to K^+ leakage from thereby creating an enhanced Na/K ratios in plants (Wu and Wang 2012). Excessive Na entry triggers membrane depolarization and causes nutrient substitution under salt stress (Nahar et al. 2016; Rahman et al. 2016). High levels of salinity cause negative effects on cell division, expansion, and differentiation, which lead to stunted growth in plants (Parihar et al. 2015). Salt stress causes low osmotic pressure that confers low water potential. Furthermore, plants suffer from reduced stomatal conductance, transpiration, and CO_2 assimilation, which impairs plant photosynthesis (Hasegawa et al. 2000; Kumar et al. 2017). It was reported that salt stress reduces plant photosynthetic apparatus such as Chl a Chl b in *Vicia faba*, *Brassica juncea*, and *Morus alba* plants (Ahmad et al. 2010; Azooz et al. 2011; Ahmad et al. 2012). The salinity stress decreases crop yield up to 62% by negatively effecting the physio-biochemical traits. Diminished yield was also reported in different crops such as *Solanum lycopersicum*, *L. V. radiata,* and *B. oleracea var.italica* (Nahar and Hasanuzzamn 2009; Parvin et al. 2015; Parvin and Haque 2017).

Drought

Drought is known as severe abiotic stress that hampers plant progress and yields mainly in the arid and semiarid areas (Kunz et al. 2017; Kasim et al. 2021). One-third of world agricultural land is impacted by drought stress that poses grave threat to crop growth and food security (Kaur and Asthir 2017; Singh et al. 2020). It has been anticipated that by about 50% of arable lands will be disturbed by severe drought by 2050 (Jaleel et al. 2009; Kour et al. 2019). Due to drought stress, plant germination, growth, and development suffer greatly (Li et al. 2017). The lack of precipitation, lack of groundwater triggers energy loss, turgor loss, impairs enzymatic machinery, and hampers cell division and differentiation. Damaging consequences of drought vary with different plant genotypes, growth stage, and severity of drought (Raheem et al. 2018; Saud et al. 2020). Drought stress triggers diminished transpiration rate, nutrient absorption through roots, and thereby causes nutrient insufficiency in plants (Toor et al. 2020). Water scarcities disturb plant enzymatic activities, leaf relative water level, and osmotic capacity in plants. Drought stress induces turgidity loss and subsequently stomatal closure causes decrease transpiration rate and CO_2 intake in *Phaseolus vulgaris* and *H. annuus* plants (Miyashita et al. 2005). Drought stress confers damage to the activity of chlorophyll synthase enzyme whereas reduction in the activities of key photosynthetic enzymes such as pyruvate phosphate dikinase, ribulose-1,5 biphosphate carboxylase/oxygenase and NADP-malate dehydrogenase were evidently decreased under drought stress conditions (He et al. 2017; Hosseini et al. 2018). Drought stress abridge plant growth by negatively effecting the photosynthetic pigments in tomato seedlings (Li et al. 2018). In another study drought stress had adverse effects on plant reproductive stage and inhibited pollen sink strength in drought sensitive wheat cultivar leading to lower crop yield (Dong et al. 2017). Results from prolong drought stress decreased photosynthetic capacity, activities of glutathione, ascorbate, ascorbate peroxidase enzymes, and increased malondialdehyde, reactive oxygen species in alfalfa plants (Zhang et al. 2019). Drought stress drastically reduced the leaf water capacity, stomatal regulation through increased hydrogen peroxide, carbonyl groups, and superoxide radical in four distinct *Argania spinosa* plants (Chakhchar et al. 2016).

Toxic Metals

Plant are constantly facing different abiotic stresses due to rapid urbanization and industrialization (Wani et al. 2018; Abbas et al. 2020). Different anthropogenic events such as industrial waste, smelting and sewage disposal, and fertilizers usage are contaminating agricultural soils with different metal/metalloids (Berni et al. 2019; Ghori et al. 2019). These toxic heavy metals effect plant growth and metabolism, and the intensity of negative effects alters with metal, plant genotype, and soil types (Pandey et al. 2019; Wei et al. 2020). Metals that may be harmful to plants include iron, nickel, chromium, cobalt, cadmium, copper, manganese, zinc, aluminum, arsenic, chromium, manganese, and lead (Javed et al. 2021; Tanwir et al. 2021). Heavy metals are nonbiodegradable, mobile, and easily taken up by plants roots (Li et al. 2019). Some metals like cadmium and lead are present in pesticide that do not have any known functions in plants and can cause damage if present in large amount in soil (Singh et al. 2018; Shukla and Jain 2020). Heavy metal toxicity results in visible signs such as chlorosis and curtailed growth leading to plant death (Kumar et al. 2019). Heavy metal/metalloids stress induce reactive oxygen species including hydroxyl radical, hydrogen peroxide, and superoxide radical, which resulted in oxidative stress in plants. Moreover, higher levels of metal caused oxidative stress denature or inactivate enzymatic activities and other proteins (Tanwir et al. 2021). Majority of heavy metals hinder the uptake of essential nutrients from soil, thereby disrupting the membrane integrity and cause alteration in the plant metabolic reactions including photosynthesis, homeostasis, and respiration (Huihui et al. 2020). Heavy metal-generated reactive oxygen species stimulate lipid peroxidation of cell membrane initiating them to leak, hampers the biomolecules, and triggers DNA strands cleavage due to oxidative injury (Shahzad et al. 2018; Syed et al. 2018).

Other Stresses

Likewise, the plants face the most widespread abiotic stresses, viz. drought, salinity, and heavy metal stress, and also face other abiotic stress conditions such as high temperature, chilling, nutrient shortage, radiation, and ozone.

Temperature fluctuation(extreme/high temperature stress) impart drastic effect on plants and may lead to death. Extreme heat due to rising greenhouse gasses causing extreme rise by $0.89 \pm 0.2\ ^{o}C$ and is expected to rise upto $2.6 \pm 4.8\ ^{\circ}C$ by 2100 (Sage et al. 2015). Each rise in Celsius temperature might reduce crop yield upto 17%. Higher temperature stress impairs physio-biochemical traits such as seed propagation, transpiration, photosynthesis, and dry matter growth leading to lower crop yield (Bita and Gerats 2013). The high temperature stress denature enzyme activities via generation ROS species in various crops. Higher plants face various deformities under high temperature stress including burning of plant parts, abscission hence reduced growth rate (Wang et al. 2010).

Chilling stress or cold stress is among abiotic stress, which limits plant growth, harvest, and geographical distribution (Djanaguiraman et al. 2010; Hu et al. 2019). Chilling-sensitive plants may suffer from cold injury and can be categorized into two kinds: chilling stress (0–20 °C) and freezing stress (<°C) (Zhuang et al. 2019; Wang et al. 2020). Many plant types originated from tropical and subtropical areas could face death of severe injury due to chilling temperature below 10 °C (Zhu et al. 2013). Plant could suffer from lower photosynthesis by chilling stress, which disrupt the entire plant physiology at the molecular level (Ray et al. 2015). Lower temperature significantly damages sugar metabolism and plant endogenous hormone levels (Hajihashemi et al. 2018). The chilling stress induces oxidative damage through the generation of free radicals inside plant cells by negatively influencing lipid peroxidation, proline, and other biomolecules (Hussain et al. 2016; Zhang et al. 2017). Plants are negatively influenced by radiation emission that decreases plant weight, photosynthesis and hampers chloroplast following negative impact of growth and harvest (Shen et al. 2010). The UV-B radiations trigger detrimental effects on plants with the generation of ROS in higher plants, which causes protein denaturation, lipid peroxidation, disruption of Photosystem I, II and light harvesting complex (Liu et al. 2013), gamma radiation cause oxidative damage to DNA, and damage macromolecules through excess ROS bioproduction (Esnault et al. 2010).

Plant nutrients that include macronutrients and micronutrients are essential structural components of plants and are implicated in metabolic, developmental, and physiological functions for plant growth (Parvin et al. 2019; Abbas et al. 2020). Deficiency of these nutrients severely damage plant growth and yield. Several studies reported that nutrients deficiency decreases root elongation, disrupt PSII working, limit glycolysis, TCA cycle working, and decrease efficiency of Rubisco as well all limit plant growth and development through increase ROS, lipid peroxidation, and electrolyte leakage (Nobori et al. 2016; Karthika et al. 2018; Parvin et al. 2019).

Role of Nutrients in Stress Tolerance

The macro and micro nutrients play a key role in mainstreaming and typical performance of plant growth and advancement. The crop fertilized with macro and micronutrients has higher ability to cope with environmental stresses.

Nitrogen

Nitrogen (N) in the periodic table that belongs to group V, which exists as the diatomic inert gas form (N2) and constitutes 78% of in the atmosphere (Karthika et al. 2018). Nitrogen is a crucial constituent of numerous cell components and is an important part of protein, nucleic acids, phospholipids, and secondary metabolites in plants. Nitrogen holds a key role in transferal of genetic code to the progeny as it is vital component of deoxyribonucleic acid or ribonucleic acid (Kim et al. 2005). Nitrogen comprises the chlorophyll pigments and is responsible for imparting green color to the leave and increase quality of leaves in fodder and vegetables (Akram et al. 2011). The chief source of N phytoavailability is industrially formed urea $(NH_2)_2CO$ that is decomposed into ammonium (NH_4^+) through urease enzyme. Recent study reported that nitrogen in plants is absorbed either as NO_3^- or as ammonium NH_4^+ reliant upon plant genotypes and soil types. Nitrate reductase and nitrite reductase are enzyme that participate in conversion of nitrate (NO_3^-) to ammonium (NH_4^+). It is reported that high affinity H+- coupled symporters arbitrate NO_3^- uptake, whereas NH_3/H^+ MT transporters facilitate NH_4^+ uptake in plants (Mon et al. 2016).

Nitrogen Role in Stress Tolerance

Nitrogen accessibility profusely contributes to plant biomass increase by increasing photosynthesis and protein synthesis (Yang et al. 2020). Application of N-fertilizer (0.3 and 0.6 N/Kg soil) induced higher growth and yield due to increased photosynthetic apparatus, soluble carbohydrate, relative water level, and lower membrane permeability in wheat plants (Agami et al. 2018). Recent study demonstrated that the addition of nitrogen fertilizer improved the biomass and grain yield in durum wheat (Mon et al. 2016). In a study performed by (Akram et al. 2011), it was informed that water use efficiency was increased and photosynthetic pigments showed enhanced levels in maize hybrids under salinity stress conditions. The nitrogen fertilizer improved photosynthetic traits, chlorophyll, stomatal conductance, water use efficiency, and photochemistry of PS II in two fescue cultivars under high temperature stress (Zhao et al. 2008). Supply of N fertilizer decreased growth inhibition and yield of rice plants under Cd stress (Jalloh et al. 2009). In another study, nitrogen treatment restored pakchoi growth by relieving Cd-stimulated oxidative injury through decreased MDA under Cd stress (Zhang et al. 2016).

Cadmium stress prompts significant reduction in grain per panicle and plant biomass; however, application of N fertilizer form caused significant progress in growth and yield response in rice plants when applied with Cd stress treatments (Alpha et al. 2009). Under salt stress, the use of nitrogen regulates the proline level and improves the photosynthesis in mustard plant (Iqbal et al. 2015). The presence of Cd metal increases the ROS accumulation, and Cd-stimulated oxidative injury is decreased by exogenous application of N, which reflects the diminished malondialdehyde and hydrogen peroxide in popular plants (Zhang et al. 2014). The exogenous treatment of N augmented the activities of catalase, superoxide dismutase, and glutathione peroxidase in *Brassica juncea* plants cultivated in contaminated segments (Giansoldati et al. 2012).

Potassium

Potassium is one of the crucial macronutrients necessary in large amount in plants for various regulatory function in several biochemical processes in plants (Waraich et al. 2012). Potassium has vital role in protein synthesis, enzyme activation, carbohydrate metabolism, and maintenance of turgor pressure (Wang et al. 2013). The foremost part of K is enzyme activation, and it contributes to the activation of more than 60 enzymes and implies directly in the pyruvate kinase activation (Karthika et al. 2018). Potassium facilitates the efficacy of photosynthesis and CO_2 assimilation; it maintains electrical charge stability at the site of ATP biosynthesis (Kafkafi et al. 2001; Fageria 2016). Potassium is well known as quality nutrient; it has an essential role in improving crop quality and provides resistance against disease and drought in plants (Masood and Bano 2016; Meena et al. 2016). Potassium-deficient plants reported decreased CO_2 fixation, assimilation, chlorophyll degradation, and they exhibit chlorotic and necrotic spots in stress conditions (Ling and Silberbush 2002; Hasanuzzamn et al. 2018b).

Potassium content varies from 1 to 5% in healthy plant tissue and is seventh highly rich element in earth's crust. Main available form of potassium in soil, i.e. exchangeable *k* 1%, noninterchangeable *k* (98.99%), and solution/dissolved form (K^+). Plants absorb K as K^+, and different sources of available K in soil are provided as potassium fertilizer. Potassium availability in soil depends upon soil texture, soil pH, soil moisture, and other factors (Dhillon et al. 2019; Pathak et al. 2020). Different proteins appear in the cell membrane known as transporters and channels, which have affinity for K^+ termed as high affinity transporters expedite K^+ uptake from external environment into different cellular compartment in plants (Hasanuzzamn et al. 2018a; Karthika et al. 2018).

Role of Potassium in Stress Tolerance

In addition to growth, potassium plays an important role in osmotic adjustment, aquaporins function, and turgor maintenance in drought stress condition (Zahoor et al. 2017). The exogenous K treatment enhanced dry accumulation, leaf area, and improved water uptake status in plants under drought conditions (Egilla et al. 2005). In an experimental study, it was reported that ample supply of K decreased ROS production by downregulating lipid peroxidation, electrolyte leakage, and hydrogen peroxide content in (*Helianthus annuus* L.) under water-stressed condition (Zain et al. 2014). Under drought condition, proline accumulation plays a protective roles in plants and facilitate in osmotic adjustments; numerous studies have reported that the use of K significantly promoted growth in *O. sativa*, *B. napus*, and *T. aestivum* plants when faced with drought stress condition (Fazeli et al. 2007; Soleimanzadeh et al. 2010)

Studies confirmed that under saline condition, the K supply significantly improves the K content in plant cell under critical salt stress conditions and increased the plant ability to tolerate salinity stress. Application of KH_2PO_4 significantly promoted shoot and root fresh dry biomass accumulation in *L. esculentum* plants under salinity stress. It was reported that *H. vulgare* plants under 150 mM NaCL stress demonstrated significant improvement in plant height and shoot fresh weight in barley crop (Fayez and Bazaid 2014) Similarly, the use of 9 mM K augmented the root and shoot dry weight in *S. lycopersicum* plants under 75 mM NaCl stress (Amjad et al. 2016). In a recent study, Chakraborty et al. (2016) reported that the addition of K could reduce the Na uptake through adapting the tissue ionic stability in *Arachis hypogea* L (Hasanuzzamn et al. 2018a).

Foliar application of KH_2PO_4 enhanced heat resistantance in wheat plants by increasing photosynthetic capacity in wheat (Dias and Lidon 2010). Plants facing chilling stress showed decreased plant growth; however, application of K regulated the photosynthesis, carbon assimilation, and metabolism in cold-sensitive maize verity due to upregulated ROS defense mechanism (Farooq et al. 2008). The addition of 6 mM K ameliorates Cd toxicity damage in *Vicia faba* L. under high levels of 200 µM Cd stress treatment (Farooq et al. 2008). In another experiment, Cd toxicity reduced the plant (shoot and root length), chlorophyll content, and disrupt the enzyme activity through increased lipid peroxidation; however, the application of K (10 mM) balanced the nutritional demand and upregulated the physiological processes in peach plants though improved antioxidant system and photosynthesis (Song et al. 2015). One of the main consequences of abiotic stress is the overproduction of ROS in plants which triggers oxidation of protein, enzyme inhibition, lipid peroxidation, and leads to cell death. Hence, it is suggested that the application of exogenous K could relieve ROS formation by maintaining electron transport chain in photosynthesis and diminishing the activity of NADPH (Song et al. 2015).

Phosphorus

Phosphorus is a crucial plant nutrient for growth and yield. Phosphorus content in plants varies from 0.05 to 0.5% of plant dry biomass. Phosphorus is considered as immobile in soils and mobile element in plants. Phosphorus performs a key role in different cellular processes that include synthesis of biomolecules and integrity of membrane structure. Phosphorus in plants helps in enzyme activation, cell division, and carbohydrate metabolism (Gupta et al. 2014). It stimulates seed germination, root development, flower as well as seed formation, and enhance crop quality. Phosphorus is a constituent of adenosine triphosphate, cytidine triphosphate, phosphoenol pyruvate, as well as uridine triphosphate and other phosphorylated compounds. Phosphorus is part of nucleic acid and lipid membrane structure (Van de Wiel et al. 2016). Phosphorus has a role in photosynthesis, sugar and starch translocation, and transfer of genetic characters into next generation as it is component of DNA and RNA. Phosphorus is found in abundance in the form of orthophosphate ion $(PO_4)^{3-}$ in the soil. Phosphorus in found in rocks and minerals in earth's crust and is released via weathering of rocks and minerals (Malhotra et al. 2018). The bioavailability of phosphorus in soil is constrained as it occurs in stable and insoluble form. Plant takes up phosphorus in the form of H_2PO4^{-2} and HPO_4^{-2} through roots. Phosphorus uptake takes place through transporters found in root plasma membrane of plants due to difference in concentration between plant and soil. Plant roots and different microbes exude different organic acids and other compounds thus acidifying rhizosphere and triggers smooth movement of P inside plant root (Ramaekers et al. 2010; Karthika et al. 2018).

Role of Phosphorus in Stress Tolerance

Exogenous treatment of phosphate fertilizers enhances the stress resistance in plants that leads in adjustments of different physio biochemical processes in plant, which help in plant growth. Increase in root surface area, water relation, antioxidant activities, and chlorophyll content are reported when phosphorus is applied under harsh drought stress (Ahmad et al. 2017). Application of superphosphate fertilizers ($ZnSO_4H_2O$) significantly increases essential oil yield components of (*Matricaria recutita* L.) (Jeshni et al. 2017; Karthika et al. 2018). The addition of phosphate fertilizers improved tolerance to drought at reproductive stage resulting in enhanced yield and improved quality of grain in soyabean P-deficient soils. The P application augmented the fresh biomass, photosynthetic efficiency, leaf area, and mineral content of plants with decreased Cd uptake and hydrogen peroxide level in wheat plants (Arshad et al. 2016). In a study, the exogenous addition of phosphorus to plants decreased the chloroplast damage and increased the chlorophyll content in maize plants when treated with Cd and Zn stress (Jiang et al. 2007).The root zone application of P improved the maize growth through root proliferation and acidification of rhizosphere (Shen et al. 2011). In another experiment, the addition of P nduced Cd tolerance in Mangrove species through decreased malonaldehyde levels and positively regulating the photosynthetic content and phytochelatins biosynthesis as adaptive strategy to cope Cd-generated stress (Dai et al. 2017).

Calcium

Calcium (Ca) belongs to one of the indispensable macronutrients for plant growth. Calcium plays several key roles in different processes such as functional integrity of cell membrane and control of ion exchange (Marschner and Rengel 2012). Calcium is known as intracellular secondary messenger that facilitates extracellular stimuli for cellular response and synchronize different endogenous actions in plants. Calcium is involved in cell division, enzyme activity, cell wall development, starch metabolism, and nitrate metabolism. Calcium has a regulatory role in seed germination, development, and water relation (Rehman et al. 2016). Calcium pectate is the chief component of cell wall, and it helps to sustain the structure of chromosome. Calcium plays a significant role in nitrogen assimilation and proteins synthesis in plants. Calcium has a role in the activation of phospholipase enzyme, amylase, arginine kinase, and ATPase (Karthika et al. 2018). Calcium concentration is found in about 3.6% of earth's crust and is the third most vital accessible nutrient in the soil. The average soil level of Ca is about 1.37% and it occurs in the type of common minerals and rocks such as apatite, dolomite, feldspar, gypsum, and gluconate (Ray and George 2010). Due to weathering of rocks, Ca enters into soil and is the component of many compounds like calcium gluconate, calcium chloride, calcium carbonate, and calcium sulfate and is available to plants in the form of Ca^{2+}. Calcium permeable channels allow the entry toward the root (Kaya et al. 2002; Nedjimi and Daoud 2009).

Role of Calcium Under Stress

Calcium supply can upregulate the plant defense and yield improvement against different environmental stresses (Karthika et al. 2018). Calcium addition could significantly improve morpho-physio-biochemical features of plants but also augments the yield attribute under saline soil stress conditions (Parvin et al. 2015). In an experiment, 75 mM salinity stress caused harmful impact on yield by decreasing the fruit number and fruit biomass in *S. lycopersicum*; however, the application of Ca 5 mM significantly increased the yield of the plant under salt stress.(Tuna et al. 2007). In another experiment, calcium addition showed higher flower numbers and fruit weight in sweet pepper plant under given drought stress (Azam et al. 2016).To counter salinity stress, the applied calcium in plants demonstrated increased fruit biomass per plant, which yielded in higher productivity in tomato plant under NaCl stress (Shabala et al. 2006). Calcium supplies augment the pod number and seed yield by 29.43, 52.45% in *C. arietinum* under Cd-induced phytotoxicity. Under various prevailing stress conditions, net increase in cytosolic Ca^{2+} concentration promotes different biological responses that facilitate plants to adjust to detrimental environment by regulating antioxidant system and reduced lipid peroxidation for plant survival. Under salt stress application, os 10 mM $CaCl_2$ at seedling stage improved plant biomass through increased chlorophyll content and oppressing the damage of reactive oxygen specie in *Oryza sativa* L. (Roy et al. 2019). Under metal stress regimes, Ca application obstructs heavy metal bioaccumulation by controlling the membrane integrity and divalent cation movement across plant cell membrane (Farzadfar et al. 2013). Under cobalt stress exogenous addition of Ca significantly increased the antioxidant system naturally to decrease the oxidative stressed in *H. vulgare* plants (Lwalaba et al. 2017). In another study, applied Ca declined H_2O_2 and MDA level by 20.4, 23.3% in *Sesamum indicium* plant-protecting plant from Cd-induced oxidative injury and upregulating the antioxidant enzyme system (Abd_Allah et al. 2017).

Sulfur

Sulfur is an essential nonmetallic element for plants and has a role in various metabolic functions. Sulfur is called as fourth major nutrient and is taken up by plants as SO_4^{2-} form, which is portable in soil. Sulfur content in plants differs between 0.1 and 0.4% analogous to Mg and P (Karthika et al. 2018). Sulfur is a crucial constituent in oilseed crops, and these crops need high amount of S than other cereal crops. Sulfur is an important component of biomolecules such as cysteine and methionine. Sulfur-containing amino acids act as precursor for different co-enzymes, and metabolites have a role in the synthesis of protein, chlorophyll, and vitamins involved in stress tolerance (Rausch and Wachter 2005). Recent data suggested that eminent supply of S significantly enhanced photosynthetic capacity in certain plants under stress conditions. Sulfur is taken up by plants in the form of sulfate and known as extremely mobile in plants (Kumaran et al. 2008). The uptake and assimilation of sulfate are assisted by H^+ gradient, and transport of sulfate ions occurs via xylem in the shoot (Ernst et al. 2008). Subsequent reduction of sulfate ions SO_4^{2-} to SO_3^{2-} and S^{2-} in shoot chloroplast and root plants take place, which is important for assimilation of sulfur (Bouranis et al. 2008). Sulfur deficiency symptoms appear in mature leave and then younger leaves and interveinal chlorosis observed in the older leaves. Typical deficiency symptoms of S include yellowish green leaflets and weakening of stem in plants (Karthika et al. 2018).

Role of Sulfur in Stress Tolerance

Some studies have reported that SO_4^{2-} transport in vascular system of plants, its assimilation in leaves, and recycling of S-compounds are linked to drought stress response (Hawkesford et al. 2012). Higher plants required S mainly in the anionic sulfate form from the soil that gets transformed into nutritionally valuable S comprising compounds like cystine, GSH, co-enzymes, and sulpholipids (Kopriva and Koprivova 2003; Rao and Reddy 2008). The preliminary product of S assimilation cysteine (Cys) is considered as rate restricting factor for GSH (γ-glutamyl-cysteinyl-glycine) biosynthesis, which is important S-constituting antioxidant and play an vital role in metabolism and stress resistance against ROS (Szalai et al. 2009). Sulfur has a considerable role in crop tolerance against metal stress through providing tripeptide thiol GSH, a phytochelatins precursor in plants, has a major role in Cd tolerance (Zechmann et al. 2008). In a study on *A. thaliana* HPLC analysis showed that Cd induced the expression of S assimilation by increasing the transcription of particular genes that augments the GSH supply for PCs biosynthesis, which improve the plant defense against Cd (Gallego et al. 2012).

Magnesium

Magnesium (Mg) is an essential biological cation, constituent of chlorophyll, and an activator of >300 enzymes like RNA polymerase, ATPases, phosphatase, protein kinases carboxylases, and glutathione synthase in plants, which are involved in ion transport regulation and cationic balance in plants (Bose et al. 2011). The Mg range for optimal growth is 1.5–3.5 g kg^{-1} in plants, and its concentration range lies amid 125 µmol l^{-1} and 8.5 mmol l^{-1}in soil. Magnesium has various physiological function in plants; it has vital role in phloem loading and photo assimilate transport in to plant sink organs such as root and shoot (Guo et al. 2016). Magnesium has positive influence on photosynthesis, ATP formation, and utilization and enzyme activation. Magnesium (Mg) is an important part of polyribosome, and Mg levels determine the gene transcription as well as gene translation in plants. It also contribute to improve phosphorus uptake and translocation in plants. Cellular Mg functions as enzyme cofactor and facilitates nucleotide stabilization in plants (Karthika et al. 2018). Magnesium is found in ionic form in soils and is carried up by plants as Mg^{2+}, whereas Mg deficiency can trigger chlorosis leading to necrosis in plants as well as exert a negative impact on the partitioning of dry matter and carbohydrate in root and shoots. Under Mg deficiency symptoms appear first in older leave and then move to younger leaves. Magnesium deficiency occurs when the Mg concentration falls below than 0.1%. Maintenance of optimum Mg status in plants is essential in the defense versus ROS-generated oxidative stress injury in plants, which occurs at the expense of limited phloem sugar export and disruption in CO_2 fixation in stress condition (Maathuis et al. 2009).

Role of Mg in Stress Tolerance

The application of $Mg(NO_3)_2$ improved photosynthetic attributes in strawberry plants facing salinity stress (Yildiri et al. 2009). Foliar application of Mg fertilizers ameliorates the nutrient deficiency stress in crops and enhanced the chlorophyll content along with yield status of plants. Under drought stress nutrient intake from leaves is restricted due to impassable stomata and alterations in leaf morphology; however, application of Mg fertilizers in combination of other nutrients increased plant growth (Römheld and Kirkby 2010). In nutrient-limited environment, the application of Mg fertilizer enhanced the crop yield and nutrient uptake in sugar beet plants (Grzebisz et al. 2010). New insights about Mg uptake and utilization specifically into the role of Mg in enhancing crop tolerance to different stresses indicate Mg demand under adverse environmental conditions in plants (Senbayram et al. 2016).

Boron

Boron is unique micronutrient found in the earth crust at low levels. Boron levels in soil range from 0.5 to 2 ppm and accessible as water-soluble borate mineral, which further utilized to generate borax and kernite (Naeem et al. 2018). Boric acid (H_3BO_3) in the form B is permeable to plant cells, and B ranges from 10 to 200 mg kg^{-1} of B in normal plants. Boron haa a role in different biological processes in plants, which include synthesis of proteins, respiration, sugar and RNA metabolism, carbohydrate, and indole acetic acid hormone (Shireen et al. 2018). Boron is involved in vital activities of plants such as cell division, meristematic tissue elongation, cell wall formation, membrane integrity, leaf expansion, and sugar translocation (Marschner 1995). Boron has a role in inducing floral organs, pollen tube growth, and flower male fertility. Boron facilitates phosphorus and chlorine transport as well as Ca nutrition in plants (Aydin et al. 2019). Boron in considered crucial for expanding tips of plants and new leaf as well as bud development in plants. Deficiency of boron ranges from 5 to 10 mg kg^{-1} in graminaceous and 20 to 70 mg kg dicotyledonous plants. Boron deficiency first appears in younger leaves and internode develops shorter giving a rosette look (Shireen et al. 2018). In boron -deficient plant, flower retention, pollen formation, pollen tube growth, and nitrogen fixation are badly effected due to B deficiency (Karthika et al. 2018).

Role of Boron Under Stress

Boron facilitates plant resistance against environmental stresses by preserving the cell wall and cell membrane integrity. Boron availability under salinity could alleviate negative impacts on plant photosynthesis (Mohamed et al. 2016). Addition of boron ameliorates the toxicity symptoms of aluminum (Al) in citrus plants as evident from lowering metabolite accumulation (Yan et al. 2019). In hydroponics grown seedlings, drought stress caused significant decrease in chlorophyll content and soluble protein content, whereas decreased MDA level was reported in tomato plants with the addition of boron (Uluisik et al. 2018). Boron supply alleviates the drought caused damage in Eucalyptus plants, whereas boron application increased drought tolerance in sunflower and maize (Hassan et al. 2011; Pita-Barbosa et al. 2016; Naeem et al. 2018). In a recent study, boron application decreased the Al bioaccumulation in root cell and compartmentalizing it into vacuoles thereby reducing the Al deposition in cytoplasm (Yan et al. 2021). Application B significantly altered the Cd-induced oxidative stress triggered by ROS accumulation in shoot of *Brassica napus* though increased pectin, protein, cellulose, and carbohydrate levels in leaf cell wall thus improved growth by immobilizing Cd (Wu et al. 2020).

Iron

Iron (Fe) is the fourth copious essential microelement found on earth's crust and is available to plant owing to neutral and alkaline soil as it occurs in an insoluble form. In plants, Fe is considered as essential nutrient in regulating major sustaining process, which include respiration, chloroplast enlargement, photosynthesis, and biosynthesis of chlorophyll. Iron is a major component of cell redox system and acts as cofactor for different antioxidants like peroxidase, ascorbate peroxidase, and catalase (Marschner 1995; Kumar et al. 2010). Iron is present in cytochrome and has a role in electron transfer in redox reaction. Plants uptake iron as ferrous ion Fe^{2+}, and its concentration ranges from 100 to 500 mg kg^{-1} of dry biomass in plants. Iron occurs in two Fe^{2+} and Fe^{3+} oxidation states in plants. Plants uptake iron as Fe^{2+} via plasma membrane-specific channels. The uptake and bioaccumulation of Fe is increased by three possible reactions: the exudation of protons via plasmalemma P-type ATPase, second by decrease of Fe^{3+} to more soluble form Fe^{2+}, and third by plasmalemma transport of Fe^{2+} via Fe transporters (Lan et al. 2011). Iron deficiency symptoms are interveinal chlorosis that appear on younger leave later turning the entire leaf white. Iron deficiencies occur when its content falls below 50 mg kg^{-1} in plants leading to severe growth reduction in plants (Kobayashi et al. 2003; Broschat 2006).

Role of Iron in Stress

To survive under stress, the nutritional status of plants has a major role in enhancing plant tolerance to stresses (Tripathi et al. 2018). Under drought stress, the application of Fe nutrition helps plants to increase its tolerance through improved assimilates production (Khan and Gulzar 2003; Pourgholam et al. 2013). Foliar application of Fe increased drought tolerance in sunflower and soybean that showed enhanced oil production (Kobraee et al. 2011). In another study, foliar application of Fe with Zn mitigates oxidative stress caused by increased hydrogen peroxide content, lipid peroxidation, and increased antioxidant system. In a recent study, it was demonstrated that the higher expression of Fe-SOD contributes to decrease in injury symptoms, which leads to increased drought resistance in plants (Samis et al. 2002). Iron supplementation showed significant effect to reduced salinity stress by improving the plant antioxidant enzymes (Sharma et al. 2012).The Fe^{2+} amino acid chelates increase the plant resistance against salinity stress in tomato cultivars (Ghasemi et al. 2014). In an experimental study, Fe plaque plays a major role in regulating Cu, Cd toxicity in plants (Huang et al. 2009). The accumulation and Cd uptake inside the rice plants were diminished with the application of Fe (Shao et al. 2007). The foliar application of Fe fertilizer EDRA-Na2Fe and FeSO4 eliminated the adverse impact of cd in rice shoot, root, and grains (Shao et al. 2008).

Zinc

Zinc is an essential plant micronutrient holding atomic number 30, Zn is considered as 23rd most abundant element on earth with five different stable isotopes (Tripathi et al. 2015). The level of Zn in plants ranges from 25 to 150 mg kg^{-1} of dry matter. Zinc is an essential component of enzymes for protein biosynthesis, whereas estimated 300 enzymes contain Zn as catalytic element (Chattha et al. 2017). The main enzymes having Zn comprise carbonic anhydrase as well as superoxide dismutase, which performs different functions. Zinc is the integral part of different biomolecules including lipids and protein. The fundamental character of Zn is being the constituent of six different enzyme classes, oxidoreductase, transferases, lyases, hydrolases, ligases, and isomerases. Zinc also has a participatory role in the Rubisco structure (Tsone and Cebola Lidon 2012). Most plants absorb zinc as Zn^{++} ion. The uptake and translocation of Zn from soil medium to plant shoot are concentration reliant, while zinc deficiency appears when its concentration is below 15 mg Zn kg^{-1} of dry matter (Mousavi et al. 2007). Zinc deficiency causes restricted growth and decreases in crop yield, whereas general Zn deficiency symptoms include white bud disease in maize, little leaf of cotton, and rosette disease of apple (Ul Hassan et al. 2017)

Role of Zn Under Stress

The drought stress decreases dry biomass and plumule length due to remobilization of nutrient resources to embryo; however, Zn priming increases the biosynthesis of IAA and GA3 that increase the plumule length and weight under stress (Cakmak 2008). Drought stress impairs the membrane permeability, antioxidant activity, and photosynthetic efficiency; however, application of Zn enhances chlorophyll content and remarkably improves the growth and yield (Karim et al. 2012). In another study, the addition of Zn increases the seed and oil yield as well as chlorophyll content and SOD expression in sunflower under drought stress at seed filling stage (Zafar et al. 2014). Addition of Zn considerably increases the quantum yield of photosystem II and activities of SOD, POD, and Cat by reducing the electrical conductivity in Triticale plants (Arough et al. 2016). Cadmium contamination negatively effects the grain quality and crop productivity in wheat seed, but the application of Zn 5 g kg^{-1} with biochar ameliorates the Cd effects and increased the grain yield upto 9.51% in comparison to Cd-stressed plants (Farooq et al. 2020). In an hydroponic experiment, the application of Zn (1 μM) significantly increased the plant biomass, height, and photosynthetic attributes with marked decreased in MDA content in two rice cultivars under Cd stress (Hassan et al. 2005).

Copper

Copper is considered as pivotal micronutrient in plants as it performs a role in controlling many biochemical responses in plants. Copper triggers reduction of one O_2 atom to water and another O_2 atom to hydroxylation of substrate like phenolase complex (Tripathi et al. 2015; Karthika et al. 2018). Copper acts as cofactor for multiple enzymes in plants like Cu-Zn-SOD as well as in cytochrome C oxidase (Rout et al. 2013). Some copper-containing enzymes include ascorbate oxidizer, cytochrome, and superoxide dismutase. Copper contributes to oxidative phosphorylation, Fe utilization, and cell wall metabolism (Ravet et al. 2011). Copper induces systematic resistance to plant against diseases, and fertility in male flower isincreased by copper (Tripathi et al. 2015). Plant absorbs copper in the Cu^{2+} form by root. Different studies have reported that Cu^{2+} and Cu^{+} are most abundant forms of Cu for plant uptake and plants absorbs Cu in the Cu^{2+} form via root. Cu concentration beyond threshold levels inhibits plant growth and cause cytotoxicity. In food crops recommended value of Cu is 30 mg kg^{-1}. Copper levels below 5 μg g^{-1} cause deficiency symptoms, and it first appear as necrosis extending along leaf margins. Leaf malfunctioning is detected leading to premature leaf abscission due to Cu deficiency (Tripathi et al. 2015). Copper deficiency caused reduction in cell wall formation, tissue lignification, leaf margins curling, and damage to apical meristem, which obstructs the fruit development and pollen growth (Burkhead et al. 2009).

Role of Copper in Stress Tolerance

Copper contributes to disproportion of superoxide ions synthesized due to different environmental stress because Cu is the component of Cu-Zn-SOD fort the H_2O_2 conversion and molecular oxygen. Optimum copper supply protects plants against pathogen and contributes toward lignin synthesis. Copper induced the ROS formation and activates copper diamine oxidase, which initiates defense mechanisms in plants (Tripathi et al. 2015).

Manganese

Manganese (Mn) is considered as essential and ubiquitous microelement on earth crust. Manganese occurs in different oxidation states like Mn^{2+}, Mn^{3+}, Mn^{4+}, and Mn^{5+} (Hebbern et al. 2009). Mn acts as an enzymes activator and have a catalytic functioning in cellular metabolism and in CAM and C_4 function in CO_2 fixation (Marschner 1995). Mn is involved in enzyme activation and play a role in photosynthesis, nitrogen metabolism, carbohydrate metabolism, and amino acid biosynthesis as well synthesis of secondary products in plants (Millaleo et al. 2010). The foremost role of Mn is in oxygen evolution as it is component of oxygen evolving complex (OEC) correlated with PSII. Mn plays a critical role in decarboxylation of C4 acids to release CO_2 that is fixed by Rubisco and added into carbohydrate pool via NAD^+ malic enzyme and phosphoenolpyruvate. Moreover, Mn is involved in ATP biosynthesis, protein biosynthesis, and fatty acid biosynthesis (Ducic and Polle 2005). Mn is a cofactor of first enzymes involved in shikimate pathway that facilitates the synthesis of aromatic amino acid and tyrosine phenyl alanine. Manganese contributes to phenolic metabolism and synthesis of lignin and provides resistance against pathogens (Millaleo et al. 2010; Tripathi et al. 2015).

Role of Mn in Stress Tolerance

Under stress condition, Mn SOD found in mitochondria plays an important role as antioxidant enzyme in stress against ROS-induced oxidative stress. It also seen in superoxide ions accumulation through their dismutation and converts them to hydrogen peroxide. The Mn-SOD induced defense in wheat seedlings under drought stress. Mn provide protection and defense against fungus infection via association in phenolics and lignin biosynthesis (Morab et al. 2003).

Molybdenum

Molybdenum (Mo) is a second transition series metal, and it has several oxidation states Mo (II), Mo (III), Mo(IV), Mo(V), and Mo(VI) of which Mo(VI) hexavalent is the most stable form. Molybdenum has high affinity for oxygen and sulfur comprising groups. Mo acts as a cofactor for several enzymes and contributes to dinitrogen fixation and nitrate assimilation in plants. It is important for free living as well as symbiotic bacteria in nitrogen fixation process (Manuel et al. 2018). Plants absorb MoO_4^{2-} stable at 4 pH and follows same transporter path as phosphate transfer in plants. Mo-containing nitrate reductase and xanthine oxidase promotes the nitrogen reactions, and its compounds and aldehyde oxidase catalyze the biosynthesis pathway of abscisic acid and auxins (Koshiba et al. 1996).

Molybdenum Role Under Stress

Application of molybdenum fertilizer 0.3 mg kg^{-1} significantly increases shoot fresh biomass and activities of antioxidant systems under salt stress in Chinese cabbage plants. The increase in plant growth was correlated with enhanced potassium ion content and decreased sodium ion concentration in Chinese cabbage plants (Taylor 1991; Zhang et al. 2012). In a recent study, Mo allocation significantly ameliorated the Cd-induced phytotoxicity in fragrant rice through increased antioxidant defense system and lowered hydrogen peroxide content, malondialdehyde content, and electrolyte leakage in fragrant rice cultivars (Imran et al. 2020). In another study, the addition of Mo in hydroponics experiment improved the activities of superoxide dismutase, catalase, ascorbate peroxidase, and ascorbic acid content; moreover, the osmotic adjustment products like proline and soluble sugar were also enhanced under Mo application in wheat plants under drought stress (Wu et al. 2014).

Chlorine Chlorine ((Cl) has single oxidation state as Cl^- and is engaged in chlorine comprising organic complexes in plants. Chlorine is a constituent of PS II manganese cluster that catalyzes the H_2O_2 oxidation. Chlorine contributes in the activity of auxin, and in living cell it facilitates turgor and osmotic adjustments. Chlorine is involved in stomata functioning and stimulates asparagine synthetase activity, which catalyzes the glutamate-dependent biosynthesis of asparagine. Chlorine transport across plasma membrane involves proton pump (H^+ATPase) and a passive element facilitated via anion channels (White and Broadley 2001). The chlorine transport occurs through xylem from root to shoot, which is affected by rate of transpiration and shoot growth (Xu et al. 1999).

Role of Chlorine Under Stress Under abiotic stress condition, Cl has an important role in osmoregulatory processes in plants. In plant roots, Cl acts as an osmotically active solute. The chlorine concentration for osmoregulatory process varies from plant to plant, and osmotic responses also vary among plants under abiotic stress regimes.

Nickel Nickel is a recently discovered micronutrient that was recognized as essential nutrient in 1987. Plants require Ni in low quantity $<0.5\,mg\ kg^{-1}$ and is taken up by plants as Ni^{2+} that exist in three oxidation stresses I, II, and III. Nickel is a key constituent of enzymes involved in nitrogen metabolism and N fixation in crops. Nickel is easily transported in plants, and two transport systems (low affinity and high affinity) absorb and transport Ni in plants. Nickel is present in prokaryotic enzyme including hydrogenase, dehydrogenase (Hänsch and Mendel 2009), whereas urease enzyme contains Ni that is essential for N metabolism in plants. In certain legumes, Ni is required in the growth of root nodules and hydrogenase enzyme activation. Nickel deficiency causes reduction in urease that resulted in toxic urea accumulation in plants.

Conclusion

The essential nutrients have significant role in plants biochemistry, particularly during abiotic stress condition due to their possible role in regulation of enzymatic reactions and synthesis of organic molecules. Insufficient amount or low phytoavailbililty of essential elements would result in restricted crop growth and productivity worldwide. Reactive oxygen species are generated in plant cell due to abiotic stresses which ultimately hamper plant cells particularly biochemical components and negatively impact physiological activities in plants. Therefore, optimum supply of essential nutrients not only promotes plant growth but also safeguards cellular metabolism against different environmental stresses. In addition, exogenous application of mineral fertilizer or by farming of genetically modified crop could also be recognized as most efficient and reliable technique to improve the mineral status of soil and plants. Furthermore, studies should be conducted to explore the more precise role of nutrients, their role in defense mechanism, and pathways under abiotic stress which may assist in the management of environmental stresses to achieve optimum growth and yield of agricultural crop for sustainable food production.

References

Abbas, S., Javed, M.T., Shahid, M. et al. (2020). Acinetobacter sp. SG-5 inoculation alleviates cadmium toxicity in differentially Cd tolerant maize cultivars as deciphered by improved physio-biochemical attributes, antioxidants and nutrient physiology. *Plant Physiol Biochem 155*: 815–827.

Abbas, S., Javed, M.T., Ali, Q. et al. (2021). Elucidating Cd-mediated distinct rhizospheric and in planta ionomic and physio-biochemical responses of two contrasting Zea mays L. cultivars. *Physiol Mol Biol Plants 27* (2): 297–312.

Abd_Allah, E.F., Hashem, A., Alqarawi, A. et al. (2017). Calcium application enhances growth and alleviates the damaging effects induced by Cd stress in sesame (Sesamum indicum L.). *J Plant Interact 12* (1): 237–243.

Agami, R.A., Alamri, S.A., Abd El-Mageed, T.A. et al. (2018). Role of exogenous nitrogen supply in alleviating the deficit irrigation stress in wheat plants. *Agri Water Manage 210*: 261–270.

Ahanger, M.A., Tyagi, S.R., Wani, M.R., and Ahmad, P. (2014). Drought tolerance: role of organic osmolytes, growth regulators, and mineral nutrients. In: *Physiological Mechanisms and Adaptation Strategies in Plants Under Changing Environment* (ed. P. Ahmad and M. Wani), 25–55. New York, NY: Springer.

Ahmad, P., Jaleel, C.A., and Sharma, S. (2010). Antioxidant defense system, lipid peroxidation, proline-metabolizing enzymes, and biochemical activities in two Morus alba genotypes subjected to NaCl stress. *Russ J Plant Physiol 57* (4): 509–517.

Ahmad, P., Kumar, A., Ashraf, M., and Akram, N.A. (2012). Salt-induced changes in photosynthetic activity and oxidative defense system of three cultivars of mustard (*Brassica juncea L.*). *Afr J Biotechnol 11* (11): 2694–2703.

Ahmad, Z., Waraich, E.A., Ahmad, R., and Shahbaz, M. (2017). Modulation in water relations, chlorophyll contents and antioxidants activity of maize by foliar phosphorus application under drought stress. *Pak J Bot 49* (1): 11–19.

Ahmed, M., Hasanuzzaman, M., Raza, M.A. et al. (2020). Plant nutrients for crop growth, development and stress tolerance. In: *Sustainable Agriculture in the Era of Climate Change* (ed. R. Roychowdhury, S. Choudhury, M. Hasanuzzaman and S. Srivastava), 43–92. Cham: Springer.

Akram, M., Ashraf, M.Y., Jamil, M. et al. (2011). Nitrogen application improves gas exchange characteristics and chlorophyll fluorescence in maize hybrids under salinity conditions. *Russ J Plant Physiol 58* (3): 394–401.

Ali, R., Hassan, S., Shah, D. et al. (2020). Role of polyamines in mitigating abiotic stress. *Protective Chemical Agents in the Amelioration of Plant Abiotic Stress: Biochemical and Molecular Perspectives* 291–305.

Alpha, J.M., Chen, J., and Zhang, G. (2009). Effect of nitrogen fertilizer forms on growth, photosynthesis, and yield of rice under cadmium stress. *J Plant Nutr 32* (2): 306–317.

Amjad, M., Akhtar, J., Murtaza, B. et al. (2016). Differential accumulation of potassium results in varied salt-tolerance response in tomato (*Solanum lycopersicum L.*) cultivars. *Hortic Environ Biotechnol 57* (3): 248–258.

Arnon, D.I. and Stout, P.R. (1939). The essentiality of certain elements in minute quantity for plants with special reference to copper. *Plant Physiol 14* (2): 371.

Arough, Y., Seyed Sharifi, R., and Seyed, S.R. (2016). Bio fertilizers and zinc effects on some physiological parameters of triticale under water-limitation condition. *J Plant Interact 11* (1): 167–177.

Arshad, M., Ali, S., Noman, A. et al. (2016). Phosphorus amendment decreased cadmium (Cd) uptake and ameliorates chlorophyll contents, gas exchange attributes, antioxidants, and mineral nutrients in wheat (*Triticum aestivum L.*) under Cd stress. *Arch Agron Soil Sci 62* (4): 533–546.

Aydin, M., Tombuloglu, G., Sakcali, M.S. et al. (2019). Boron alleviates drought stress by enhancing gene expression and antioxidant enzyme activity. *J Soil Sci Plant Nutr 19* (3): 545–555.

Azam, M., Norman, M., Abbasi, N.A. et al. (2016). Effect of foliar application of micro-nutrient and soil condition on growth and yield of sweet pepper (*Capsicum annuum L.*). *Sci Technol Dev 35*: 75–81.

Azooz, M.M., Youssef, A.M., and Ahmad, P. (2011). Evaluation of salicylic acid (SA) application on growth, osmotic solutes and antioxidant enzyme activities on broad bean seedlings grown under diluted seawater. *Int J Plant Physiol Biochem 3* (14): 253–264.

Barzana, G., Rios, J.J., Lopez-Zaplana, A. et al. (2021). Interrelations of nutrient and water transporters in plants under abiotic stress. *Physiol Plant 171* (4): 595–619.

Berni, R., Luyckx, M., Xu, X. et al. (2019). Reactive oxygen species and heavy metal stress in plants: impact on the cell wall and secondary metabolism. *Environ Exp Bot 161*: 98–106.

Bita, C. and Gerats, T. (2013). Plant tolerance to high temperature in a changing environment: scientific fundamentals and production of heat stress-tolerant crops. *Front Plant Sci 4*: 273.

Bose, J., Babourina, O., and Rengel, Z. (2011). Role of magnesium in alleviation of aluminium toxicity in plants. *J Exp Bot 62* (7): 2251–2264.

Bouranis, D.L., Buchner, P., Chorianopoulou, S.N. et al. (2008). Responses to sulfur limitation in maize. In: *Sulfur Assimilation and Abiotic Stress in Plants* (ed. N.A. Khan, S. Singh and S. Umar), 1–19. Berlin, Heidelberg: Springer.

Broschat, T.K. (2006). Iron deficiency in palms. *EDIS 2006* (1): 1–3.

Burkhead, J.L., Gogolin Reynolds, K.A., Abdel-Ghany, E. et al. (2009). Copper homeostasis. *New Phytol 182* (4): 799–816.

Cakmak, I. (2008). Enrichment of cereal grains with zinc: agronomic or genetic biofortification? *Plant Soil 302* (1): 1–17.

Chakhchar, A., Lamaoui, M., Aissam, S. et al. (2016). Differential physiological and antioxidative responses to drought stress and recovery among four contrasting *Argania spinosa* ecotypes. *J Plant Interact 11* (1): 30–40.

Chakraborty, K., Bhaduri, D., Meena, H.N., and Kalariya, K. (2016). External potassium (K^+) application improves salinity tolerance by promoting Na^+-exclusion, K^+-accumulation and osmotic adjustment in contrasting peanut cultivars. *Plant Physiol Biochem 103*: 143–153.

Chattha, M.U., Hassan, M.U., Khan, I. et al. (2017). Biofortification of wheat cultivars to combat zinc deficiency. *Front Plant Sci 8*: 281.

Dai, M., Lu, H., Liu, W. et al. (2017). Phosphorus mediation of cadmium stress in two mangrove seedlings *Avicennia marina* and *Kandelia obovata* differing in cadmium accumulation. *Ecotoxicol Environ Saf 139*: 272–279.

Dhillon, J.S., Eickhoff, E.M., Mullen, R.W., and Raun, W.R. (2019). World potassium use efficiency in cereal crops. *Agron J 111* (2): 889–896.

Dias, A.S. and Lidon, F.C. (2010). Bread and durum wheat tolerance under heat stress: a synoptical overview. *Emi J Food Agri* 412–436.

Djanaguiraman, M., Prasad, V., and Seppanen, M. (2010). Selenium protects sorghum leaves from oxidative damage under high temperature stress by enhancing antioxidant defense system. *Plant Physiol Biochem 48* (12): 999–1007.

Dong, B., Zheng, X., Liu, H. et al. (2017). Effects of drought stress on pollen sterility, grain yield, abscisic acid and protective enzymes in two winter wheat cultivars. *Front Plant Sci 8*: 1008.

Ducic, T. and Polle, A. (2005). Transport and detoxification of manganese and copper in plants. *Braz J Plant Physiol 17* (1): 103–112.

Egilla, J.N., Davies, F.T., and Boutton, T.W. (2005). Drought stress influences leaf water content, photosynthesis, and water-use efficiency of *Hibiscus rosa-sinensis* at three potassium concentrations. *Photosynthetica 43* (1): 135–140.

Ernst, W.H., Krauss, G.J., Verkleij, J.A., and Wesenberg, D. (2008). Interaction of heavy metals with the sulphur metabolism in angiosperms from an ecological point of view. *Plant Cell Environ 31* (1): 123–143.

Esnault, M.A., Legue, F., and Chenal, C. (2010). Ionizing radiation: advances in plant response. *Environ Exp Bot 68* (3): 231–237.

Fageria, N.K. (2016). *The Use of Nutrients in Crop Plants*. CRC Press.

Farooq, M., Basra, S.M.A., Rehman, H., and Saleem, B.A. (2008). Seed priming enhances the performance of late sown wheat (*Triticum aestivum L.*) by improving chilling tolerance. *J Agron Crop Sci 194* (1): 55–60.

Farooq, M., Ullah, A., Usman, M., and Siddique, K.H. (2020) Application of zinc and biochar help to mitigate cadmium stress in bread wheat raised from seeds with high intrinsic zinc. *Chemosphere 260*: 127652.

Farzadfar, S., Zarinkamar, F., Modarres-Sanavy, S.A.M., and Hojati, M. (2013). Exogenously applied calcium alleviates cadmium toxicity in *Matricaria chamomilla L.* plants. *Environ Sci Pollut Res 20* (3): 1413–1422.

Fayez, K.A. and Bazaid, S.A. (2014). Improving drought and salinity tolerance in barley by application of salicylic acid and potassium nitrate. *J Saudi Soc Agric Sci 13* (1): 45–55.

Fazeli, F., Ghorbanli, M., and Niknam, V. (2007). Effect of drought on biomass, protein content, lipid peroxidation and antioxidant enzymes in two sesame cultivars. *Biol Plant 51* (1): 98–103.

Gallego, S.M., Pena, L.B., Barcia, R.A. et al. (2012). Unravelling cadmium toxicity and tolerance in plants: insight into regulatory mechanisms. *Environ Exp Bot 83*: 33–46.

Ghasemi, S., Khoshgoftarmanesh, A.H., Afyuni, M., and Hadadzadeh, H. (2014). Iron (II)–amino acid chelates alleviate salt-stress induced oxidative damages on tomato grown in nutrient solution culture. *Sci Hortic 165*: 91–98.

Ghori, N.H., Ghori, T., Hayat, M.Q. et al. (2019). Heavy metal stress and responses in plants. *Int J Environ Sci Technol 16* (3): 1807–1828.

Giansoldati, V., Tassi, E., Morelli, E. et al. (2012). Nitrogen fertilizer improves boron phytoextraction by *Brassica juncea* grown in contaminated sediments and alleviates plant stress. *Chemosphere 87* (10): 1119–1125.

Grzebisz, W., Przygocka-Cyna, K., Szczepaniak, W. et al. (2010). Magnesium as a nutritional tool of nitrogen efficient management – plant production and environment. *J Elem 15* (4): 771–788.

Gull, A., Lone, A.A., and Wani, N.U.I. (2019). Biotic and abiotic stresses in plants. *Abiotic and Biotic Stress in Plants* 1–19.

Guo, W., Nazim, H., Liang, Z., and Yang, D. (2016). Magnesium deficiency in plants: an urgent problem. *Crop J 4* (2): 83–91.

Gupta, D.K., Chatterjee, S., Datta, S. et al. (2014). Role of phosphate fertilizers in heavy metal uptake and detoxification of toxic metals. *Chemosphere 108*: 134–144.

Gupta, S., Schillaci, M., Walker, R. et al. (2020). Alleviation of salinity stress in plants by endophytic plant-fungal symbiosis: current knowledge, perspectives and future directions. *Plant Soil* 61 (1): 219–224.

Hajihashemi, S., Noedoost, F., Geuns, J.M.C. et al. (2018). Effect of cold stress on photosynthetic traits, carbohydrates, morphology, and anatomy in nine cultivars of *Stevia rebaudiana*. *Front Plant Sci* 9: 1430.

Hänsch, R. and Mendel, R.R. (2009). Physiological functions of mineral micronutrients (cu, Zn, Mn, Fe, Ni, Mo, B, cl). *Curr Opin Plant Biol 12* (3): 259–266.

Hasanuzzaman, M., Hossain, M.A., da Silva, J.A.T., and Fujita, M. (2012). Plant response and tolerance to abiotic oxidative stress: antioxidant defense is a key factor. In: *Crop Stress and Its Management: Perspectives and Strategies* (ed. B. Venkateswarku, A. Shanker, C. Shanker and M. Maheswari), 261–315. Dordrecht: Springer.

Hasanuzzaman, M., Bhuyan, M.H.M., Nahar, K. et al. (2018a). Potassium: a vital regulator of plant responses and tolerance to abiotic stresses. *Agronomy 8* (3): 31.

Hasanuzzaman, M., Bhuyan, M.H.M.B., Mahmud, J.A. et al. (2018b). Interaction of sulfur with phytohormones and signaling molecules in conferring abiotic stress tolerance to plants. *Plant Signal Behav 13* (5): e1477905.

Hasegawa, P.M., Bressan, R.A., Zhu, J.K., and Bohnert, H.J. (2000). Plant cellular and molecular responses to high salinity. *Annu Rev Plant Biol 51* (1): 463–499.

Hassan, M.J., Zhang, G., Wu, F. et al. (2005). Zinc alleviates growth inhibition and oxidative stress caused by cadmium in rice. *J Plant Nutr Soil Sci 168* (2): 255–261.

Hassan, N.M., El-Sayed, A.K., Ebeid, H.T., and Alla, M.N. (2011). Molecular aspects in elevation of sunflower tolerance to drought by boron and calcium foliar sprays. *Acta Physiol Plant 33* (2): 593–600.

Hawkesford, M., Horst, W., Kichey, T. et al. (2012). Functions of macronutrients. In: *Marschner's Mineral Nutrition of Higher Plants* (ed. P. Marschner), 135–189. Academic Press.

He, F., Sheng, M., and Tang, M. (2017). Effects of Rhizophagus irregularis on photosynthesis and antioxidative enzymatic system in *Robinia pseudoacacia L.* under drought stress. *Front Plant Sci 8*: 183.

Hebbern, C.A., Laursen, K.H., Ladegaard, A.H. et al. (2009). Latent manganese deficiency increases transpiration in barley (*Hordeum vulgare*). *Physiol Plant 135* (3): 307–316.

Hosseini, M.S., Samsampour, D., Ebrahimi, M. et al. (2018). Effect of drought stress on growth parameters, osmolyte contents, antioxidant enzymes and glycyrrhizin synthesis in licorice (*Glycyrrhiza glabra L.*) grown in the field. *Phytochemistry 156*: 124–134.

Hu, T., Wang, Y., Wang, Q. et al. (2019). The tomato 2-oxoglutarate-dependent dioxygenase gene SlF3HL is critical for chilling stress tolerance. *Horti Res 6* (1): 1–12.

Huang, Y., Hu, Y., and Liu, Y. (2009). Heavy metal accumulation in iron plaque and growth of rice plants upon exposure to single and combined contamination by copper, cadmium and lead. *Acta Ecol Sin 29* (6): 320–326.

Huihui, Z., Xin, L., Zisong, X. et al. (2020). Toxic effects of heavy metals Pb and Cd on mulberry (*Morus alba L.*) seedling leaves: photosynthetic function and reactive oxygen species (ROS) metabolism responses. *Ecotoxicol Environ Saf 195*: 110469.

Hussain, S., Khan, F., Cao, W. et al. (2016). Seed priming alters the production and detoxification of reactive oxygen intermediates in rice seedlings grown under sub-optimal temperature and nutrient supply. *Front Plant Sci* 7: 439.

Imran, M., Hussain, S., El-Esawi, M.A. et al. (2020). Molybdenum supply alleviates the cadmium toxicity in fragrant rice by modulating oxidative stress and antioxidant gene expression. *Biomol Ther 10* (11): 1582.

Iqbal, N., Umar, S., and Khan, N.A. (2015). Nitrogen availability regulates proline and ethylene production and alleviates salinity stress in mustard (*Brassica juncea*). *J Plant Physiol 178*: 84–91.

Jaleel, C.A., Manivannan, P., Wahid, A. et al. (2009). Drought stress in plants: a review on morphological characteristics and pigments composition. *Int J Agric Biol 11* (1): 100–105.

Jalloh, M.A., Chen, J., Zhen, F., and Zhang, G. (2009). Effect of different N fertilizer forms on antioxidant capacity and grain yield of rice growing under Cd stress. *J Hazard Mater 162* (2-3): 1081–1085.

Javed, M.T., Akram, M.S., Tanwir, K. et al. (2017). Cadmium spiked soil modulates root organic acids exudation and ionic contents of two differentially Cd tolerant maize (*Zea mays L.*) cultivars. *Ecotoxicol Environ Saf 141*: 216–225.

Jeshni, M.G., Mousavinik, M., Khammari, I., and Rahimi, M. (2017). The changes of yield and essential oil components of German Chamomile (*Matricaria recutita L.*) under application of phosphorus and zinc fertilizers and drought stress conditions. *J Saudi Soc Agric Sci 16* (1): 60–65.

Jiang, H.M., Yang, J.C., and Zhang, J.F. (2007). Effects of external phosphorus on the cell ultrastructure and the chlorophyll content of maize under cadmium and zinc stress. *Environ Pollut 147* (3): 750–756.

Kafkafi, U., Xu, G., Imas, P. et al. (2001). *Potassium and chloride in crops and soils: the role of potassium chloride fertilizer in crop nutrition*. Basel: International Potash Institute.

Kanwal, S., Ilyas, N., Shabir, S. et al. (2018). Application of biochar in mitigation of negative effects of salinity stress in wheat (*Triticum aestivum L.*). *J Plant Nutr 41* (4): 526–538.

Karim, M.R., Zhang, Y.Q., Zhao, R. et al. (2012). Alleviation of drought stress in winter wheat by late foliar application of zinc, boron, and manganese. *J Plant Nutr Soil Sci 175* (1): 142–151.

Karthika, K.S., Rashmi, I., and Parvathi, M.S. (2018). Biological functions, uptake and transport of essential nutrients in relation to plant growth. In: *Plant Nutrients and Abiotic Stress Tolerance* (ed. M. Hasanuzzaman, M. Fujita, H. Oku, et al.). 1–49. Singapore: Springer.

Kasim, W.A., Osman, M.E., Omar, M.N., and Salama, S. (2021). Enhancement of drought tolerance in *Triticum aestivum L.* seedlings using *Azospirillum brasilense* NO40 and Stenotrophomonas maltophilia B11. *Bullet Natl Res Cent 45* (1): 1–14.

Kaur, G. and Asthir, B. (2017). Molecular responses to drought stress in plants. *Biol Plant 61* (2): 201–209.

Kaya, C., Kirnak, H., Higgs, D., and Saltali, K. (2002). Supplementary calcium enhances plant growth and fruit yield in strawberry cultivars grown at high (NaCl) salinity. *Sci Hortic 93* (1): 65–74.

Khan, M.A. and Gulzar, S. (2003). Germination responses of *Sporobolus ioclados*: a saline desert grass. *J Arid Environ 53* (3): 387–394.

Kim, J.H., Chung, B.Y., Kim, J.S., and Wi, S.G. (2005). Effects ofin Planta gamma-irradiation on growth, photosynthesis, and antioxidative capacity of red pepper (*Capsicum annuum L.*) plants. *J Plant Biol 48* (1): 47–56.

Kobayashi, T., Nakayama, Y., Itai, R.N. et al. (2003). Identification of novel cis-acting elements, IDE1 and IDE2, of the barley IDS2 gene promoter conferring iron-deficiency-inducible, root-specific expression in heterogeneous tobacco plants. *Plant J 36* (6): 780–793.

Kobraee, S., Shamsi, K., and Rasekhi, B. (2011). Effect of micronutrients application on yield and yield components of soybean. *Ann Biol Res 2* (2): 476–482.

Kopriva, S. and Koprivova, A. (2003). Sulphate assimilation: a pathway which likes to surprise. In: *Sulphur in Plants* (ed. Y.P. Abrol and A. Ahmad), 87–112. Dordrecht: Springer.

Koshiba, T., Saito, E., Ono, N. et al. (1996). Purification and properties of flavin-and molybdenum-containing aldehyde oxidase from coleoptiles of maize. *Plant Physiol 110* (3): 781–789.

Kour, D., Rana, K.L., Yadav, A.N. et al. (2019). Drought-tolerant phosphorus-solubilizing microbes: biodiversity and biotechnological applications for alleviation of drought stress in plants. In: *Plant Growth Promoting Rhizobacteria for Sustainable Stress Management* (ed. R. Sayyed, N. Arora and M. Reddy), 255–308. Singapore: Springer.

Kumar, N. and Mallick, S. (2019). Ameliorative mechanisms of polyamines against abiotic stress in the rice plants. In: *Advances in Rice Research for Abiotic Stress Tolerance* (ed. M. Hasanuzzaman, K. Nahar, M. Fujita and J.K. Biwas), 725–735. Woodhead Publishing.

Kumar, P., Tewari, R.K., and Sharma, P.N. (2010). Sodium nitroprusside-mediated alleviation of iron deficiency and modulation of antioxidant responses in maize plants. *AoB Plants 2010*: plq002.

Kumar, J., Singh, S., Singh, M. et al. (2017). Transcriptional regulation of salinity stress in plants: a short review. *Plant Gene 11*: 160–169.

Kumar, V., Singh, J., and Kumar, P. (2019). Heavy metals accumulation in crop plants: sources, response mechanisms, stress tolerance and their effects. In: *Contaminants in Agriculture and Environment: Health Risks and Remediation*, vol. *1* (ed. V. Kumar, R. Kumar, J. Singh and P. Kumar), 38.

Kumaran, S., Francois, J.A., Krishnan, H.B., and Jez, J.M. (2008). Regulatory protein-protein interactions in primary metabolism: the case of the cysteine synthase complex. In: *Sulfur Assimilation and Abiotic Stress in Plants* (ed. N.A. Khan, S. Singh and S. Umar), 97–109. Berlin, Heidelberg: Springer.

Kunz, J.N., Voronine, D.V., Lee, H.W.H. et al. (2017). Rapid detection of drought stress in plants using femtosecond laser-induced breakdown spectroscopy. *Opt Express 25* (7): 7251–7262.

Lan, P., Li, W., Wen, T.N. et al. (2011). iTRAQ protein profile analysis of *Arabidopsis* roots reveals new aspects critical for iron homeostasis. *Plant Physiol 155* (2): 821–834.

Li, J., Cang, Z., Jiao, F. et al. (2017). Influence of drought stress on photosynthetic characteristics and protective enzymes of potato at seedling stage. *J Saudi Soc Agri Sci 16* (1): 82–88.

Li, C., Zhou, K., Qin, W. et al. (2019). A review on heavy metals contamination in soil: effects, sources, and remediation techniques. *Soil Sed Cont: An Inter J 28* (4): 380–394.

Ling, F. and Silberbush, M. (2002). Response of maize to foliar vs. soil application of nitrogen–phosphorus–potassium fertilizers. *J Plant Nutr 25* (11): 2333–2342.

Liu, W., Yu, K., He, T. et al. (2013). The low temperature induced physiological responses of *Avena nuda L.*, a cold-tolerant plant species. *Sci World J* 2013: 658793.

Lwalaba, J.L.W., Zvobgo, G., Fu, L. et al. (2017). Alleviating effects of calcium on cobalt toxicity in two barley genotypes differing in cobalt tolerance. *Ecotoxicol Environ Saf 139*: 488–495.

Malhotra, H., Sharma, S., and Pandey, R. (2018). Phosphorus nutrition: plant growth in response to deficiency and excess. In: *Plant nutrients and abiotic stress tolerance* (ed. M. Hasanuzzaman, M. Fujita, H. Oku, et al.), 171–190. Singapore: Springer.

Manuel, T.J., Alejandro, C.A., Angel, L. et al. (2018). Roles of molybdenum in plants and improvement of its acquisition and use efficiency. In: *Plant Micronutrient Use Efficiency* (ed. M.A. Hossain, T. Kamiya, D.J. Burrit, et al.), 137–159. Academic Press.

Marschner, H. (1995). *Mineral Nutrition of Higher Plants*, 153–168. London, UK: Academic Press.

Marschner, H. (2011). *Marschner's Mineral Nutrition of Higher Plants*. Academic Press.

Marschner, P. and Rengel, Z. (2012). Nutrient availability in soils. In: *Marschner's Mineral Nutrition of Higher Plants* (ed. P. Marschner), 315–330. Academic Press.

Masood, S. and Bano, A. (2016). Mechanism of potassium solubilization in the agricultural soils by the help of soil microorganisms. In: *Potassium Solubilizing Microorganisms for Sustainable Agriculture* (ed. V. Meena, B. Maurya, J. Verma and R. Meena), 137–147. New Delhi: Springer.

McCauley, A., Jones, C., and Jacobsen, J. (2009). Plant nutrient functions and deficiency and toxicity symptoms. *Nutrient Management Module 9*: 1–16.

Meena, V.S., Maurya, B.R., Verma, J.P., and Meena, R.S. (ed.) (2016). *Potassium Solubilizing Microorganisms for Sustainable Agriculture*, 1–20. New Delhi: Springer.

Mengel, K. and Kirkby, E.A. (1987). *Principles of Plant Nutrition*, 687–695. Bern: *International Potash Institute*.

Millaleo, R., Reyes-Díaz, M., Ivanov, A.G. et al. (2010). Manganese as essential and toxic element for plants: transport, accumulation and resistance mechanisms. *J Soil Sci Plant Nutr 10* (4): 470–481.

Miyashita, K., Tanakamaru, S., Maitani, T., and Kimura, K. (2005). Recovery responses of photosynthesis, transpiration, and stomatal conductance in kidney bean following drought stress. *Environ Exp Bot 53* (2): 205–214.

Mohamed, A.K.S., Qayyum, M.F., Shahzad, A.N. et al. (2016). Interactive effect of boron and salinity on growth, physiological and biochemical attributes of wheat (*Triticum aestivum*). *Int J Agric Biol 18* (2).

Mon, J., Bronson, K.F., Hunsaker, D.J. et al. (2016). Interactive effects of nitrogen fertilization and irrigation on grain yield, canopy temperature, and nitrogen use efficiency in overhead sprinkler-irrigated durum wheat. *Field Crop Res 191*: 54–65.

Morab, H., Koti, R.V., Chetti, M.B. et al. (2003). Role of nutrients in inducing rust resistance in soybean. *Ind J Plant Physiol (India)* 8: 85–88.

Mousavi, S.R., Galavi, M., and Ahmadvand, G. (2007). Effect of zinc and manganese foliar application on yield, quality and enrichment on potato (*Solanum tuberosum L.*). *Asian J Plant Sci* 6: 1256–1260.

Naeem, M., Naeem, M.S., Ahmad, R. et al. (2018). Improving drought tolerance in maize by foliar application of boron: water status, antioxidative defense and photosynthetic capacity. *Arch Agron Soil Sci* 64 (5): 626–639.

Nahar, K. and Hasanuzzaman, M. (2009). Germination, growth, nodulation and yield performance of three mungbean varieties under different levels of salinity stress. *Green Farming* 2 (12): 825–829.

Nahar, K., Hasanuzzaman, M., Rahman, A. et al. (2016). Polyamines confer salt tolerance in mung bean (Vigna radiata L.) by reducing sodium uptake, improving nutrient homeostasis, antioxidant defense, and methylglyoxal detoxification systems. *Front Plant Sci* 7: 1104.

Nedjimi, B. and Daoud, Y. (2009). Ameliorative effect of CaCl2 on growth, membrane permeability and nutrient uptake in Atriplex halimus subsp. schweinfurthii grown at high (NaCl) salinity. *Desalination* 249 (1): 163–166.

Nobori, T., Kobayashi, N.I., Tanoi, K., and Nakanishi, T.M. (2016). Alteration in caesium behavior in rice caused by the potassium, phosphorous, and nitrogen deficiency. *J Radioanal Nucl Chem* 307 (3): 1941–1943.

Nxele, X., Klein, A., and Ndimba, B.K. (2017). Drought and salinity stress alters ROS accumulation, water retention, and osmolyte content in sorghum plants. *South Afri J Bot* 108: 261–266.

Pandey, J., Verma, R.K., and Singh, S. (2019). Suitability of aromatic plants for phytoremediation of heavy metal contaminated areas: a review. *Intern J Phyto* 21 (5): 405–418.

Parihar, P., Singh, S., Singh, R. et al. (2015). Effect of salinity stress on plants and its tolerance strategies: a review. *Environ Sci Pollut Res* 22 (6): 4056–4075.

Parvin, K. and Haque, M.N. (2017). Protective role of salicylic acid on salt affected broccoli plants. *J Agri Ecol Res Intern* 10 (2): 1–10.

Parvin, K., Ahamed, K.U., Islam, M.M. et al. (2015). Reproductive behavior of tomato plant under saline condition with exogenous application of calcium. *Middle-East J Sci Res* 23: 2920–2926.

Parvin, K., Nahar, K., Hasanuzzaman, M. et al. (2019). Calcium-mediated growth regulation and abiotic stress tolerance in plants. In: *Plant Abiotic Stress Tolerance* (ed. M. Hasanuzzaman, K. Hakeem, K. Naha and H. Alharby), 291–331. Springer.

Pathak, J., Ahmed, H., Kumari, N. et al. (2020). Role of calcium and potassium in amelioration of environmental stress in plants. In: *Protective Chemical Agents in the Amelioration of Plant Abiotic Stress: Biochemical and Molecular Perspectives* (ed. A. Roychoudhury and D.K. Tripathi), 535–562. https://doi.org/10.1002/9781119552154.ch27.

Pita-Barbosa, A., Hodecker, B.E.R., and de Barros, N.F. (2016). Boron as mitigator of drought damage in Eucalyptus: a genotype-dependent mechanism? *Sci For* 44 (112): 851–861.

Pourgholam, M.O.H.S.E.N., Nemati, N.A.B.I.O.L.A.H., and Oveysi, M.E.Y.S.A.M. (2013). Effect of zinc and iron under the influence of drought on yield and yield components of rapeseed (Brassica napus). *Annals Biol Res* 4 (4): 186–189.

Raheem, A., Shaposhnikov, A., Belimov, A.A. et al. (2018). Auxin production by rhizobacteria was associated with improved yield of wheat (Triticum aestivum L.) under drought stress. *Arch Agron Soil Sci* 64 (4): 574–587.

Ramaekers, L., Remans, R., Rao, I.M. et al. (2010). Strategies for improving phosphorus acquisition efficiency of crop plants. *Field Crop Res* 117 (2-3): 169–176.

Rao, A.C. and Reddy, A.R. (2008). Glutathione reductase: a putative redox regulatory system in plant cells. In: *Sulfur Assimilation and Abiotic Stress in Plants*, 111–147. Berlin: Springer.

Rausch, T. and Wachter, A. (2005). Sulfur metabolism: a versatile platform for launching defence operations. *Trends Plant Sci* 10 (10): 503–509.

Ravet, K., Danford, F.L., Dihle, A. et al. (2011). Spatiotemporal analysis of copper homeostasis in Populus trichocarpa reveals an integrated molecular remodeling for a preferential allocation of copper to plastocyanin in the chloroplasts of developing leaves. *Plant Physiol* 157 (3): 1300–1312.

Ray, J.G. and George, K.J. (2010). Calcium accumulation in grasses in relation to their root cation exchange capacity. *J Agron* 9: 70–74.

Ray, D.K., Gerber, J.S., MacDonald, G.K., and West, P.C. (2015). Climate variation explains a third of global crop yield variability. *Nat Commun* 6: 5989.

Römheld, V. and Kirkby, E.A. (2010). Research on potassium in agriculture: needs and prospects. *Plant Soil* 335 (1): 155–180.

Rout, J.R., Ram, S.S., Das, R. et al. (2013). Copper-stress induced alterations in protein profile and antioxidant enzymes activities in the in vitro grown Withania somnifera L. *Physiol Mol Biol Plants* 19 (3): 353–361.

Roy, P.R., Tahjib-Ul-Arif, M., Polash, M.A.S. et al. (2019). Physiological mechanisms of exogenous calcium on alleviating salinity-induced stress in rice (Oryza sativa L.). *Physiol Mol Biol Plants* 25 (3): 611–624.

Sage, T.L., Bagha, S., Lundsgaard-Nielsen, V. et al. (2015). The effect of high temperature stress on male and female reproduction in plants. *Field Crop Res* 182: 30–42.

Samis, K., Bowley, S., and McKersie, B. (2002). Pyramiding Mn-superoxide dismutase transgenes to improve persistence and biomass production in alfalfa. *J Exp Bot* 53 (372): 1343–1350.

Saud, S., Fahad, S., Cui, G. et al. (2020). Determining nitrogen isotopes discrimination under drought stress on enzymatic activities, nitrogen isotope abundance and water contents of Kentucky bluegrass. *Sci Rep* 10 (1): 1–16.

Senbayram, M., Gransee, A., Wahle, V., and Thiel, H. (2016). Role of magnesium fertilisers in agriculture: plant–soil continuum. *Crop Past Sci* 66 (12): 1219–1229.

Shabala, S., Demidchik, V., Shabala, L. et al. (2006). Extracellular $Ca2^+$ ameliorates NaCl-induced K^+ loss from Arabidopsis root and leaf cells by controlling plasma membrane K^+-permeable channels. *Plant Physiol* 141 (4): 1653–1665.

Shahzad, B., Tanveer, M., Che, Z. et al. (2018). Role of 24-epibrassinolide (EBL) in mediating heavy metal and pesticide induced oxidative stress in plants: a review. *Ecotoxicol Environ Saf* 147: 935–944.

Shao, G., Chen, M., Wang, D. et al. (2008). Using iron fertilizer to control Cd accumulation in rice plants: A new promising technology. *Sci China C Life Sci* 51 (3): 245–253.

Sharma, P., Jha, A.B., Dubey, R.S., and Pessarakli, M. (2012). Reactive oxygen species, oxidative damage, and antioxidative defense mechanism in plants under stressful conditions. *Aust J Bot* 2012 (1): 1–26.

Shen, X., Li, X., Li, Z. et al. (2010). Growth, physiological attributes and antioxidant enzyme activities in soybean seedlings treated with or without silicon under UV-B radiation stress. *J Agron Crop Sci* 196 (6): 431–439.

Shen, J., Yuan, L., Zhang, J. et al. (2011). Phosphorus dynamics: from soil to plant. *Plant Physiol* 156 (3): 997–1005.

Shireen, F., Nawaz, M.A., Chen, C. et al. (2018). Boron: functions and approaches to enhance its availability in plants for sustainable agriculture. *Int J Mol Sci* 19 (7): 1856.

Shu, K., Qi, Y., Chen, F. et al. (2017). Salt stress represses soybean seed germination by negatively regulating GA biosynthesis while positively mediating ABA biosynthesis. *Front Plant Sci* 8: 1372.

Shukla, L. and Jain, N. (2020). A review on soil heavy metals contamination: effects, sources, and remediation techniques. *J Soil Contamination* 13 (4): 299–304.

Singh, R., Ahirwar, N.K., Tiwari, J., and Pathak, J. (2018). Review on sources and effect of heavy metal in soil: its bioremediation. *Inter J of Res in Appl, Nat Soc Sci* 2018: 1–22.

Singh, D.P., Singh, V., Gupta, V.K. et al. (2020). Microbial inoculation in rice regulates antioxidative reactions and defense related genes to mitigate drought stress. *Sci Rep* 10 (1): 1–17.

Soleimanzadeh, H., Habibi, D., Ardakani, M.R. et al. (2010). Effect of potassium levels on antioxidant enzymes and malondialdehyde content under drought stress in sunflower (Helianthus annuus L.). *Ameri J Agric Bioll Sci* 5 (1): 56–61.

Song, Z., Duan, C.L., Guo, S.L. et al. (2015). Potassium contributes to zinc stress tolerance in peach (Prunus persica) seedlings by enhancing photosynthesis and the antioxidant defense system. *Genet Mol Res* 14 (3): 8338–8351.

Sunita, K., Mishra, I., Mishra, J. et al. (2020). Secondary metabolites from halotolerant plant growth promoting rhizobacteria for ameliorating salinity stress in plants. *Front Microbiol* 11: 2619.

Syed, R., Kapoor, D., and Bhat, A. (2018). Heavy metal toxicity in plants: a review. *Plant Arch* 18 (2): 1229–1238.

Szalai, G., Kellős, T., Galiba, G., and Kocsy, G. (2009). Glutathione as an antioxidant and regulatory molecule in plants under abiotic stress conditions. *J Plant Growth Regul* 28 (1): 66–80.

Taiz, L. and Zeiger, E. (2002). *Plant Physiology*, 782. Sunderland: Sinauer Associates Inc.

Tanwir, K., Javed, M.T., Abbas, S. et al. (2021). Serratia sp. CP-13 alleviates Cd toxicity by morpho-physio-biochemical improvements, antioxidative potential and diminished Cd uptake in Zea mays L. cultivars differing in Cd tolerance. *Ecotoxicol Environ Saf* 208: 111584.

Taylor, I.B. (1991). Genetics of ABA synthesis. In: *Physiology and Biochemistry Abscisic Acid*, 23–38.

Thakur, P., Kumar, S., Malik, J.A. et al. (2010). Cold stress effects on reproductive development in grain crops: an overview. *Environ Exp Bot* 67 (3): 429–443.

Toor, M.D., Adnan, M., Javed, M.S. et al. (2020). Foliar application of Zn: Best way to mitigate drought stress in plants; a review. *Int J Appl Res* 6 (8): 16–20.

Tripathi, D.K., Singh, S., Singh, S. et al. (2015). Micronutrients and their diverse role in agricultural crops: advances and future prospective. *Acta Physiol Plant* 37 (7): 1–14.

Tripathi, D.K., Singh, S., Gaur, S. et al. (2018). Acquisition and homeostasis of iron in higher plants and their probable role in abiotic stress tolerance. *Front Environ Sci* 5: 86.

Tsonev, T. and Cebola Lidon, F.J. (2012). Zinc in plants-an overview. *Emi J Food Agri (EJFA)* 24 (4): 322–333.

Tuna, A.L., Kaya, C., Ashraf, M. et al. (2007). The effects of calcium sulphate on growth, membrane stability and nutrient uptake of tomato plants grown under salt stress. *Environ Exp Bot* 59 (2): 173–178.

Ul Hassan, Z., Ali, S., Rizwan, M. et al. (2017). Role of zinc in alleviating heavy metal stress. In: *Essential Plant Nutrients*, 351–366. Springer.

Uluisik, I., Karaka, H.C., and Koc, A. (2018). The importance of boron in biological systems. *J Tra Element Med Biol* 45: 156–162.

Van de Wiel, C.C., van der Linden, C.G., and Scholten, O.E. (2016). Improving phosphorus use efficiency in agriculture: opportunities for breeding. *Euphytica* 207 (1): 1–22.

Wang, L.J., Fan, L., Loescher, W. et al. (2010). Salicylic acid alleviates decreases in photosynthesis under heat stress and accelerates recovery in grapevine leaves. *BMC Plant Biol* 10 (1): 1–10.

Wang, M., Zheng, Q., Shen, Q., and Guo, S. (2013). The critical role of potassium in plant stress response. *Int J Mol Sci* 14 (4): 7370–7390.

Wang, M., Zhang, S., and Ding, F. (2020). Melatonin mitigates chilling-induced oxidative stress and photosynthesis inhibition in tomato plants. *Antioxidants* 9 (3): 218.

Wani, W., Masoodi, K.Z., Zaid, A. et al. (2018). Engineering plants for heavy metal stress tolerance. *Rend Lincei Sci Fis Nat* 29 (3): 709–723.

Waraich, E.A., Ahmad, R., Halim, A., and Aziz, T. (2012). Alleviation of temperature stress by nutrient management in crop plants: a review. *J Soil Sci Plant Nutr* 12 (2): 221–244.

Wei, L., Zhang, J., Wang, C., and Liao, W. (2020). Recent progress in the knowledge on the alleviating effect of nitric oxide on heavy metal stress in plants. *Plant Physiol Biochem* 147: 161–171.

White, P.J. and Broadley, M.R. (2001). Chloride in soils and its uptake and movement within the plant: a review. *Ann Bot* 88 (6): 967–988.

Wu, G.Q. and Wang, S.M. (2012). Calcium regulates K^+/Na^+ homeostasis in rice (Oryza sativa L.) under saline conditions. *Plant Soil Environ 58* (3): 121–127.

Wu, S., Hu, C., Tan, Q., and Nie Sun, X. (2014). Effects of molybdenum on water utilization, antioxidative defense system and osmotic-adjustment ability in winter wheat (Triticum aestivum) under drought stress. *Plant Physiol Biochem* 83: 365–374.

Wu, X., Song, H., Guan, C., and Zhang, Z. (2020). Boron mitigates cadmium toxicity to rapeseed (Brassica napus) shoots by relieving oxidative stress and enhancing cadmium chelation onto cell walls. *Environ Pollut* 263: 114546.

Xu, G., Mage, H., Tarchitzky, J., and Kafkafi, U. (1999). Advances in chloride nutrition of plants. *Adv Agron* 68: 97–150.

Yan, L., Riaz, M., Du, C. et al. (2019). Ameliorative role of boron to toxicity of aluminum in trifoliate orange roots. *Ecotoxicol Environ Saf* 179: 212–221.

Yan, L., Riaz, M., Liu, J. et al. (2021). Boron reduces aluminum deposition in alkali-soluble pectin and cytoplasm to release aluminum toxicity. *J Hazard Mater* 401: 123388.

Yang, Y., Xiong, J., Tao, L. et al. (2020). Regulatory mechanisms of nitrogen (N) on cadmium (Cd) uptake and accumulation in plants: a review. *Sci Total Environ* 708: 135186.

Yildiri, E., Karlidag, H.Ü.S.E.Y.İ.N., and Turan, M. (2009). Mitigation of salt stress in strawberry by foliar K, Ca and Mg nutrient supply. *Plant Soil Environ* 55 (5): 213–221.

Zafar, S., Nasri, M., Moghadam, H.R.T., and Zahedi, H. (2014). Effect of zinc and sulfur foliar applications on physiological characteristics of sunflower (Helianthus annuus L.) under water deficit stress. *Int J Biosci* 5 (12): 87–96.

Zahoor, R., Dong, H., Abid, M. et al. (2017). Potassium fertilizer improves drought stress alleviation potential in cotton by enhancing photosynthesis and carbohydrate metabolism. *Environ Exp Bot* 137: 73–83.

Zain, N.M., Ismail, M.R., Puteh, A. et al. (2014). Drought tolerance and ion accumulation of rice following application of additional potassium fertilizer. *Commun Soil Sci Plant Anal* 45: 2502–2514.

Zechmann, B., Müller, M., and Zellnig, G. (2008). Modified levels of cysteine affect glutathione metabolism in plant cells. In: *Sulfur Assimilation and Abiotic Stress in Plants* (ed. A.K. Nafees, S. Singh and S. Umar), 193. Berlin: Springer.

Zhang, M., Hu, C., Zhao, X. et al. (2012). Molybdenum improves antioxidant and osmotic-adjustment ability against salt stress in Chinese cabbage (Brassica campestris L. ssp. Pekinensis). *Plant Soil* 355 (1): 375–383.

Zhang, F., Wan, X., and Zhong, Y. (2014). Nitrogen as an important detoxification factor to cadmium stress in poplar plants. *J Plant Interact* 9 (1): 249–258.

Zhang, R., Liu, Y., Xue, W.L. et al. (2016). Slow-release nitrogen fertilizers can improve yield and reduce Cd concentration in pakchoi (Brassica chinensis L.) grown in Cd-contaminated soil. *Environ Sci Pollut Res* 23 (24): 25074–25083.

Zhang, X., da Silva, J.A.T., Niu, M. et al. (2017). Physiological and transcriptomic analyses reveal a response mechanism to cold stress in *Santalum album* L. leaves. *Sci Rep* 7: 42165.

Zhang, C., Shi, S., Liu, Z. et al. (2019). Drought tolerance in alfalfa (Medicago sativa L.) varieties is associated with enhanced antioxidative protection and declined lipid peroxidation. *J Plant Physiol* 232: 226–240.

Zhao, W.Y., Xu, S., Li, J.L. et al. (2008). Effects of foliar application of nitrogen on the photosynthetic performance and growth of two fescue cultivars under heat stress. *Biol Plant* 52 (1): 113–116.

Zhu, J.J., Li, Y.R., and Liao, J.X. (2013). Involvement of anthocyanins in the resistance to chilling-induced oxidative stress in *Saccharum officinarum* L. leaves. *Plant Physiol Biochem* 73: 427–433.

Zhuang, K., Kong, F., Zhang, S. et al. (2019). Whirly1 enhances tolerance to chilling stress in tomato via protection of photosystem II and regulation of starch degradation. *New Phytol* 221 (4): 1998–2012.

Index

Plant Ionomics: Sensing, Signaling, and Regulation, First Edition.
Edited by Vijay Pratap Singh and Manzer H. Siddiqui.